Photoelectrochemical Solar Conversion Systems

Molecular and Electronic Aspects

Photoelectrochemical Solar Conversion Systems

Molecular and Electronic Aspects

Andrés G. Muñoz

CRC Press
Taylor & Francis Group
Boca Raton London New York

CRC Press is an imprint of the
Taylor & Francis Group, an **informa** business

CRC Press
Taylor & Francis Group
6000 Broken Sound Parkway NW, Suite 300
Boca Raton, FL 33487-2742

First issued in paperback 2019

© 2013 by George Koenigsaecker
CRC Press is an imprint of Taylor & Francis Group, an Informa business

No claim to original U.S. Government works

ISBN-13: 978-1-4398-6925-3 (hbk)
ISBN-13: 978-0-367-38068-7 (pbk)

Visit the Taylor & Francis Web site at
http://www.taylorandfrancis.com

and the CRC Press Web site at
http://www.crcpress.com

Contents

Preface

The continuous increase of collective consciousness of the world's habitants in recent decades about global climate changes and the alarming increase of greenhouse gas emissions [1] have resulted in renewed interest in renewable and environmental-friendly energy sources, particularly in industrialized countries. The new impulse received by nonemitting technologies emerged as an answer to the need for mitigation of the alarming increase of the CO_2 levels, to which more frequent world catastrophes are attributed. Recently, the sad news about the nuclear disaster in Japan has forced governments in several countries to review their energy programs, exhorted to an accelerated repudiation of atomic energy. Nuclear energy was profiled as a clean energy source. Nevertheless, in spite of considerable progress achieved concerning plant security, it has received a considerable negative image. The conversion of solar energy into electricity or chemical energy, together with the wind and geothermal energy technologies, are promising substitutes for some fossil-based nonrenewable sources of energy, which nowadays are mainly provided by coal thermal and nuclear energy power stations. Within the solar solution, the energy storage arises as a complementary issue to be resolved. The storage of electrical energy requires the use of high-density batteries, which limits the implementation of solar energy systems to private consumers' houses and cars. The conversion of solar energy in a transportable fuel is an issue that is gaining more and more attention. In this sense, the generation of hydrogen by means of water splitting in electrolyzers fed with current from renewable sources is pursued to replace the prevailing production method based on fossil fuel reforming. This would complete what we could call a clean energy cycle: water–energy–water. The storage and combustion of hydrogen does not seem to be a hindrance for further development of the clean cycle, considering the large advances achieved in the last decade in this field: safe high-pressure bottles, fuels cells, and combustion engines for cars. On the other hand, the CO_2 reduction by means of electrolysis has recently received attention as an alternative to produce transportable fuels. The development of the technology of electrical current-generating or photovoltaic solar cells is closely connected with the progress made in the research area of aerospace between 1954 and 1973. In 1954, the first reliable solar cell based on silicon technology with a conversion efficiency of 6% was set up for the provision of electricity in the American satellite Vanguard I [2]. It is not an accident that silicon is the most used material for the construction of solar cells. This can be considered as a result of longstanding experience and good knowledge of the structural and electronic properties of this material inherited from the research and industry of electronic devices, thus ensuring a continuous provision of absorber material of good electronic quality. For this reason, the silicon-based solar cells, consisting in n^+p-junctions, are considered first-generation cells. In spite of the simultaneous development of other more efficient light absorber materials, such as GaAs, $CuInSe_2$ and $Cu(InGa)Se_2$, and InP, providing a broader palette of organic compounds and polymers, silicon still prevails in the large spectrum of designed conversion devices available today.

Although silicon is one of the most plentiful elements in the earth's crust with an abundance of 27.7% [3], its extraction is extremely energy consuming and demands practically 60% of total costs for the production of crystal wafers with the necessary electronic quality. This fact was generally used as a critical argument to discourage the advance of the solar cell technology in favor of the continuity of traditional energy technologies. On the other hand, this criticism has served as a positive argument for the promotion of alternative solar energy conversion technologies. This situation has, however, been reversed from the 1990s on, since the foregoing amortization time of 8 to 10 years was able to be continuously reduced up to periods of 1 to 4 years as reported in the last decade [4]. The still-high energy costs for the production of high-quality silicon wafers 20 years ago motivated the simultaneous development of new light absorber materials, for instance ternary chalcopyrites [5,6], organic layers, and semiconducting oxides sensitized with organic dyes [7–10], among others. Because of the high sunlight absorption capability of materials such as $CuInS_2$, $CuInSe_2$, or InP, the utilization of thin films of these materials is evidently advantageous. In fact, efficiencies up to 19.9% were reached with solar cells consisting of heterostructures of $Cu(In_xGa_{1-x})Se_2$ and CdS [11] using polycrystalline materials. In contrast to this achievement, low conversion efficiencies have been reached in solar cells constructed on the basis of organic semiconductors, in spite of their low production costs. This has limited its application to low-power-consuming devices such as electronic devices and thermal jackets. The realization of the dye-sensitized cell concept has been technically accomplished in the form of photoelectrochemical cells, where an inorganic oxide semiconductor soaked with an organic dye is connected with a carbon or platinum counter-electrode through a redox-electrolyte. Conversion efficiencies up to 10.4% have been reached with this concept [12], and large efforts are being presently made to improve it, as shown by the surprisingly large amount of research reports on this device concept. An efficient improvement requires a spreading of the spectral absorption region of the dyes to cover the whole solar spectrum; the suppression of recombination processes, for example, by modification of the energy levels of the molecular orbital (the so-called inverse Marcus process [13]); and a better anchoring of the organic molecules onto the semiconducting support to ensure more efficient transfer of photogenerated excitons. Together with the development of dye-sensitized systems, innumerable cell concepts based on nanodimensioned systems have been proposed, thus largely extending the available palette. The modification of the semiconductor band structure by quantum size effects on the one hand, and the enhancing of light absorption by the generation of surface plasmons on the other hand, are the main phenomena motivating the development of those systems. In addition, better knowledge of the natural photosynthetic processes has motivated some researchers to mimic the natural mechanism in a more profitable way with artificial photosynthesis.

This book aims to introduce the reader to fundamental aspects concerning the conversion of sunlight into electricity or chemical energy by photoelectrochemical systems with an emphasis on the inorganic Schottky-type electrolyte–metal–semiconductor contacts. This type of system consists of a semiconducting light absorber, which is protected from the aggressive electrolyte by an interfacial passive layer, through which the transport of light-generated electrons from the light absorber to the electrochemical reaction sites should occur. The interfacial layer, which protects the underlying substrate against chemical reaction with the electrolyte, should not only preserve good electronic quality of the interface (i.e., a reduced density of recombination centers), but also guarantee a high electronic conductivity. The formation of an efficient conversion absorber–passive layer–metal–electrolyte interface involves several physical, chemical, and electrochemical steps, which are referred to as surface conditioning steps. Anodic oxide formation and metal electrodeposition belong to these processes.

The first chapter introduces general concepts of the conversion of sunlight into chemical energy and a comprehensive overview of the different actual conversion concepts. Chapters 2 and 3 present an introduction to the electrochemical methods for the construction and characterization of electrolyte–metal–oxide–semiconductor contacts (EMOS) in the nanodimensions, the so-called nanoemitter concept, including the electrochemical formation of metal clusters of catalytic metals and the formation of passivating layers by anodization. Chapter 4 discusses the fundamentals of electrocatalysis with emphasis on the hydrogen evolution reaction and the electrochemical CO_2 reduction. A critical revision of the classical and quantum mechanical theories of electron transfer reactions in metal–electrolyte interfaces and their relation with surface electronics is pursued. The physicochemical characterization of the model system Si-SiO$_x$–metal–electrolyte by means of modern electrochemical, surface, and spectroscopic methods is given in Chapter 5. Improvements in conversion efficiency by means of optical effects, such as the generation of surface plasmons by nanodimensioned arrangements of optically active metals or surface absorption enhancement by construction of absorber nanostructures, is complementarily discussed in Chapter 6.

Finally, I want to express my gratitude to Prof. Hans-Joachim Lewerenz and Prof. Dieter Schmeißer for their invaluable support and fruitful scientific discussions. Not least, I acknowledge my wife, Ina, for her infinite patience and spiritual support during the writing of this book.

Andrés G. Muñoz

About the Author

Dr. Andrés G. Muñoz is a senior scientist dedicated to the investigation of fundamental questions of applied and basic electrochemistry of metal and semiconductors. He made his first steps in the electrochemical science at the Institute of Electrochemistry and Corrosion of the University del Sur, Argentina. After receiving his PhD in materials science, he worked under renowned scientists from the German electrochemistry school at the Heinrich-Heine University of Düsseldorf and the Research Centre Jülich and was distinguished with an Alexander von Humboldt fellowship. He combined his background in electrochemistry science with surface science during his stay at the Institute for Solar Fuels and Energy Storage Materials at the Helmholtz Center in Berlin to resolve fundamental questions of interfacial photo-electrochemical processes in solar energy conversion systems based on inorganic semiconductors. Presently, his research activities at the Company Global Research for Safety focus on the application of electro-analytical methods in the area of thermodynamic of electrolytes connected with different questions of nuclear, geothermal and renewable energy sources. He also continues with his lectures at the Brandenburg Technical University at Cottbus.

1 Principles and Systems in Light-Induced Energy Conversion

1.1 BASIC ASPECTS OF LIGHT-INDUCED GENERATION OF FUELS

The general principle of the conversion of sunlight into chemical energy or electricity consists of the generation of electron-hole pairs by the absorption of incident light and their subsequent spatial separation. Charge carriers are consumed by an electrochemical reaction (photoelectrochemical cell) or in an external circuit (photovoltaic cell). In the particular case of photoelectrochemical conversion, electrochemical reactions such as hydrogen evolution and CO_2-reduction are mostly catalyzed by metals and alloys. Thus, this book will usually refer to the metal phase to signify a catalyst surface.

The charge separation in photoelectrochemical systems is driven by a gradient of electrochemical potential of the electrons. This quantity is a thermodynamic parameter defined as the Gibbs free energy of charged particles, which for the case of electrons is given by

$$\tilde{\mu}_e = \left(\frac{\partial G}{\partial n_e} \right)_{P,T} = \mu_e - \varphi \tag{1.1}$$

where μ_e is the chemical potential and φ is the electrostatic potential. In some systems, the gradient of the electrochemical potential appears in the form of an electric field, as in the case of semiconducting absorbers, or as a difference of chemical potential between two contacting phases as found in dye-sensitized and bio-inspired systems.

In the following we analyze general aspects of photoelectrocatalytic processes. The scheme in Figure 1.1 represents a light conversion system as a black box without regard to the internal electron transfer mechanisms involved. Under stationary conditions and without energy accumulation, the incident flux of photon energy (J_{ph}) is partially converted into chemical energy by generation of new chemical products ($J_{\Delta G}$), while the remaining energy is dissipated by reflection or transmission or in the form of phonons or irradiated light (J_{dis}). The energy balance can be expressed as follows:

$$J_{ph} = J_{\Delta G} + J_{dis} \tag{1.2}$$

1

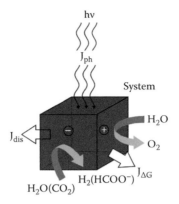

FIGURE 1.1 Schematic of the light energy conversion into chemical energy. The picture represents the conversion process as a black box where light-induced charges are consumed by electrochemical reactions occurring at the system boundaries.

The flux of chemical energy delivered by the photoconversion system can be calculated from the change of Gibbs free energy for the total reaction, ΔG_r, as

$$J_{\Delta G} = \frac{d\Delta G_r}{dt} = \frac{dn}{dt}\sum_i v_i \mu_i \qquad (1.3)$$

where v_i is the stoichiometric factor, μ_i is the chemical potential of the species i, and dn/dt represents the molar flux of a reference species. In the case of an electrochemical reaction, the Gibbs free energy is calculated from the electrochemical potential of the participating charged species as

$$\Delta G_r = \sum_i v_i \tilde{\mu}_i \qquad (1.4)$$

where μ_i is the electrochemical potential of the charged species i, given by $\tilde{\mu}_i = \mu_i + z_i \phi$. In the particular case of a reaction in equilibrium, we have that

$$\sum_i v_i \tilde{\mu}_i = 0.$$

This equation gives a useful relation between the change of Gibbs free energy of the reaction and the potential established at the electrode–electrolyte interface:

$$V^0 = -\sum_i v_i \mu_i / ze = -\frac{\Delta G_r}{ze} \qquad (1.5)$$

where z is the number of electrons involved in the reaction. Equation (1.5) shows that the reaction rate and the course of the electrochemical reaction can be controlled by changing the electrode potential beyond its equilibrium value. In the case of water splitting, for instance, the whole reaction is the result of two electrochemical reactions energetically and spatially separated: the reduction of water at the cathode (reaction 1.6a) by photogenerated electrons and the oxidation of water by photogenerated holes (reaction 1.6b) at the anode.

$$4 \, H_2O(l) + 4 \, e^-_{(hv)} \rightarrow 2 \, H_2(g) + 4 \, OH^-(aq) \tag{1.6a}$$

$$2 \, H_2O \, (l) + 4 \, h^+_{(hv)} \rightarrow O_2 \, (g) + 4 \, H^+ \, (aq) \tag{1.6b}$$

$$2 \, H_2O \, (l) \rightarrow 2 \, H_2 \, (g) + O_2 \, (g) \tag{1.6c}$$

The change of free energy for reaction (1.6a) and (1.6b) is given by

$$\Delta G_1 = \Delta \mu^0_1 + (kT/e) \, [\ln f_{H2}^{1/2} + \ln a_{OH^-}] \tag{1.7}$$

and

$$\Delta G_2 = \Delta \mu^0_2 + (kT/e) \, [\ln f_{O2}^{1/2} + kT \ln a_{H+}] \tag{1.8}$$

respectively, where f_i and a_i are the fugacities and activities of gaseous and electrolyte components, respectively. $\Delta \mu^0$ represents the change of Gibbs free energy under standard conditions, that is, $T = 298$, 15 K, $f_i = 1$, and ideal solution with $a_i = 1$. The sum of Eqs. (1.7) and (1.8) gives the change of Gibbs free energy for the water-splitting reaction:

$$\Delta G_r = 1.23 + (kT/e) \, [\ln f_{H2}^{1/2} + kT \ln f_{O2}^{1/2}] \, eV_/mol \, H_2O \tag{1.9}$$

It can be noted that the total reaction does not introduce change of pH in the electrolyte. A similar analysis can be made for the CO_2 reduction (reactions 1.10a and b).

$$2 \, CO_2 \, (g) + 2 \, H_2O \, (l) + 4 \, e^-_{(hv)} \rightarrow 2 \, HCOO^- \, (aq) + 2 \, OH^- \, (aq) \tag{1.10a}$$

$$2 \, H_2O \, (l) + 4 \, h^+_{(hv)} \rightarrow O_2 \, (g) + 4 \, H^+ \, (aq) \tag{1.10b}$$

$$2 \, CO_2 \, (g) + 2 \, H_2O \, (l) \rightarrow 2 \, HCOO^- \, (aq) + O_2 \, (g) + 2 \, H^+ \, (aq) \tag{1.10c}$$

Unlike the water splitting, the change of Gibbs free energy for reduction of CO_2 to formic acid as the final product gives the free energy relation, which changes with solution pH as

$$\Delta G_r: 1.243 - 0.0295 \times pH + (kT/e) \tag{1.11}$$
$$[\ln f_{O2}^{1/2} + kT \ln a_{HCOO^-} - 1/2 \ln f_{CO2}] \, eV_/mol \, H_2O$$

The flux of chemical energy can be measured by taking hydrogen as the reference substance. Thus, we have

$$J_{\Delta G} = \dot{m}_{H_2} \Delta G_R \tag{1.12}$$

where $\dot{m}_{H_2}$ is the mass flux of evolving hydrogen and ΔG_R is the total change of free energy per mol of evolved hydrogen. According to this scheme, the conversion efficiency can be defined as

$$\eta_{h\nu} = \frac{J_{\Delta G}}{\int_0^\infty \left(\dfrac{\partial J_{ph}}{\partial h\nu}\right) dh\nu} \tag{1.13}$$

where the integration is performed over the whole incident light spectrum.

Two routes can be presently regarded as viable for light-induced generation of hydrogen. One way consists of the use of existent photovoltaic cells in connection with electrolyzers [1,2]. The total efficiency of this type of system is given by the product of the photovoltaic cell efficiency times that of the electrolyzer. Maximal efficiencies are typically around 19% for the best commercially available solar modules, while efficiencies around 70% are found for electrolyzers. Hence, this combination is expected to operate with efficiencies of about 13%. The other way ponders the direct conversion in monolithic integrated structures [3]. This concept can be realized by coupling two photoactive semiconductors by means of an intermediate ohmic contact (*pn*-heterojunctions). The structure is immersed in the electrolyte and the electrochemical reactions occur under illumination of both surfaces of the semiconductors. This concept requires the semiconducting materials to fulfill determined conditions concerning their electron affinities and band gaps, as further discussed in Section 1.4. The investigation of the performance of each photoelectrode constituting this closed device is carried out separately in three electrode-cell systems with the introduction of an auxiliary counter- and a reference electrode. There are, in principle, two approaches to be considered for the construction of fuel electrodes based on semiconducting absorbers, as shown schematically in Figure 1.2. The first approach (i) utilizes a light absorber with an electrocatalytic surface, which can be eventually covered with an electrochemical co-catalyst. In the other approach (ii), an electrochemical active catalyst is provided. The catalytic phase is usually separated from the light absorber by insertion of a stabilization film formed in between by several surface preconditioning chemical and/or electrochemical steps. The purpose of this film is to reduce recombination centers at the absorber surface and to protect against (photo)-corrosion when brought into contact with the electrolyte. In a third concept (iii), the light absorption occurs on dye molecules attached to a wide band-gap semiconductor. The charge separation and transport mechanism distinguishes approaches (i) and (ii) from (iii). In the former case, the charge separation is driven by drift and diffusion of photogenerated charges by an electric field arising at the interface of the semiconductor in contact with a second phase. The contacting phase may be (i) a semiconductor of

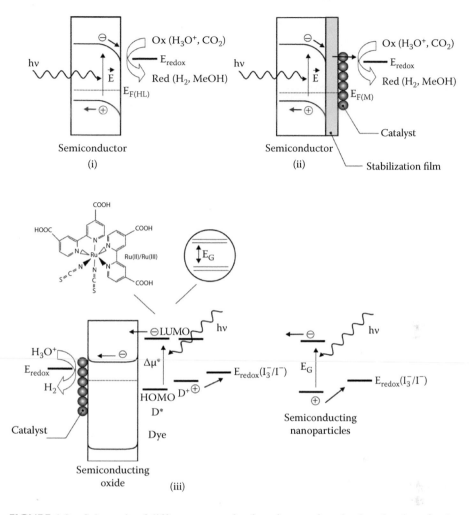

FIGURE 1.2 Schematic of different approaches based on semiconducting absorbers for the construction of photoelectrodes to be used for light-induced generation of fuels: (i) electrocatalytic light absorber; (ii) light absorber covered with a stabilization layer and an electrocatalyst; (iii) dye-sensitized semiconducting system. The dye-absorber can be replaced by nanoparticles of semiconducting materials.

the same (*pn*-homojunction) or of a different material (*pn*-heterojunction), (ii) a metal (Schottky-type junction) [4–6], or (iii) an electrolyte (photoelectrodes) [7,8]. The third approach (iii in Figure 1.2) belongs to the category of excitonic solar cells [9,10]. Here, the charge separation is driven by a chemical potential gradient between the dye and the supporting wide band-gap semiconductor.

In the following, an overview of the operation principles of a photoelectrode based on a semiconducting absorber covered with an electrocatalyst will be presented. The analysis is based on the particular case of a photocathode, but it can be readily extended to the case of a photoanode. The photoelectrode in contact with

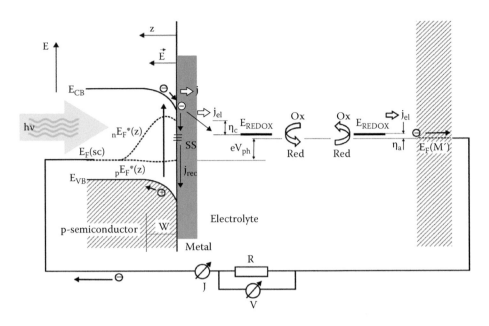

FIGURE 1.3 Working principle of a photoelectrochemical cell.

a redox-electrolyte is connected by an external circuit with an auxiliary electrode, generally consisting of an electroactive material such as Pt or graphite. Figure 1.3 shows the energetic diagram of a photoelectrochemical cell working under stationary conditions. Under illumination, the excess of photogenerated electrons in the semiconductor leads to stationary nonequilibrium conditions, and hence to different free energies of the charge carriers. For further discussion, the reader should accept without proofs that the free energy of charge carriers is given by their quasi-Fermi level: $_nE^*$ for electrons and $_pE^*$ for holes. The decreasing intensity of incident light toward the semiconductor bulk, because of absorption as well as the diffusion and the recombination in bulk, and at the interface create a spatial distribution of excess photogenerated charge carriers and hence of the quasi-Fermi levels. The recombination at the interface takes places via localized energy levels in the band gap known as surface states (ss). The electron current toward the semiconductor–electrolyte interface, arising from a field and a diffusion contribution, is given by [11]

$$j = e \times \left[n_e \upsilon_e \nabla E^*{}_e - n_h \upsilon_h \nabla E^*{}_h \right]$$ (1.14)

where $n_{e,h}$ is the concentration of electrons or holes, $\upsilon_{e,h}$ is the corresponding motility, and $\nabla E^*_{e,h}$ is the gradient of quasi-Fermi levels for electrons or holes. Photoelectrons that reach the interface are then consumed by the electrochemical reduction reaction, which can be written in the general form

$$Ox^{z+} + n\ e^- \rightarrow Red^{z-n}$$ (1.15)

Depending on the relative generation rate of photoelectrons and the electron transfer rate at the semiconductor–electrolyte interface, the electrochemical reaction can be the limiting step in the whole conversion path. In this case, the observed photocurrent corresponds to the maximal achievable electrocatalytic current density on the metal phase, that is, $j_{ph} = j_{el}$. In the absence of mass transport limitations in the electrolyte, the electron transfer rate at the surface of the photocathode is phenomenologically described by the cathodic part of the Butler–Volmer equation [12], the derivation of which will be discussed in Chapter 4.

$$j_{el} = F \times c_{ox} \times k_{et} = F \times c_{ox} \times k_0 \exp\left[\frac{\alpha_c}{kT}\left(_pE^*_{(0)} - E_{Redox}\right)\right] = j_0 \times \exp\left[\frac{e}{kT}\alpha_c\eta_c\right] \quad (1.16)$$

According to this expression, which was adapted from the classical equation for metals, the electron transfer rate is proportional to the concentration of reducing species c_{ox} by a transfer rate constant k_{et}. The latter depends exponentially on the potential difference given by $(_pE^*_{(z=0)} - E_{Redox})/e$, or reaction overvoltage η_c, by factor α_c, known as the charge transfer coefficient. In contrast to classical electrochemical reactions on metals, it is here assumed that photoelectrons are discharged at their Fermi level at the interface $_pE^*_{(z=0)}$, different from the bulk Fermi level of the semiconductor, which is assumed to be equal to the Fermi level in the metal phase. Furthermore, the electron energy in the solution is given by the redox energy level, which can be regarded as the work function of the solution. The value for the hydrogen evolution reaction in standard conditions (i.e., $T = 278, 15$ K, $p = 1$ atm, and $a_{H3O^+} = 1$) was calculated to be -4.5 ± 0.1 eV. Thus, the values for all other electrochemical reactions are calculated taking this value as a reference in the electrochemical scale.

The pre-exponential Butler–Volmer factor j_0 is the reaction rate at the equilibrium potential (i.e., $\eta_c = 0$), and it is known as the exchange current density. In equilibrium, the cathodic partial reaction equals the anodic one, and $j_{el} = 0$. The exchange current density is a measure of the catalytic activity of the surface and is a strong function of the electronic and structural properties of the metal surface, such as the work function, orientation, potential of zero charge, and the presence of d-orbitals, among others. The exponential factor α_c can be interpreted in some special cases as the fraction of the potential drop in the electrochemical double layer contributing the activation energy [13]. It adopts in general a value of about 0.5 for a single electron step reaction.

The current circuit is closed by an external electrical connection to an auxiliary electrode, whereon an oxidation reaction (anodic) takes place. In this case, the current can also be expressed by a Butler–Volmer-type equation as

$$j_{el} = F \times c_{red} \times k_0 \exp\left[\frac{\alpha_a}{kT}\left(E_{Redox} - E_{F(M'')}\right)\right] = j'_0 \times \exp\left[\frac{e}{kT}\alpha_a\eta_a\right] \quad (1.17)$$

where c_{red} is the concentration of the reduced species and $E_{F(M')}$ is the Fermi energy of the auxiliary metal. The difference between the bulk Fermi levels of the photoelectrode (semiconductor) and the auxiliary electrode $(E_{(M')} - E_{(sc)})$ is eV_{ph}, that is, the cell voltage. As $E_{(M')} - E_{(sc)} = 0$, the difference between the quasi-Fermi level for the minority charge carriers, $_nE^*$ for a p-semiconductor, and the bulk Fermi level of the semiconductor and thus the photocurrent ($j_{ph} = j_{el}$), reach their maximal values (see Eq. 1.16) (short circuit current j_{sc}). In addition, the band bending in the semiconductor and hence the electric field also increase, contributing to a more efficient charge carrier separation.

Should the circuit be opened, the system evolves toward an electrochemical equilibrium and the band bending of the semiconductor reduces until the rate of charge-carrier photogeneration equals the recombination rate at the interfacial states. At this point the maximal photopotential is achieved.

An efficient energy conversion requires an adequate electrocatalytic surface so that the electrochemical reaction does not introduce additional limitations to the conversion process, that is, with an as-small-as-possible reaction overpotential (η_c). Since the most common semiconducting light absorbers—such as Si, InP, or chalcopyrites—have bad catalytic properties for the hydrogen evolution, it is necessary to cover their surfaces with a catalytic metal in the form of a thin film or as nanoparticles. This electrocatalytic phase is, as a rule, separated from the semiconductor surface by an in-between formed stabilizing film consisting of an ultra thin insulating or conducting oxide. Figure 1.4 presents an energy band diagram describing in more detail the electronic processes at the working photoelectrode. Photogenerated minority charge carriers, first thermalized in the semiconductor, are further transported by tunneling from the semiconductor surface to the metal catalyst at the quasi-Fermi level. Thermalization of electrons in metals occurs typically

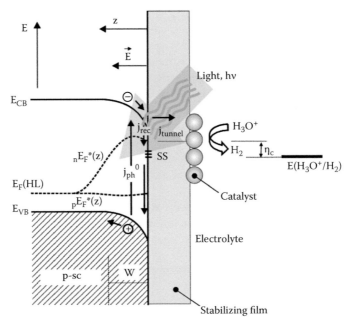

FIGURE 1.4 Energy band diagram of an electrocatalytic activated photoelectrode for hydrogen evolution.

within some hundred fs [14–16]. This period of time is much shorter than that for the electron exchange in electrochemical reactions, which proceeds within a time of 10^{-8} s to 10^{-12} s. Therefore, one would expect that injected electrons thermalize to the bulk semiconductor Fermi level ($E_{F(sc)}$) before reaching the metal–electrolyte interface (reaction without illumination). In the case of thin metal films, however, it is plausible that electrons are ballistically spread and consumed by the electrochemical reaction at their quasi-Fermi level ($_nE^*$). Experimental evidence for electrochemical reactions with injected hot electrons was reported by Sung and Bard [30]. These authors studied the reduction of thianthrene radical cations by injection of hot electrons at Ta/Ta_2O_5 electrodes covered with thin Pt films. There, it was shown that electron emission is not hindered up to metal thickness of 40 nm, in agreement with the expected mean free path of hot electrons in the Pt.

The heterogeneous transfer rate constant (k_0) of typical redox systems such as $Ru(NH_3)^{2+/3+}$ and $Fe(CN)_6^{3-/4-}$ shows values of ~ 1 cm s^{-1} [31] and ~ 0.3 cm s^{-1} [32], respectively. Accordingly, on supposing that the electron transition takes place over a distance of 1 Å from the metal surface (double layer), a transition time of 10^{-8} s can be estimated. It was experimentally observed that the electron transfer rate increases exponentially up to three orders of magnitude with the gap between the transition energy and the redox level: $(E - E_{redox})/e$. A saturation is reached for $(E - E_{redox})/e > 0.5$ V. These findings are in agreement with the behavior predicted by the Marcus theory [33] (see also Section 1.5). Hence, it is expected that the transition time drops exponentially with increasing transition energy up to values of about 1 ps.

For an optimal cell operation, the material of the auxiliary electrode should be selected in a way that $\eta_a \approx 0$, so that only the photoelectrode controls the cell response under illumination. Figure 1.5 shows a typical current–voltage curve of a photoelectrochemical cell as shown in Figure 1.3. The cell consists of a photoelectrode and an auxiliary electrode immersed in the same redox electrolyte. Thus, $V = 0$ corresponds to the situation in which $E_{F(sc)} = E_{redox}$. The photocurrent rises up at voltages

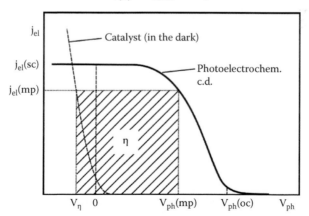

FIGURE 1.5 Typical current-voltage curve for a photoelectrochemical cell. $j_{el}(sc)$: short circuit current; $V_{ph}(oc)$: voltage at open circuit; $j_{el}(mp)$: current at the maximal power point; $V_{ph}(mp)$: voltage at the maximal power point.

$V_{ph} > V_{redox}$ ($E_{F(sc)} < E_{redox}$) as the bulk Fermi level of the semiconductor enters the depletion region. The steepness of the curve is related to the rate of recombination of electron-hole pairs at the interface and the rapidity of the electrochemical reaction as well, that is, the electrocatalytic activity. The photocurrent attains a saturation current, the short circuit current, which depends on the intensity of incident light, the absorption properties of the semiconducting material, the surface reflection, and the recombination rate in the semiconductor bulk. In photoelectrochemical systems based on light absorbers covered with an electrocatalytic material, as for instance that for light-induced hydrogen evolution, the conversion efficiency is commonly defined in reference to the overpotential necessary to get the same current density of the photoelectrode on the surface of the electrocatalytic metal (Eq. 1.18) [34,36].

$$\eta_\% = \frac{j_{el(mp)} \times (V_{ph(mp)} - V_\eta)}{P_{abs}} \times 100 \tag{1.18}$$

Here, $j_{el(mp)}$ [A cm^{-2}] and $V_{ph(mp)}$ [V] are the current density and the photovoltage at the maximal power point, P_{abs} [W cm^{-2}] is the power density of the incident light, and V_η is the overvoltage for comparable current densities at the surface of the electrocatalytic metal.

One simple manner to analyze the performance of a photoelectrochemical system is by adapting the classical equivalent circuit for solid-state solar cells to the photoelectrochemical case, shown in Figure 1.6. It consists of a current source connected in parallel with a diode. A parallel resistance represents the recombination effects, while the in-series connected resistance accounts for ohmic losses and the rate of electron consumption controlled by the electrochemical reaction. Disregarding losses at ohmic contacts, the nonlinear series reaction resistance can be expressed as

$$R_r = \left(\frac{\partial j_{el}}{\partial \eta_c}\right)^{-1} = \left(j_{el}\frac{e}{kT}\alpha_c\right)^{-1} \tag{1.19}$$

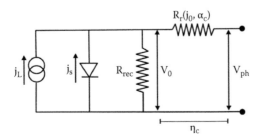

FIGURE 1.6 Equivalent circuit representing the current-voltage behavior of a photoelectrochemical cell.

After this scheme, the voltage output is reduced by $j_{el} \times R_r$ because of the electrochemical reaction. The voltage drop is here identified with the reaction overpotential η_c, and thus the current-voltage relation is given by

$$j_{el} = -j_L + j_s \left[\exp\left(\frac{e}{kT}(V_{ph} + \eta_c) \right) - 1 \right] \tag{1.20}$$

where j_L is the saturation photocurrent and j_s is the limiting reverse current of the diode in the dark. V_{ph} is given by $(E_{redox} - E_{F(sc)})/e$. For $V_{ph} = 0$ the bulk Fermi level of the semiconductor reaches the redox energy level, given by $E_{Redox} = E_H - e\, V_{redox} = -4.5 \pm 0.1$ eV $+ e\, V_{redox}$, with V_{Redox} as the redox potential in the electrochemical scale. The introduction of Eq. (1.19) into Eq. (1.20) gives, after some rearrangement,

$$V_{ph} = \frac{kT}{e} \left[\ln\left(\frac{j_{el} + j_L}{j_s} \right) - \frac{1}{\alpha_c} \ln \frac{j_{el}}{j_0} \right] \tag{1.21}$$

This expression presents a first relationship between the electrochemical parameters and the performance of the photoelectrochemical conversion device. Figure 1.7 exemplifies the current-voltage behavior for different values of j_0 and α_c. It should be noted that a decrease of the exchange current density by two orders of magnitude results in a drop of efficiency over 30 times (denoted in the figure by the decrease of the area of dashed squares). This would be the case if Ni is used as the electrocatalyst for the hydrogen evolution ($j_0 = 10^{-2.2}$ mA cm^{-2}) instead of Pt ($j_0 = 1$ mA cm^{-2}). Furthermore, the influence of the electron transfer coefficient α_c is reflected in the modification of the steepness of the rising part of the current-voltage characteristic. Note that for $\alpha_c \to \infty$, Eq. (1.21) adopts the form expected for a solid-state photodiode.

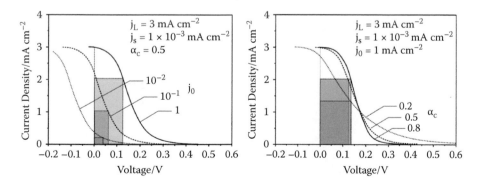

FIGURE 1.7   Current-voltage curves drawn after the modified equivalent circuit for different electrochemical parameters. The variation of the cell efficiency is represented by the shadowed squares under the curves.

1.2 DEVELOPMENT OF ENERGY CONVERSION SYSTEMS

One could say that the history of photovoltaics began with investigations into the photoelectric properties of Pt sheets covered with silver chloride in an aqueous electrolyte carried out in the first half of the seventeenth century by Becquerel [36,37]. The birth of the first solid-state solar cell, however, is attributed to the experimental device based on selenium reported by Smith in 1873 [38] and Adams in 1876 [39]. Nonetheless, the first solar cells consisting of selenium, as well as those based on copper oxide-cooper junctions [40,41] had to wait over 50 years for their commercialization. But, because of their low efficiency, ranging from 0.1% to 0.5%, they were solely used as photodetectors. The photovoltaics followed a new course after the birth of modern physics founded on the basis of the theories of Max Planck [42], relating for the first time the energy of light with its wavelength upon absorption and emission of radiation in materials.

One of the first attempts to realize an efficient solar cell was made by Ohl in 1948 at Bell Telephone Labs in the United States. Ohl developed a photodevice based on Pt-silicon Schottky-type junctions using polycrystalline silicon wafers. However, these experiments failed inasmuch as the polycrystalline nature of the material and its high level of impurities caused a large number of recombination centers. Six years later, Chapin, Fuller, and Pearson [43] were able to construct the first reliable solar cell based on a silicon *pn*-homojunction with an efficiency of 6% by AM 1.5 illumination. Similar concepts were later realized using cadmium sulfide [44] and gallium arsenide [45]. The working principle of this type of device is based on a single *pn*-junction that generates an interfacial electric field for the separation of the light-induced electron-hole pairs in the semiconductors. Cells constructed under this principle are classified under the name *first generation cells*. Limitation of the output of these cells are (i) the thermalization of the photogenerated electrons with

FIGURE 1.8 *Left:* Alexandre-Edmond Becquerel discovered the photoelectric effect in 1839 after illuminating a sheet of selenium and silver chloride. Lithography from Pierre Petit (1832–1885), published by Charles Jeremie Fuhr. *Source:* The Dibner Library Portrait Collection, Smithsonian Institution Libraries. *Right:* Calvin Fuller, Gerald Pearson, and Daryl Chapin invented the first working solar cell of silicon in 1954. *Source:* http://theenergylibrary.com.

energies higher than semiconductor band gap and (ii) the unabsorbed part of the sunlight spectrum for photon energies lower than the band gap.

Queisser and Shockley reported in 1961 a theoretical calculation of the maximal efficiency to be achieved by *pn*-junctions upon sunlight illumination [46]. Calculations were made under the assumption that the solar cell behaves as a blackbody and that recombination takes place exclusively by a radiative process. Under these assumptions, the reasoning line to calculate the yielded current density starts from the expression

$$j_{ph} = e \times (j_{\gamma,em} - j_{\gamma,abs}) \qquad (1.22)$$

where $j_{\gamma,e}$ and $j_{\gamma,abs}$ represent the emitted and the absorbed photon fluxes, respectively. The flux of emitted photons can be calculated by the following expression [47]:

$$j_{\gamma,em} = \exp\left[\frac{(E_e{}^* - E_h{}^*)}{kT}\right] \times \frac{\Omega}{4\pi^3\hbar^3 c^2} \int_{E_G}^{\infty} \frac{\alpha(h\nu) \times (h\nu)^2}{\exp\left(\dfrac{h\nu}{kT_{em}}\right) - 1} dh\nu = \qquad (1.23)$$

$$= \exp\left[\frac{e}{kT} V_{ph}\right] \times \int_{E_G}^{\infty} \alpha(h\nu) dj_{\gamma}^0(h\nu) = \exp\left[\frac{e}{kT} V_{ph}\right] \times j_{\gamma,em}^{;0}$$

where $E_e{}^*$ and $E_h{}^*$ are the quasi-Fermi levels of electrons and holes, respectively; Ω is the solid angle of incidence; and $\alpha(h\nu)$ is the spectral absorption coefficient. The first exponential multiplier of Eq. (1.23) represents the deviation of the emitted photon flux at thermal equilibrium conditions. The latter is equal to the absorption flux of a blackbody, which for the case of a semiconductor corresponds to the integration over the photon density given by the Planck's distribution upon photon energies $h\nu \geq E_G$.

On the other hand, the absorbed photon flux provided by the sunlight spectrum can be obtained by its integration according to

$$j_{\gamma,abs} = \int_{E_G}^{\infty} dj_{\gamma}^{;0} + \int_{E_G}^{\infty} dj_{\gamma,sun}(h\nu) \qquad (1.24)$$

In this expression the first integral represents the absorbed photon flux under thermal equilibrium with the surrounding environment and the second integral represents the additional absorbed radiation provided by the sun.

From the combination of Eqs. (1.22), (1.23), and (1.24) we have

$$j_{ph} = e \times \left[\exp\left(\frac{e}{kT} V_{ph}\right) - 1\right] \times \frac{\Omega}{4\pi^3\hbar^3 c^2} \int_{E_G}^{\infty} \frac{(h\nu)^2}{\exp\left(\dfrac{h\nu}{kT_{em}}\right) - 1} dh\nu - e \times \int_{E_G}^{\infty} dj_{\gamma,sun}(h\nu) \quad (1.25)$$

The last term of Eq. (1.25) is calculated by numerical integration of the sunlight spectrum. Equation (1.25) provides an expression for the ideal current-voltage characteristic from which the maximal power point can be calculated. From the condition of a maximum, $\partial(j_{ph} \times V_{ph})/\partial V_{ph} = 0$, the maximal power point (j_{mp}, V_{mp}) is obtained and hence the maximal efficiency:

$$\eta\% = \frac{j_{mp} \times V_{mp}}{\int_0^\infty (h\nu)\, dj_{\gamma, sun}(h\nu)} \times 100 \tag{1.26}$$

The emission spectrum AM0 shown at the left side of Figure 1.9 corresponds to that of the sunlight over the atmosphere. In the figure, it was replaced by its equivalent blackbody emission at a temperature of 5800 K. The AM1.5 spectrum [48], with a power of 84.4 mW cm^{-2}, represents terrestrial conditions after which solar cell efficiencies are commonly evaluated. The right side of Figure 1.9 presents a comparison of the theoretical maximal efficiencies as a function of the semiconductor band gap after the Queisser–Shockley calculations for single junctions and the real efficiencies reached to date for some semiconducting materials.

Solar cells based on crystalline (Czochralski) and polycrystalline silicon still dominate the photovoltaics market. This can be viewed as the heritage of microelectronics. The considerable progress in the fabrication technology of silicon, and the extensive knowledge of the physicochemical properties of this material impulsed by this branch, were advantageously used by the emerging photovoltaic industry. Large improvements introduced in the surface passivation process and ingenious modification of the cell optics have led to an increase in the efficiency of single *pn*-junction cells up to a current value of 25% in the case of passivated emitters rear locally

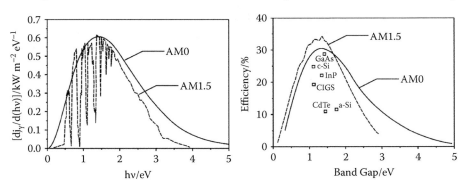

FIGURE 1.9 *Left:* Solar spectrum AM1.5 (dashed line) shown in comparison with that of a blackbody at 5800 K (comparable to the solar spectrum AM0). *Right:* Maximal theoretical achievable efficiency of a solar cell based on a single semiconductor junction as a function of the band gap for the AM0 and AM1.5 spectra (after the calculation by Shockley and Queisser. 1964). Squares represent the presently achievable efficiency for some semiconductor materials.

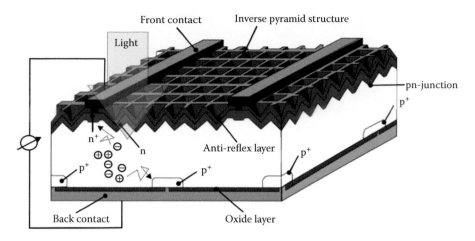

FIGURE 1.10 Schematic of a PERL-cell depicting its main features.

diffused (PERL)-cell [49,12]. This cell constitutes considerable progress in view of the theoretical maximal achievable efficiency of 33% predicted by the Queisser–Shockley limit of 33% for a *pn*-junction.

The most important features of this cell are pointed out in Figure 1.10. The front side of the cell is structured by an etching process to form an arrangement of inverse pyramids. This structure improves the absorption properties by an optimized capture of reflected light beams. The surface is additionally passivated with a Si_3N_4 layer, which diminishes surface reflection. The operating *pn*-junction is formed by inward diffusion of P into the p-type silicon wafer. The back electric contact is made by deposition of an Al-layer. High-doped p$^+$-type points are made by fire contact on the back side of the cell for efficient trapping of photogenerated holes. The localized heat treatment is performed at a temperature below the eutectic temperature of the Al-Si system. Thus, during the sintering process, the aluminum reduces the silicon dioxide layer through pinholes and contacts to the underlying silicon substrate. The p$^+$-p-contact introduces an electron barrier owing to an upward band-bending at the back side that diminishes the recombination process.

The low absorption coefficient and, until the 1980s, still high production costs of high-quality silicon have motivated the search for high-absorbing (compound) semiconductors, stressing to the fabrication of thin film structures with smaller consumption of material than for *first-generation* cells. The progress made in thin film technology provided the necessary elements for the birth of so-called *second-generation* cells. Amorphous Si, CdTe, $CuInSe_2$, GaAs, InP, and $Cu(In,Ga)Se_2$ are the most investigated materials. This technology has also offered a possibility for the construction of integrated assemblies of thin film modules, which constitutes another advantage of this type of cells. The copper indium gallium selenide (CIGS) cells, as well as those constructed on the basis of amorphous and microcrystalline silicon, with actual efficiencies of 19.4% [50] and 11.7% [51] respectively, are the best representatives of this technology. Some construction details and the operation mode of an in-series interconnected thin film CIGS cell are depicted in Figure 1.11.

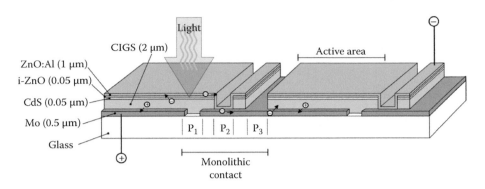

FIGURE 1.11 Schematic showing the construction details of an in series integrated CIGS-cell. P_1, P_2, and P_3 indicate the structuring steps for the realization of a series connection.

Here, the absorber consists of polycrystalline $CuIn_{1-x}Ga_xSe_2$, whose band gap can be varied according to the concentration relation $x = (at\% \ In)/(at\% \ Ga)$ from 1.03 eV for $x = 0$ to 1.7 eV for $x = 1$. The material is made p-conducting by introducing a small excess of Se in the structure. The absorber material is deposited by co-evaporation on a Mo-layer, which was previously deposited by sputtering on a glass supporting substrate. The *pn*-junction is formed by chemical bath deposition of a CdS layer, a low band gap n-semiconductor. The construction is completed by a front contact made by deposition of a transparent conductive i-ZnO/ZnO:Ga junction by RF-sputtering. In spite of the many advantages of this type of cell, further improvements are still required in the manufacturing technology to outweigh the predominance of silicon-based cells.

A more efficient absorption of the solar spectrum may be attained by assembling several semiconductor layers with a stepped configuration of band gaps. This concept, known as a *tandem* cell, enables a reduction of thermalization losses and, hence, an increase of the conversion efficiency beyond the Shockley–Queisser limit. A schematic visualization of this concept is shown in Figure 1.12. The whole sunlight spectrum could be theoretically absorbed by an in-series connection of *pn*-junctions with step-decreasing band gaps of the active p-type semiconductor, with the higher band absorber at the illuminated front.

The *tandem* concept belongs to the group of *third-generation* cells. Historically, this name was used to identify a group of concepts based on the generation of many energy levels, seeking a projection of the conversion efficiency over the Shockley–Queisser limit by a broadening of the absorbed fraction of the sunlight spectrum.

The theoretical maximal achievable efficiency of a *tandem* assembly of n thin film cells can be calculated by maximizing the cell power output, given by [52]

$$P = j_{ph} \times (\mu_1 + \mu_2 + \dots \mu_n) \tag{1.27}$$

where j_{ph} is the photocurrent density flowing through the system. In Eq. (1.27), μ_i represents the chemical potential of the absorbed or emitted radiation by or from the *i*-cell [53,54], which is also equivalent to the chemical potential of the electron-hole

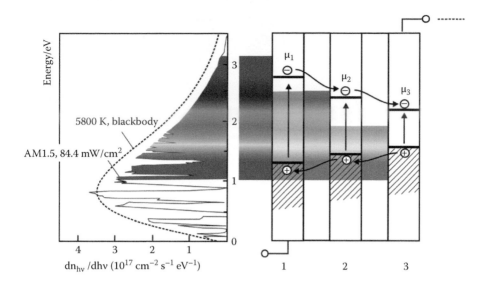

FIGURE 1.12 **(See color insert.)** Schematic showing the working principle of a tandem-cell. The solar spectrum is indicated as the energy density of photons as a function of photon energy.

pair, $\mu_e + \mu_h$, and at the same time to the energy distance between the quasi-Fermi levels. The light-induced current density at each cell component can be calculated from the difference between the absorbed and the emitted radiation, as proposed by Shockley and Queisser [46]. Let us consider a semiconductor as a blackbody at temperature T_c. Thus, the photon flux emitted by this blackbody per unit surface area into a hemisphere over the energy range $E_G \rightarrow \infty$ for the component n is given by

$$j_{\gamma,n}(E_{G,n}, \infty, T_c, \mu) = \frac{2\pi}{h^3 c^2} \int_{E_{G,n}}^{h\nu=\infty} \frac{(h\nu)^2}{\exp\left[(h\nu - \mu_n)/kT\right]-1} dh\nu \qquad (1.28)$$

Here, the integration is made upon energies higher than the band gap $E_{G,n}$. The current density can be written as the difference between the absorbed photon flux coming from the sun and the flux emitted by the semiconductor n as

$$j_{ph} = e \times \left[j_\gamma(E_{G,n}, \infty, T_s, 0) - j_\gamma(E_{G,n}, \infty, T_c, \mu_n) \right] \qquad (1.29)$$

where T_s is the sun temperature. If this expression is written for each i-cell, a series of n-equations is generated. On condition that the current is the same for each i-cell component, the maximum efficiency and the optimized band gaps can be calculated by an iterative method, as explained in Ref. [52]. Further assumptions for this calculations are that: (i) the recombination process is exclusively radiative; (ii) all the light with energies higher than the band gap is absorbed and all the light with energies below the band gap goes through; (iii) the difference in the quasi-Fermi levels at the place where photocarriers are generated equals those at the place of extraction

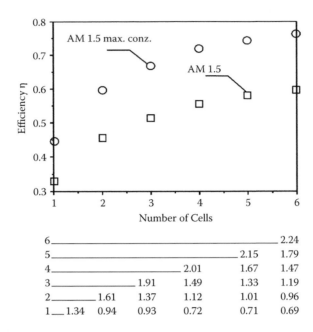

6					2.15	2.24
5					2.15	1.79
4				2.01	1.67	1.47
3			1.91	1.49	1.33	1.19
2		1.61	1.37	1.12	1.01	0.96
1	1.34	0.94	0.93	0.72	0.71	0.69

FIGURE 1.13 According to theoretic calculations, it is possible to achieve efficiencies up to 59.2% by the stacking of 6 cells for an AM1.5 spectrum. At an ideal maximal sunlight concentration, it would be possible to extend this value up to 72%. Optimal band gaps are indicated below in eV.

of those carriers (this can be ensured by infinite mobility of carriers), and (iv) each absorbed photon generates a single electron-hole pair. The results of this calculation are shown in Figure 1.13 as a function of the number of cells for an AM1.5 spectrum with and without concentration. For maximum efficiency conditions the solid angle of photoabsorption must equal the solid angle of photoemission; this corresponds to a concentration ratio of 46/200; under one sun conditions the cell is assumed to emit photons over the entire hemisphere. Note that efficiencies of about 60% are attained with the stacking of six cell-unities under one sun.

One of the successful realizations of the *tandem* concept was made by means of organometallic vapor phase deposition of III-V semiconductors. For instance the GaInP/GaInAs/Ge system ranks as one of the highest-efficiency cells with a conversion of 40.7% at 240-sun [55] (1 sun = 100 mW cm^{-2}).

Other cell concepts are grouped in the category of the *third-generation* cells based on (i) the excitation of several electron-hole pairs by photons of high energy [56] and (ii) the capture of photogenerated charge carriers before they thermalize. As already mentioned, all concepts within the *third-generation* group aim to increase the conversion efficiency by reducing thermalization losses. Figure 1.14 presents a scheme depicting the general working principles of these concepts.

Concepts based on carrier multiplication follow those based on the tandem concept. This phenomenon was observed for the first time in bulk semiconductors in the 1950s. This effect would provide increased power conversion efficiency by increasing

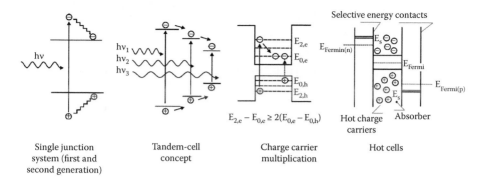

Single junction   Tandem-cell   Charge carrier   Hot cells
system (first and   concept   multiplication
second generation)

FIGURE 1.14 Schematic representation of several concepts for the reduction of thermalization losses (third-generation cells).

the solar cell photocurrent. As supported by experimental evidence, nanosized semiconductor crystals (*quantum dots*) provide a regime where carrier multiplication can be enhanced by impact ionization. One of the first reports describes the appearance of this effect in quantum dots of PbSe [57]. Afterward, observations of this phenomenon in quantum dots of II-VI semiconductors [58] and silicon [59] were reported. The multiplication process consists basically of an Auger-type process whereby a high-energy exciton (electron-hole pair), created in a semiconductor by absorbing a photon of energy $h\nu \geq 2\,E_G$, relaxes to the band edge via energy transfer of at least $1\,E_G$ to a valence band electron, which is excited above the energy gap. As a result, two electron-hole pairs are formed for one absorbed photon and a larger fraction of the high photon energy part of the sunlight spectrum is converted into electrical or chemical energy.

The last scheme in Figure 1.14 represents the working principle of the *hot carrier* cells. It consists of the slowdown of the thermalization of highly energetic photo-generated charge carriers, owing to their interaction with the phonons, so that they can be caught at energies higher than that of the semiconductor band gap [60]. This concept was realized by formation of selective energy contacts based on resonance-tunnel level, which arise from a single layer of quantum dots [61,62].

The extension of the spectrum absorption by introduction of localized energy levels within the band gap is another of the many approaches thought to overwhelm the Shockley–Queisser limit [63]. However, the concomitant increase of recombination events at defects makes it impossible to reach conversion improvements beyond of 1–2% [64–67]. Theoretical calculations have shown, on the other hand, that relevant improvements could only be reached if semiconductors with band gaps of 1.9 eV to 2.5 eV are used [68].

The *quantum-well* concept is a modified and more modern implementation of the above principle. It consists of a multi-quantum well system in the undoped region of a *p-i-n* solar cell construction. As shown schematically in Figure 1.15, the working principle of this type of cell is based on the stacking of nanodimensioned semiconductor films that lead to quantum confinement and to the formation of discrete energy levels. The light-induced electron excitation in the quantum wells with the lowest confined states determines the absorption threshold of the cell. This can be

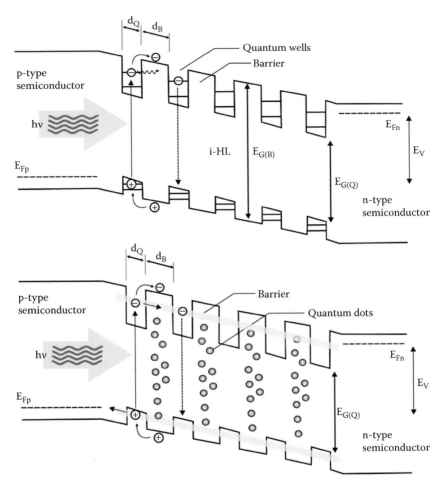

FIGURE 1.15   Working principle of a multiple-quantum-well solar cell.

simply tailored by the suitable choice of quantum well (d_Q) and barrier width (d_B) [69–71]. The potentiality of this concept resides in the collection rate of the light-induced charge carriers, which in the original concept is determined by the thermoionic emission of charge carriers across the potential barrier of the quantum-well structure.

The maximal achievable efficiency in quantum-well systems is still a controversial matter [70,72]. It has been clearly established that the short-circuit current can be enhanced over that of conventional, homojunction cells formed from the barrier material of a band gap $E_{G(B)}$. However, the question of the output voltage of an ideal quantum-well solar cell is still being discussed. This voltage is determined by the variation in the electrons and holes quasi-Fermi level separation across the cell, and it is precisely this behavior that is still not known. It was shown [73], for instance, that under the supposition of exclusively radiative recombination, an efficiency enhancement above the limit for an ideal single band-gap cell was possible only for narrow wells. Calculations based on radiation-coupled well-barrier transitions, however, predict maximal efficiencies of 63.2% [74,75].

The reduction of the thickness of the barrier material (d_B) greatly increases the tunnel probability of photogenerated carriers between adjacent wells, and the carriers are no longer localized in individual wells. Hence, carriers are easily spread throughout the whole superlattice by formation of continuous mini-bands, providing high conductivity and improved performance of the superlattice solar cell. This concept was realized in a tandem system consisting of a quantum-well superlattice of AlGaAs/GaAs coupled with a superlattice of InGaAsP/GaAs [76].

A similar effect can be achieved by embedding nanodimensioned semiconductor crystals (quantum dots) in the barrier material. Though this concept is difficult to realize because of the significantly high quantum dot density to be inserted ($\sim 10^{19}$ cm^{-2}), some attempts have been reported [77,78]. This is surely motivated by the high theoretical maximal achievable efficiency of around 87% [79,80].

1.3 ADVANCED CONCEPTS FOR SUNLIGHT ENERGY CONVERSION

The group of so-called third-generation cells includes a battery of other concepts such as dye-sensitized cells and systems based on organic semiconducting polymers. A common characteristic of this type of conversion system is the use of abundant and low-cost materials. On the other hand, the conversion mechanism differs from those based on inorganic semiconductors in the generation and transport of excitons. In principle, the existent cells can be classified in two main groups: (i) solid-state cells based on organic semiconductors, and (ii) photoelectrochemical cells, which use dyes supported on inorganic wide band gap semiconductors as light-absorbing materials.

Organic semiconductors consist of long molecular chains, which are held together by short-range van der Waals forces. Therefore, no continuous bands appear, as in the case of inorganic semiconductors where covalent and ionic bonds predominate, but instead, discrete energy levels result from the coupling of the highest occupied (HOMO) and low unoccupied molecular orbitals (LUMO).

When two metals of different work functions are connected by an interfacial organic semiconductor film, the resulting electronic structure resembles that of a *pin*-junction. Thus, light-induced charge carriers formed inside the semiconductor are transported by a thermal-activated hopping over the localized energy states, assisted by the induced electric field (see the left-side energy diagram depicted in Figure 1.16). Thus, organic semiconductors present low charge carrier mobility between 10^{-5} $cm^2V^{-1}s^{-1}$ and 10^0 $cm^2V^{-1}s^{-1}$. These values contrast with those of Si, which are typically between 10^3 $cm^2V^{-1}s^{-1}$ and 10^5 $cm^2V^{-1}s^{-1}$.

The first photovoltaic single junction devices made of semiconducting polymers have shown efficiencies lower than 1% [81,82]. The light absorption in organic solar conversion devices, as in the case of dyes, mainly takes place by generation of excitons. Owing to the low dielectric constant of these materials, the Coulomb binding energy of photogenerated electron-hole pairs is especially high (up to 0.4 eV). Thus,

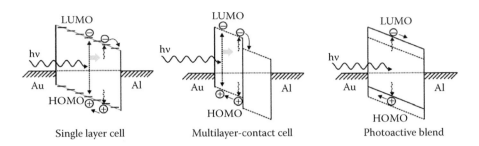

FIGURE 1.16 Schematic representation of the working principle of organic solar cells. Here, the light-induced excitons diffuse toward existent defects (*left*) or heterojunctions (*middle*) where they can be split. *Right:* The intimate mixing of two conducting polymers generates a homogeneous distribution of the energy-step, thus increasing the effectiveness of the exciton splitting.

the probability of dissociation in positive and negative polarons (mobile charge carriers), which is accomplished at chemical and structural defects, is very low. The diffusion length of excitons in polymers and disordered molecular solids varies between 1 nm and 100 nm. This introduces a limitation in the design distances between the absorption and collection sites, which should be maintained within these values to reduce losses by recombination. Efficient organic solar cells were prepared by nano-structured heterojunctions of an electron donor with an electron-acceptor material [83]. The photogenerated excitons in one of the cell components diffuse toward the contact between the two substances, whereby splitting occurs because of the jump of the HOMO and LUMO energy levels [84] (see also the schematic representation in the middle of Figure 1.16). The negative polaron is further transferred to the acceptor, while the positive one, which remains at the donator material, is driven by the electric field toward the cathode.

The present technology for the construction of heterojunction organic solar cells is based on an intimate mixing of donor and acceptor materials [85,86]. In this way, excitons can be split within their diffusion length at the hetero-contact. Both phases must be blended together to yield a dispersed continuous path for the transport of separated charges to the contacts. Isolated domains can trap charges, causing recombination. Polythiophene derivates and C_{60}-fullerenes are typically used as donor and acceptor materials, respectively [87], with which efficiencies up to 3.5% are achieved. Another frequently used system is the combination of copper-phthalocyanine (donator) with PTCBI (3,4,9,10-perylen-tetracarboxyl-bis-benzomidazol) (Figure 1.17), with an efficiency of 5% [88,89]. In this case, the introduction of a high conducting polymer layer of PEDOT:PSS (poly-3,4-(ethylendioxy)-thiophene with polystyrene-sulfonate) between the photoactive blend and the conducting ITO-glass (indium tin oxide) improves the cell efficiency because of the higher work function of the polymeric layer. On the other hand, the ultra-thin LiF film decreases the work function of the cathode and also improves the electrical contact, avoiding the direct contact of C_{60} with the metal, which might lead to quenching of photogenerated excitons.

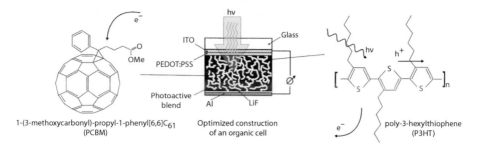

FIGURE 1.17 Schematic of an organic solar cell based on a blend of an electron donor and acceptor material.

The working principle of dye-sensitized cells involves the light-induced generation of excitons as well. Here, however, the exciton splitting succeeds at the contact of an inorganic wide band gap semiconductor, such as TiO_2, ZnO [90], or Nb_2O_5 [91] with a redox-electrolyte; this system may be considered the first practical realization of a photoelectrochemical solar cell. One of the most discussed cases was named after its inventor: the Grätzel-cell [92,93]. As shown schematically on the right side of Figure 1.18, the active part of this cell consists of a layer constituted by sintered nanoparticles of TiO_2 deposited on a glass plate covered with transparent conducting oxide (TCO). After adsorption of the dye on the oxide particles by a dipping step, the photoelectrode is next impregnated with the redox electrolyte and pressed together with a Pt- or a TCO-glass covered with graphite to form a thin electrochemical cell.

The porous layer offers a conducting support with a large active dye-acceptor area. The N3-ruthenium complexes and their derivates are usually used as dye [94].

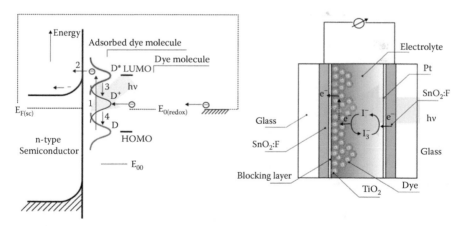

FIGURE 1.18 **(See color insert.)** Basic electronic processes taking place at a semiconductor–electrolyte interface in a dye-sensitized photoelectrochemical cell. The energy band diagram (*left*) represents a cell in the short circuit case. The photon absorption brings the dye-molecule (D) to its excited state (D*). The electron transfer to the semiconductor conduction band (2) leaves an oxidized dye-molecule (D+) (3), which is returned to its original state by reaction with the redox-electrolyte (I_3^-/I^-) (4). *Right:* Construction details of a dye-sensitized cell.

Photons of lower energy than the band gap of the supporting semiconductor may excite electrons from the HOMO level of the dye-molecule up to the LUMO level. It must be remarked that the localized levels of a single molecule spread out following a Gaussian distribution after the adsorption on the semiconductor. The distribution width depends on the strength of the bond involved in the attachment. In some cases, the adsorption process also induces a shift of the energy levels. In particular, the broadening of the excitation level (LUMO) depends on the electronic interactions with the empty states of the semiconductor conduction band and may be determined experimentally by high-resolution optical spectrometry. On condition that the excited level lies over the conduction band edge the electron is transferred by tunneling within some ps; the live-time of the excited state is typically on the order of nanoseconds. The injected electron, which is a majority charge carrier, is taken out rapidly from the interface owing to the electric field. The probability of a recombination of this electron from one determined position in the semiconductor surface with the hole left on the oxidized dye-molecule is very low, because of the unfavorable Franck–Condon overlapping between the donor and acceptor states, separated by a large energy gap (higher than 1.5 eV) [95,96]. The oxidized dye-molecule (D$^+$) resulting after the electron transfer returns to its original state by a reduction reaction with the redox-electrolyte. The regeneration reaction should be sufficiently rapid in order to avoid a side reaction, which might lead to undesirable instability effects.

Since its first report, continuous improvements are being reported for this cell concept. Last reports noted a record efficiency of 10.4% [12].

In spite of the evident advantages of this cell, such as the utilization of abundant and low cost TiO_2 and its easy assembly, the toxicity of the used electrolyte (iodide/iodine solution in acetonitrile) is a matter for further work. Furthermore, efficient dyes contain Ru, an expensive and scarce material [3].

Now, we return briefly to the inorganic materials, since an alternative to replace the organic dyes for electron injection in the conduction band is the use of nanosized semiconducting particles. The reduction of the semiconductor dimensions under the de Broglie wavelength leads to a quantization of the permitted energy levels. The influence of the dimensionality on the electronic properties of nanostructures can be exemplified by solving the Schrödinger equation of an electron gas of radius R surrounded by an infinitely large potential wall. The resulting energy levels are given by

$$E_n = \frac{\hbar^2}{2m_e}\left(\frac{\pi n}{R}\right)^2 \qquad (1.30)$$

with $n = 1,2,3,\dots$. In the case of semiconductors, a similar procedure is applied for the calculation of the exciton energy levels. Assuming that the wave function of an exciton can be written as $\psi_{exc} = \psi_i(r_e)\psi_j(r_h)$, where r_e and r_h represent the radial positions of the negative and positive polarons, respectively; from the solution of the Schrödinger equation the following expression for the discrete energy levels results:

$$E_{exc} = \frac{\hbar^2}{2}\left(\frac{n\pi}{R}\right)^2\left[\frac{1}{m_e^*}+\frac{1}{m_h^*}\right] - \frac{1.786e^2}{\varepsilon\varepsilon_0 R} - 0.248E_{Ry} \qquad (1.31)$$

where E_{Ry} is the Rydberg energy, given by $e^4/[2\ \varepsilon\varepsilon_0\hbar^2(m_e^{-1} + m_h^{-1})]$. $m_{e,h}^*$, where is the effective mass of electrons or holes, and ε is the relative dielectric constant of the medium. The first term of this equation represents the quantum effects of a particle in a potential box of size R; the second term represents the screened Coulomb interaction which stabilizes the electron-hole pair; and the third term is the polarization energy, which is generally very small. Although Eq. (1.31), derived from the effective mass approximation contains the basic physics of quantum size effects, it does not reproduce exactly the experimentally measured exciton energy as a function of the dot size. This is basically a consequence of the deviation of the energy band from the parabolic form extrapolated from the effective mass near $k = 0$. According to this expression, it might be possible to tailor the size of semiconductor particles to get, for instance, an optimal value of the band gap of 1.35 eV (maximum of the theoretical achievable efficiency after the Shockley–Queisser limit) in the case of a single pn-junction solar cell.

The working principle of a solar cell based on a wide band gap semiconductor sensitized with nanosized semiconductor crystals is shown schematically in Figure 1.19 for the case of TiO_2 sensitized with CdS nanoparticles [97]. The dimension reduction leads to the formation of discrete energy levels above and below the conduction and valence band of a bulk semiconductor, respectively. Electrons excited by illumination to discrete energy levels in the conduction band may be transferred to the conduction band of the supporting semiconductor after a possible thermalization process. Holes left in the discrete valence band levels are trapped by the redox system. The rightmost part of Figure 1.19 presents the experimental determination of the increase of the energy distance between the high occupied from the lowest unoccupied energy level, taking the band gap of the bulk semiconductor as a reference. It can be observed that the gap may be doubled up at sizes lower than 2 nm. It should be noted that this example in particular does not present significant advances from the practical point of view, since CdS nanoparticles absorb principally in the high ultraviolet region. However, it constitutes an elegant way to exemplify this phenomenon.

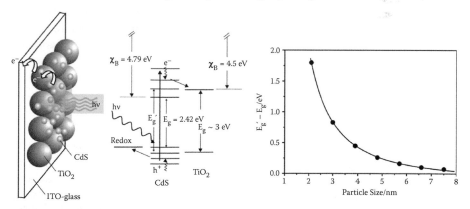

FIGURE 1.19   (**See color insert.**) *Left:* Working principles of a photoelectrochemical cell based on wide band gap semiconductor sensitized with nanosized semiconducting crystals. *Right:* Experimentally measured band gap as a function of the size of CdS particles (adapted from Ref. 97).

The application of colloidal particles of semiconductors as sensitizing material on TiO_2-based photoelectrochemical cells was proposed in 1986 in a pioneering work by Gerischer [99]. Quantum dots of other semiconducting materials such as CdSe [100,101], PbS [102], InAs [103], and InP [104] may be considered as promising for the construction of solar cells based on the sensitization concept.

The quantum confinement and the strong Coulomb electron-hole interactions in nanosized semiconductors lead to an enhancement of the Auger-processes, and consequently to multiple exciton generation. This phenomenon was found to appear in nanocrystals of PbSe, PbS, PbTe, and CdSe [104]. The multiple exciton generation has received a great deal of interest since it may lead to quantum efficiencies up to 300%, as have been predicted for particles of PbSe of 3.9 nm size at an incident radiation $h\nu = 4\,E_G$ [105].

In organic solids, on the other hand, the decay of a photogenerated exciton in two triplet states is a frequent process. This phenomenon, known as exciton fission, was discovered for first time in 1965 on illuminating anthracene with a laser [106]. However, the first reports were focused in investigation of the unique resulting magnetic properties [107,108].

Single exciton splitting was also observed in several polymers such as polydiacetylene [109], poly(p-phenylene vinylene), and poly(p-phenylene) [110] as well as in charotenoides [111] and covalent tetracene [112] and 1,3-diphenylisobenzofuran dimmers [114].

The exciton fission is represented by an energy band diagram as shown Figure 1.20. Here, it is assumed that the molecules behave as harmonic oscillators, and thus the potential energy is represented as a parabola. Electrons are excited from their ground states S_0 to a permitted energy vibration level (S_1^*) of the excited state (S_1) by illumination. As the excitation energy increases over the energy threshold $E_{(S1)} + 2E_{(T1)}$, there is certain probability that the singlet splits in two triplet states T_1. Exciton fission is spin-allowed and presents some advantages in comparison with its related inorganic multiple exciton generation. First, it can occur from the relaxed singlet state. Therefore, there is no competition with other rapid intramolecular processes like vibrational relaxation. Second, exciton fission occurs without the constraint of a quantum confinement and the newly produced triplets are able to diffuse away from each other into the bulk solid, thus reducing the probability of self-annihilation, as commonly observed in the inorganic case. The exciton fission has opened a new alternative for further development of the dye-sensitized-cell concept, on condition that each of both triplet states can inject an electron independently. The introduction of an additional dye molecule that shows single fission promises to increase the maximal theoretical achievable efficiency of 33.4% to 44.4% [103].

1.4   INORGANIC CONCEPTS

Among the numerous concepts conceived for the conversion of solar energy into chemical energy carriers, the inorganic concept based on monolithic tandem pn-junctions is highly feasible from the point of view of its stability and high efficiency. Its realization requires an adequate selection of n-type and p-type semiconductor absorbers, so that the band edges lie above and below the redox-level of the hydrogen reduction and the water oxidation reactions respectively, that is, $E^s_{CB\,(p-sc)} > E_{H2/H}{}^+$ and $E^s_{VB\,(n-sc)} < E_{H2O/}$

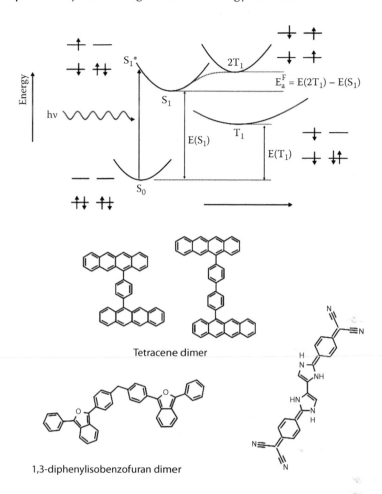

FIGURE 1.20 Exciton fission process in organic solids (top). Some compounds showing this phenomenon (bottom).

O_2. Figure 1.21 shows an energy band diagram and the main electronic processes for an illuminated idealized pn-junction. Under electronic equilibrium both phases adopt the same Fermi-level, requiring a band bending at the limit with the electrolyte. The alignment of the Fermi levels at the contact will be achieved by a tunnel p^+-n^+-junction. This type of contact ensures on the one hand a rapid annihilation of the majority charge carriers generated by separation of photo-induced electron-hole pairs. On the other hand, this contact also avoids the formation of a space-charge region, which introduces a potential barrier for the majority charge carriers and drain for the minority charge carriers. The total photovoltage generated by illumination of the junction is given by the distance between the quasi-Fermi-levels at both semiconductor–electrolytes interfaces, equal to the addition of the respective semiconductor photopotentials: $V_{ph} = V_{ph\ (p-sc)} + V_{ph(n-sc)}$. The minimum voltage required

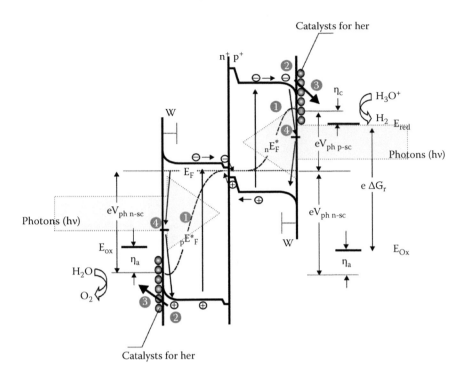

FIGURE 1.21 Inorganic concept for the conversion of solar energy into chemical energy based on a monolithic construction. The photogenerated minority charge carriers (1) diffuse toward the semiconductor surface (2), where they are caught by the electric field and further transported to the catalyst–electrolyte interface and consumed by the electrochemical reaction (3). Some of the excited minority charge carriers recombine at the interface states (4).

for the junction working at a given generation rate must be higher than that for the onset of the water splitting to overcome the electrochemical overtensions at each electrode surface and the ohmic resistance:

$$V = \Delta G_r / e + \eta_a + \eta_c + j\Delta\Omega \tag{1.32}$$

where $\Delta G_r / e$ represents, from a thermodynamic point of view, the reversible work to be made to reverse the reaction driving force; η_a and η_c are the anodic and cathodic reaction overpotentials, respectively; and $j\Delta\Omega$ is the total ohmic overvoltage. In the semiconductor, one can differentiate between the space-charge layer and a field-free zone (see Figure 1.22). Photogenerated charge carriers inside the space-charge layer will be rapidly collected by the electric field and driven to the reaction centers (drift) where they are partially consumed by the electrochemical reaction, whereas an important part will be annihilated in the recombination centers. In the field-free area, charge carriers diffuse in all directions. The maximal distance covered is given by $d_{diff} = \sqrt{D_{e,h}\tau_{e,h}}$, where $D_{e,h}$ is the diffusion coefficient of electrons (holes) and $\tau_{e,h}$ is the carrier lifetime, determined by the energy distribution and concentration of recombination centers. The concentration profile of charge carriers is essentially determined by the relation between the spatial dependence of the generation rate, the charge carrier diffusion, and the recombination rate. A simplified analysis based on the solution of the continuity equation for calculating the photocurrent generated in the field-free semiconductor is:

$$\frac{1}{e}\vec{\nabla}\cdot\vec{j}_{e,h}+G-R=0 \tag{1.33}$$

where $\vec{j}_{e,h}$ is the charge carrier flux vector and G and R are the bulk generation and annihilation rate, respectively. The generation rate is calculated from the Lambert–Beer relation $J_{ph}=J_{ph}{}^0\exp\left(-\alpha_{abs}\,z\right)$, as $G=-dJ_{ph}/dz$, where J_{ph} is the photon flux and α_{abs} is the absorption coefficient of the semiconducting material. In the field-free zone, the current is given by diffusion of the charge carrier. Thus, $\vec{j}_{e,h}=-D\vec{\nabla}(\Delta p)$, where Δp is the excess of minority charge carrier generated under illumination. Replacing the last expressions in Eq. (1.33) and solving the second order differential equation with the boundary limits $\Delta p=0$ for $z\to\infty$ and $\Delta p=\Delta p^0$ for $z=z_w$, we obtain

$$\Delta p=\Delta p^0\exp\left(\frac{z_W-z}{d_{diff}}\right)+A\left[\exp\left(-\alpha_{abs}z_W+\frac{z_W-z}{d_{diff}}\right)-\exp\left(-\alpha_{abs}z_W\right)\right] \tag{1.34}$$

where

$$A=\frac{\alpha_{abs}J_{ph}^0 d_{diff}^2}{D(d_{diff}^2\alpha_{abs}^2-1)}$$

The total current driven at the semiconductor surface is given the diffusion and the drift component. This latter is obtained by integration of the generation rate inside the space-charge layer, z_W. Thus, we arrive at

$$j_s=e\left[D\left(\frac{d\Delta p}{dz}\right)_{z=zW}+J_{ph}^0\left[1-\exp\left(-\alpha_{ads}z_W\right)\right]\right] \tag{1.35}$$

which, after the introduction of the derivative of Eq. (1.34), converts to

$$j_s=eD\left[A\left(\alpha_{ads}-\frac{1}{d_{diff}}\right)\exp(-\alpha_{ads}z_W)-\frac{\Delta p^0}{d_{diff}}\right]+eJ_{ph}^0\left[1-\exp\left(-\alpha_{ads}z_W\right)\right] \tag{1.36}$$

This expression can be further reduced to the form

$$\xi=\frac{j_s}{eJ_{ph}^0}=1-\frac{\exp(-\alpha_{ads}z_W)}{(d_{diff}\alpha+1)} \tag{1.37}$$

where ξ is the effective quantum yield. The depletion layer width of the semiconductor z_W is given by

$$z_W = \left[\frac{2\varepsilon_{sc}\varepsilon_0}{eN_A} \left(\Delta\varphi_{sc} - 2\frac{kT}{e} \right) \right]^{1/2} \approx \left(\frac{2\varepsilon_{sc}\varepsilon_0}{eN_A} \right)^{1/2} \sqrt{\Delta\varphi_{sc}} \qquad (1.38)$$

where N_A is the acceptor concentration of the p-type semiconductor. Thus, Eq. (1.37) can be rewritten as

$$-\ln(1-\xi) = \ln(1+d_{diff}\alpha) + \alpha \left(\frac{2\varepsilon_{sc}\varepsilon_0}{eN_A} \right)^{1/2} \sqrt{\Delta\varphi_{sc}} \qquad (1.39)$$

This expression is usually used to experimentally determine the parameters α and d_{diff} by plotting $-\ln(1-\xi)$ vs. $|(V_{\text{fb}}-V)|^{1/2}$ obtained by illumination of the semiconductor–electrolyte interface with monochromatic light of different wavelengths. This model offers a good approach for the description of the photoelectrochemical behavior of semiconductor–electrolyte interfaces at high band bendings. However, it fails in reproducing the experimental behavior near the flat band potential, where the effects of surface recombination and bulk recombination inside the space-charge layer cannot be disregarded. This model was originally developed for solid-state junctions and later adapted to semiconductor–electrolyte junctions. But a direct translation does not include some inherent phenomena that appear at the semiconductor–electrolyte interface, such as the interfacial charge accumulation due to a slow electron transfer to the electrolyte and the reduction of the band bending by a related increase of the interfacial potential. Reichman [114] has revised the application of Gärtner's theory for semiconductor–electrolyte interfaces, taking into account the rate of hole and electron transfer from the band edges of the semiconductor to the energy states of the electrolyte (electrons to the oxidized states and holes to the reduced states). According to this theory, the total photocurrent is given by

$$j = \frac{j_G - j_0 \exp(eV/kT)}{1 + \dfrac{j_0 \exp(eV/kT)}{j_n^0}} - j_p^0 \left[\exp(eV/kT) - 1 \right] \qquad (1.40)$$

where j_G is the Gärtner's current density given by Eq. (1.37), and eV is the difference between the Fermi level of the semiconductor and the work function of the redox reaction (see scheme in Figure 1.22). j_0 is the saturation current density in the dark, given by: $j_0 = e\, n_0\, d_{\text{diff}}/\tau_n$ with n_0 as the equilibrium concentration of minority charge carriers with a lifetime of τ_n. j_n^0 and j_p^0 represent the exchange currents for electrons and holes, respectively. Applying the method of Sah, Noyce, and Shockley [115,116] to calculate the recombination rate in the space-charge region, Reichmann derived the following expression for the total current:

$$j = j_n^0 \left[\left(\frac{-c + (c^2 + 4ab)^{1/2}}{2a} \right)^2 - 1 \right] - j_p^0 \left[\exp(eV/kT) - 1 \right] \qquad (1.41)$$

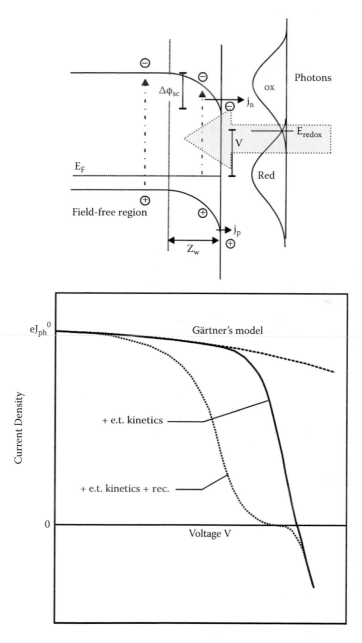

FIGURE 1.22 Energy band diagram for a p-type semiconductor-electrolyte junction under illumination showing the transfer of a photogenerated charge carrier to the energy states of the electrolyte (top). The bottom diagram shows a qualitative representation of the current voltage characteristics of the p-type semiconductor-electrolyte junction after Gärtner's equation and its extensions, considering the electron transfer kinetics and the recombination in the space-charge region.

where $c = \pi \, k \, T \, n_i \, z_W \exp(e \, V/2kT)/4\tau \, \Delta\varphi_{sc}$, $a = j_n^0 + j_0 \exp(eV/kT)$, and $b = j_n^0 + j_G$ (n_i is the intrinsic donor concentration of the semiconductor). Figure 1.22 presents a comparison of the photocurrent response predicted by Gärtner's model and Eqs. (1.40) and (1.41). The Gärtner's current decreases with voltage V due to a reduction of the length of the space-charge layer and a larger recombination in the field-free region. The inclusion of the limiting transfer kinetics in Eq. (1.40) enlarges the recombination in the field-free region due to a buildup of minority charge carriers. The recombination inside the space-charge layer enhances the current decrease near the flat band potential (Eq. 41). It can be noted that the course of the current-voltage characteristic, considering the limiting charge transfer rate and the recombination in the space-charge layer, is similar to that shown in Figure 1.7 derived from adapting the classical equivalent circuit of a solar cell to the photoelectrochemical case.

The electrocatalytic centers in the type of systems shown in Figure 1.21 consist generally of nanodimensioned particles of noble metals or transition metal alloys, which are good electrocatalysts for the hydrogen evolution and the CO_2 reduction. In the latter case, as discussed in more detail in Chapter 4, the reduction process occurs by a mechanism involving the co-adsorption of H; this mechanism is essentially different from that operating at high overpotentials, involving the formation of adsorbed CO_2^- radicals. On the other hand, metal oxides [117–119] and noble metals are commonly used as electrocatalyst materials for the construction of photoanodes (water oxidation on the n-type semiconductor).

The water oxidation follows an especially complex reaction mechanism, involving the transfer of four electrons per evolved oxygen molecule. From the point of view of quantum mechanics, there is no probability of a simultaneous transfer of more than one electron. Thus, the oxidation process implies the formation of adsorbed and aqueous intermediate species formed by successively one-electron transfer steps. Inspired by the natural photosynthetic process, some researchers tried to design new catalyst structures based on Mn [120,121]. The anodic reaction center contained in photosystem II consists of a Mn_4O_xCa-complex, which will be successively oxidized after the absorption of one light photon. The oxidation states, which are traditionally named as S_i (i: 0,1,2,3,4), form a cycle known as the Kok-cycle at the end of which an oxygen molecule is released and the cycle starts again [122]. The final state, S_4, represents a transition state which evolves to the initial S_0 state by liberating O_2 and 2 H^+ with a simultaneous binding of one or two water molecules. The final process takes place within 1–2 ms as determined by time-resolved experiments [123,124]. Therefore, the photosystem II must be able to stabilize, accumulate, and synchronize four holes, which are created within some picoseconds.

The development of an electrocatalytic system for water reduction based on local precipitation of catalytic oxides by anodization of inert electrodes immersed in neutral phosphate solutions containing Co^{2+} represents significant progress in this subject considering the utilization of abundant and environmental friendly substances [125].

The construction of inorganic high efficiency photoelectrodes based on the semiconductor–metal–electrolyte interfaces requires the following:

1. Effective light collection, including the energy excess
2. Low recombination losses
3. Rapid electron transfer of the photogenerated minority charge carriers toward the metal–electrolyte interface

These preconditions are basically fulfilled by a photoelectrode consisting of collector centers made of catalytic metal particles embedded in or deposited on a good passivated surface of semiconductors with a high extinction coefficient in the frequency range of the solar spectrum. Thereby, structures consisting fundamentally in Schottky-type emitters of reduced dimensions in a passivated matrix result. The term *passivated* is used in this case to mean a stable surface free from recombination centers. The passivation process in the case of treatment of semiconductor surfaces is closely related to the formation of an anodic oxide, as researchers from corrosion science are used to. The covering of a semiconductor surface with a stabilizing film, which in most cases consists of an oxide, aims to protect the attack of the semiconductor by the aggressive electrolyte (corrosion) and to maintain a reduced level of the density of recombination states. The formation of small Schottky contacts can be achieved, apart from physical evaporation methods using masks, by photoelectrodeposition of catalytic metals and alloys using previous structured surfaces. In the case of Si, two methods can be considered for surface stabilization: the formation of a thin oxide by thermal [126] or chemical [127,128] treatment, or functionalization by impregnation with organic compounds. The formation of stable adsorbed films of $Si-CH_3$ [129], Si-phenyl [130], and other sylanol-bonds [131] was reported, most of them with recombination rates similar to that found in hydrogenated surfaces.

1.5   ENERGY AND ELECTRON TRANSFER PROCESSES

The energy absorption in semiconductor-based conversion systems occurs by separation of the photogenerated electron-hole pairs by an electric field arising at the interface upon contact of the semiconductor with another phase. The contacting phase can be another semiconductor of the same or a different material, a metal, or an electrolyte. The driving force for the charge motion is the gradient of its electrochemical potential. The excess of photogenerated charge carriers leads to a stationary state with splitting of the Fermi levels for electrons and holes. The basic electronic processes occurring in an illuminated interface will be illustrated for the special case of a *pn*-junction. Figure 1.23 shows an energy band diagram for this contact. Photogenerated electrons in the p-type semiconductor migrate to the n-type semiconductor and vice versa. The concentration of photogenerated minority charge carriers vanishes toward the semiconductor bulk due to recombination. The decay distance is given in each case by the diffusion length L_e for electrons and L_h for holes. The circulating current can be calculated from the longitudinal component of the divergence of the space current density vector j_p of holes (the same analysis is also valid for electrons).

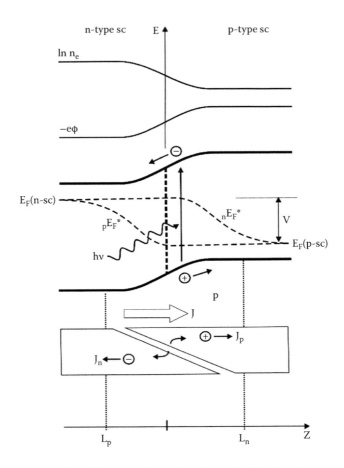

FIGURE 1.23   Energy band diagram of a *pn*-junction under illumination.

$$j = -e \times \int_{-L_p}^{L_n} \nabla \cdot j_p \big|_z \, dz \tag{1.42}$$

The current density vector can be calculated from the continuity equation for holes in steady-state conditions as follows:

$$\frac{\partial n_h}{\partial t} = G_h - R_h - \operatorname{div} j_p = 0 \tag{1.43}$$

where G_h and R_h are the bulk generation and the recombination rates, respectively. The generation rate can be expressed as the sum of the rate under equilibrium without illumination G_h^0 and the additional term generated under illumination:

$$G_h = G_h^0 + \Delta G_h \tag{1.44}$$

On the other hand, the recombination rate can be written as

$$R_h = R_h^0 \times \frac{n_e n_h}{n_i^2} = R_h^0 \exp\left(\frac{\tilde{\mu}_e + \tilde{\mu}_h}{kT}\right) \qquad (1.45)$$

In the dark, phonons and photons are in equilibrium, so that $\tilde{\mu}_e + \tilde{\mu}_h = 0$ and $G_h^0 = R_h^0$. Introducing Eq. (1.44) and (1.45) into Eq. (1.37), we have [132]

$$j = -e \times \int_{-L_p}^{L_e} \left\{ G_h^0 \left[1 - \exp\left(\frac{\tilde{\mu}_e + \tilde{\mu}_h}{kT}\right) \right] + \Delta G_h \right\} dz \qquad (1.46)$$

The sum of electrochemical potentials, $\tilde{\mu}_e + \tilde{\mu}_h$, is independent from the space coordinates and is given by the contact tension V. Thus, the following expression results:

$$j = -e \times G_h^0 (L_e + L_p) \left[\exp\left(\frac{V}{kT}\right) - 1 \right] - e \int_{-L_p}^{L_e} \Delta G_h dz = j_s \left[\exp\left(\frac{V}{kT}\right) - 1 \right] - j_L \qquad (1.47)$$

Equation (1.47) is the well-known expression of the current density for an illuminated diode, where j_L and j_s represent the short circuit and the saturation current of a diode in the reverse-bias region. Under electronic equilibrium in the dark, the recombination rate is determined by the livetime and concentration of the minority charge carriers. Thus,

$$R_{e,h}^0 = G_{e,h}^0 = \frac{n_e^p}{\tau_e} = \frac{n_h^n}{\tau_h} \qquad (1.48)$$

where n_e^p and n_h^n are the electron concentration in the p-type region and the hole concentration in the n-type region, respectively. τ_e and τ_h are the lifetimes of electrons and holes, respectively, which are determined by their diffusion lengths as $\tau_e = L_e^2 / D_e$ and $\tau_h = L_p^2 / D_h$. Therefore, we have

$$j_s = e n_i^2 \left(\frac{D_e}{n_A L_e} + \frac{D_h}{n_D L_p} \right) \qquad (1.49)$$

where n_D and n_A are the donor and acceptor concentrations of the n-type and p-type semiconductors, respectively. The short circuit current density j_L is calculated by integration of the additional generation rate under illumination, $\Delta G_{e,h}(z)$, which can be calculated as a function of the distance z by integrating the absorbed solar light upon the frequency:

$$\Delta G_{e,h}(z) = \int_{h\nu=0}^{\infty} \alpha(h\nu) \left[1 - r(h\nu) \right] \times e^{-\alpha(h\nu)z} dj_\gamma (h\nu, z = 0), \qquad (1.50)$$

with $\alpha(h\nu)$ and $r(h\nu)$ as the spectral absorption and reflexion coefficients, respectively. Thus, it follows that

$$j_L = -e \times \int\limits_{h\nu=0}^{\infty} \int\limits_0^{L_e} \alpha(h\nu)\big[1 - r(h\nu)\big] \times e^{-\alpha(h\nu)z} dz dj_\gamma(h\nu, z = 0) \qquad (1.51)$$

$$= -e \times \int\limits_{h\nu=0}^{\infty} \big[1 - r(h\nu)\big] \times \big[1 - e^{-\alpha(h\nu)L_e}\big] dj_\gamma(h\nu, z = 0)$$

The absorption of photons in the natural and artificial photosynthetic [133] systems, as well as in organic semiconductor devices [134], takes place by formation of excitons [135,136] and their subsequent transport to stabilization centers, whereby they are split into electrons and holes that are further driven to the reaction centers. Excitons are bounded electron-hole states stabilized by attractive coulombic forces. They can be classified in Wannier- or Frenkel-type excitons, depending on whether the exciton radius is larger or smaller than the atomic distance in the crystal lattice.

Figure 1.24 presents a simplified scheme of the absorption process in natural systems. Here, photons are captured by the chlorophyll complex, formed by a bound of molecules of chlorophyll a and b. Chlorophyll molecules are formed by a group of porphyrin units coordinated by a central Mn atom. Hereby, the electronic π-system makes it possible to absorb light in the visible range. The excitons generated in this manner are transported by bridge structures to the stabilization centers, which by means of a screening mechanism involving carotenoids units hinder that electron-hole pairs' recombinant. The exciton transport to the stabilization centers takes place within some picoseconds, a time several orders of magnitude shorter than the typical lifetime of excitons, which is measured in nanoseconds. The transport of the separated charges occurs by a train of acceptors (for electrons) and donators (for holes). In the particular case of green algae, electrons are transported by a combined electron–proton mechanism to the enzymatic cathodic unit hydrogenase, where hydrogen evolution occurs. Holes, on the other hand, are transported to the MnCa-complex, where oxygen evolution takes place. The particularity of some microorganisms to generation hydrogen [137,138] by a photosynthetic process constitutes a potential source for the design of bio-inspired conversion systems [139].

The light absorption in photosynthetic and organic conversion systems involves excitation of organic molecules. The decay to the original ground state can follow different paths, which are represented in the Jablonski's diagram (Figure 1.25) [140]. The ground and excited states of a molecule are characterized by discrete energy levels delimited by Morse's curves of potential energy. These allowed energy levels of one molecule, the behavior of which is described as a harmonic oscillator, arises from the solution of the Schrödinger equation by introducing Morse's function for the potential term. Additionally, vibration and rotation energies are considered independently, using the Born–Oppenheimer approach. Thus, the energy term can be written as

$$E = E_e + E_{vib} = E_e + (n_\nu + \tfrac{1}{2})h\nu_{vib} - C(n_\nu + \tfrac{1}{2})^2 h\nu_{vib} + E_{rot} \qquad (1.52)$$

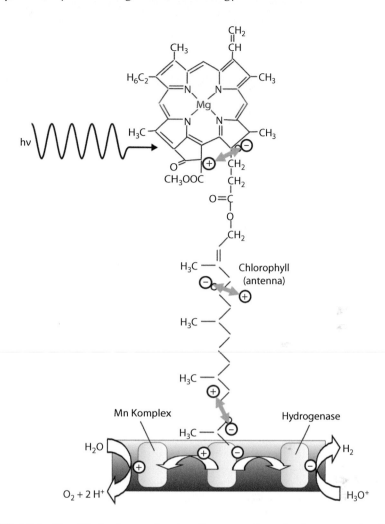

FIGURE 1.24 Simplified scheme showing the conversion of solar energy in chemical energy in natural systems.

where n_v is the quantum number of the vibration state, E_e is the electronic energy, C is an inharmonicity constant, and v_{vib} is the vibration frequency of the molecule. The position of the energy minimum for the ground and excited state depends on the distances of the atomic nuclei of the molecule. According to the Franck–Condon principle, the atomic positions do not change during the transfer process, the time constant of which is several orders of magnitude larger than that of the electron exchange. It means that the transition can be represented as a vertical shift in this type of diagram. Therefore, electrons are generally transferred from the ground state to higher vibrations levels of the excited singlet state, $v' \neq 0$. Within picoseconds, the excited molecule relaxes to the first quantum level of the excited state. Thereafter, the excited molecule decays to its original state. This might occur by emission of a photon upon conditions as required by the Franck–Condon principle.

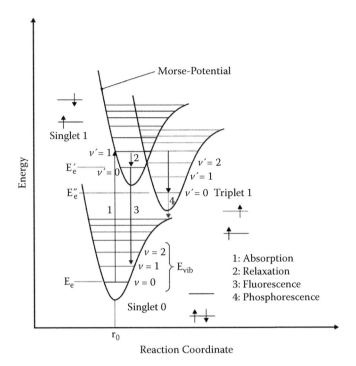

FIGURE 1.25 Jablonski diagram showing the molecular excitation by incident photons and the different decaying mechanisms.

This type of transition is better known as fluorescence and takes place in a time span of some nanoseconds. Alternatively, the excited state can decay to a triplet state, characterized by a pair of electrons with the same spin. The probability for the formation of this state will be determined by a series of factors, such as the selection rule, spin-orbit coupling, or participation of higher electronic states, as well as by the energetic changes involved (only exergonic processes are permitted). Due to the complexity of the deactivation of triplet states, known as phosphorescence, their lifetime is several orders of magnitude larger than singlet-states; triplet-state of organic molecules show a typical value of microseconds to milliseconds. For this reason, many biomolecular reactions take place via triplet states. This has important consequences for the energy transport in photosynthetic systems, since here it is necessary that the energy transfer process occurs within a time much shorter than the lifetime of the excited states of the metal complexes in the light-collecting antennas. As already mentioned, the energy transfer in organic systems occurs via transport of excitons from the antenna complexes to the electrocatalytic centers [141]. This process is described using quantum mechanics by applying the time-dependent perturbation theory. Thus, the transfer rate constant can be written in the following general form [142,143]:

$$k_{en} = \frac{2\pi}{\hbar} \left\langle \psi'_e \middle| H \middle| \psi_e \right\rangle^2 \left\langle \psi'_s \middle| \psi_s \right\rangle^2 \left\langle \psi'_{vib} \middle| \psi_{vib} \right\rangle^2 \tag{1.53}$$

where H represents the operator, which mixes the wave functions of the donator and acceptor. ψ_s and ψ_s' are the related spin-wave functions and ψ_{vib} and ψ_{vib}' represent the total vibration-wave functions. The final form adopted by Eq. (1.53) depends on the particular energy transfer mechanism. In general, two mechanisms are considered—the Förster [144] and the Dexter [145] transfer modes—the main characteristics of which are schematically shown in Figure 1.26. The former is adequate for describing the energy transfer among molecules that are separated by relatively long distances, much larger than that indicated by the van der Waals radius. This mechanism is based on electrostatic dipole interactions between the donor and acceptor molecules. According to this model, the transfer rate is given by

$$k_{en} = \frac{1}{\tau_f}\left(\frac{z_0}{z}\right)^6 \quad ; \quad z_0^6 = \frac{9 \times \ln 20 \times f_{dip}^2 \Phi_f}{128\pi^5 n^4 N_A} \times \int \frac{\alpha_A(h\nu)\varepsilon_D(h\nu)}{\nu^4}d\nu \quad (1.54)$$

In this equation z_0 is the so-called Förster radius, τ_f is the fluorescence lifetime, Φ_f is the fluorescence quantum yield, n is the refractive index, f_{dip} is a factor accounting for effects of dipole orientation, and $\alpha_A(\nu)$ and $\varepsilon(\nu)$ are the normalized spectral emission function of the donor and the normalized spectral absorption of the acceptor, respectively. The integral in Eq. (1.54) represents the spectral overlap of donor and acceptor [146]. In the Dexter mechanism, the electron exchange takes place by a mixing of the electron wave functions of the donor and acceptor. Therefore, the term $\left\langle\psi_e'|H|\psi_e\right\rangle^2$ results proportional to $\exp(-z/r_B)$, where r_B is the mean Bohr radius. The spin wave functions ψ_s and ψ_s' are included in the formulation of both

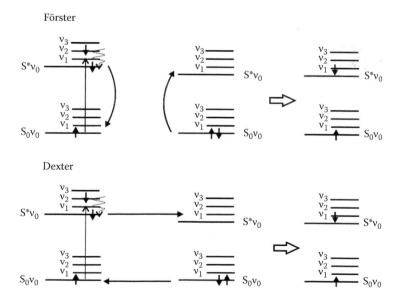

FIGURE 1.26 Schematic representation of the exciton transfer after Förster and Dexter mechanisms.

mechanisms. The energy transfer operator is, however, spin-independent. This means that in the absence of spin-orbit coupling the energy transfer takes place between states with the same spin multiplicity only. The spin-orbit coupling makes the change of spin character possible, but only under a considerable drop of the term $\langle \psi'_e | H | \psi_e \rangle$.

In the following text, the electron transfer between a donor and acceptor species will be discussed. For the description of this process, the classical approach after the Marcus theory is usually applied [33], for instance in the quantification of electrochemical reactions. This theory assumes that the donor and acceptor particles can be described as classical harmonic oscillators (see Figure 1.27). We can distinguish the vibrations of the inner structure of the reactant molecules and that of the surrounding medium, which are characterized by different frequencies. According to this model, the potential energy is represented by a paraboloid surface $U_{pot}(z_i, z_a)$, where the coordinates indicate the displacements of the molecule from the equilibrium due to inner and outer vibrations. Thus, the potential surfaces of the oxidized (acceptor) and reduced (donor) particles are expressed as

$$U_{ox} = G_{ox}^0 + \sum_j \tfrac{1}{2} m_{ox,j} \omega_j^2 (z_{ij} - z_{aj}^0)^2; \quad U_{red} = G_{red}^0 + \sum_j \tfrac{1}{2} m_{red,j} \omega_j^2 (z_{ij} - z_{aj}^0)^2 \quad (1.55)$$

where $G^0_{ox,red}$ is the energy of the particle in the equilibrium position, $m_{ox,red,j}$ is the effective mass of the oscillator j, and ω_j is its characteristic vibration frequency. The crossing of both paraboloids defines a reaction surface with a minimum upon which the reaction should occur. The minimum must fulfill the condition $[\partial U_s(z)/\partial z] = 0$, where the function $U_s(z)$ defines a line in the reaction surface for which the condition

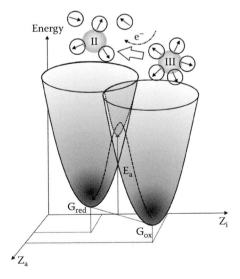

FIGURE 1.27 Schematic representation of the potential energy surfaces of the particles participating in the electron transfer process.

$U_{red}(z) = U_{ox}(z)$ applies. The minimum of this line can be found by applying the Lagrange method. Thus, we have

$$z_i^{min} = z_a^{ox} + \beta(z_a^{ox} - z_a^{red}) \tag{1.56}$$

where β is the Lagrange operator, given by

$$\beta = \frac{\lambda + G_{red}^0 - G_{ox}^0}{2\lambda}; \qquad \lambda = \frac{1}{2} \sum_j m_j v_j^2 (z_{aj}^{ox} - z_{aj}^{red})^2 \tag{1.57}$$

The Lagrange operator contains the parameter λ known as reorganization energy, related with the energy changes derived from structural changes of the solvation and inner structure during the transfer process. The particle in its equilibrium position needs to overcome the energy barrier or activation energy to reach the minimum of the reaction surface. It is given by

$$E_{a,ox} = \frac{(\lambda + G_{ox}^0 - G_{red}^0)^2}{4\lambda} \tag{1.58}$$

Basically, this theory considers that the transfer process is controlled by the energetic of the reacting particles, whereas details on the electron behavior are not considered. Here, it is assumed that the electron transfer occurs adiabatically as the particle overcomes the energy barrier. This holds whenever the time constant for the electron transfer is several orders of magnitude shorter than that for the molecular oscillations, what applies for the most common electrochemical reactions. The Marcus theory was completed and validated by Gerischer using a semiclassical treatment of the process [147]. The theoretical basis of Gerischer's theory will be explained in more detail in Chapter 4 in the frame of the description of electrochemical reactions at metal-electrolyte interfaces.

From the classical description of the electron transfer between two molecules, the following expression for the electron transfer rate results:

$$k_{et} = v \exp\left[-\frac{(\lambda + \Delta G^0)^2}{4\lambda kT} \right] \tag{1.59}$$

where λ contains the inner and outer reorganization energies, ΔG^0 is the change of the free Gibbs energy of the reaction, and the pre-exponential factor v is given by the product of the probability of the electron transfer and the attempt frequency of the reacting particle. The classical description of the electron transfer process suits for the particular case of a strong electronic coupling between donor and acceptor. This case does not apply, however, for systems with weak electronic interactions, and thus a quantum mechanical description is necessary here.

The formulation of an expression for the rate constant of the electron transfer, based on quantum mechanics, starts from the application of the golden rule of quantum mechanics. Thus, the following expression, as originally formulated by Landau [148] and Zehner [149] independently of each other [148]:

$$k_{et} = \frac{2\pi}{\hbar} H_{AD}^2 f_{FC} \tag{1.60}$$

In this expression, H_{AD} is the matrix element, which represents the resonance energy from the coupling of the acceptor and donor orbitals. f_{FC} is the Franck–Condon factor given by the overlap integral of the vibration wave functions of the end and initial states. As mentioned before, the electron transfer process can be treated in terms of one of two limit situations, defined by the strength of the electronic coupling between donor and acceptor. The strength of the electronic coupling is given by the matrix element. This later defines a characteristic time: $t_e = \hbar / |H_{AD}|$. On the other hand, a characteristic time for the nuclear movements can be defined as $t_{vib} = 2\pi / \omega_{vib}$. If $t_e \ll t_{vib}$, we are in the adiabatic case. If $t_e \gg t_{vib}$, we speak of a nonadiabatic transfer. The former case can be interpreted as an electronic delocalization with frozen molecular coordinates at the transfer point. This case is generally described by a classical treatment of the molecular oscillation of the solvation shell of the electroactive particles. The nonadiabatic transfer requires a more complex analysis and is particularly suited for the description of long-range electron transfer processes. For electronic transfers where $kT \gg \hbar\omega$, the following expression applies:

$$k_{et} = |H_{AD}|^2 \sqrt{\frac{\pi}{\hbar^2 kT\lambda}} \exp\left[-\frac{(\Delta G^0 - \lambda)^2}{4\lambda kT}\right] \tag{1.61}$$

where

$$\lambda = \frac{v^2}{2}(z^A - z^D)^2 \tag{1.62}$$

with z^A and z^D as the coordinates of the acceptor and donor, respectively. In the case of a coupling of the electron transfer with high frequency states of the inner molecular structure and low frequency states of the surrounding solvent, the whole process can be described by a quantum mechanical–classical treatment, where the high frequency modes are quantum mechanical and the low frequency modes are classically described. In accordance with this analysis, we have

$$k_{et} = |H_{AD}|^2 \sqrt{\frac{\pi}{\hbar^2 kT\lambda}} \sum_{N=0}^{\infty} \left|\langle \Psi_{D(0)} | \Psi_{AN} \rangle\right|^2 \exp\left[-\frac{(\Delta G^0 - \hbar v_i N - \lambda)^2}{4\lambda kT}\right] \tag{1.63}$$

where

$$\sum_{N=0}^{\infty} \left|\langle \Psi_{D(0)} | \Psi_{AN} \rangle\right|^2$$

is the sum of Franck–Condon factors, which correspond to the coupling of the donor ground states with N electron levels of the acceptor.

The aforementioned theories are based on a traditional quantum mechanical treatment of the electron transfer in one medium regarded as a bath of harmonic oscillators: the solvent. This theoretical description can be unsatisfactory in some biological systems though, where the electronic coupling is comparable to the energy fluctuations of the excited states. This means that if the electronic coupling is weak, the transfer probability over the crossing point is small, and many attempts will occur before an electronic state change can occur. If the dephasing rates in the reactant and product manifolds become comparable to the time scale of the electron tunneling, the semi-classical description is no longer valid and the nuclear motion must be treated quantum mechanically. In this situation, the initially excited neutral state does not equilibrate before reaction occurs.

The Redfield theory offers the theoretical frame for the description of systems with complex quantum dynamics [150]. This theory is based on the matrix density formalism used for the solution of the time-dependent Schrödinger equation. The formalism consists of the division of the total system in a relevant system coupled with its environment, both described in terms of a given set of coordinates (s) and (z), respectively. According to this concept, the total Hamiltonian can be formulated as

$$H = H_S + H_{S-B} + H_B \qquad (1.64)$$

where H_S and H_B are the Hamilton-operators of the system and the environment, respectively. H_{S-B} represents the interactions between both systems. In this context, an expectation value or *observable* can be calculated by introduction of the Hermitian operator, defined as

$$\left\langle \hat{f} \right\rangle = \int ds\, dz\, \psi^*(s,z) f(s) \psi(s,z) \qquad (1.65)$$

or as

$$\left\langle \hat{f} \right\rangle = \int ds \left[f(s') \rho(s,s') \right]_{s=s'} \qquad (1.66)$$

after introducing the matrix density $\rho(s,s') = \int dz\, \psi^*(s,z) f(s) \psi(s',z)$ into Eq. (1.65). The trace of this quantity gives the reduced matrix density $\rho_r = Tr_s\{\rho\}$. This mathematical operation is physically interpreted as the reduction of the total probability density to the space of the relevant system. The Redfield approach for the description of the transfer between two energy states, α and β, is mathematically expressed by the following differential equation:

$$\frac{\partial \rho_{r(\alpha,\beta)}(t)}{\partial t} = -i\omega_{(\alpha,\beta)}\rho_{r(\alpha,\beta)}(t) + \sum_{\gamma,\delta} R_{\alpha\beta\gamma\delta}\rho_{r(\lambda,\delta)} \qquad (1.67)$$

where $\rho_{r(\alpha,\beta)} = \left\langle \alpha | \rho_r(t) | \beta \right\rangle$ is the matrix element of the reduced matrix density, $\omega_{(\alpha,\beta)} = E_\alpha - E_\beta$ is the eigenfrequency of the system, and $R_{\alpha\beta\gamma\delta}$ is the Redfield tensor.

In spite of the mathematical complexity, it is important here to show the qualitative interpretation of Eq. (1.67). The first term represents the dynamic of the isolated system. The interactions of the system with its environment are expressed by the dissipative tensors in the second term. After the computational solving of the Redfield dynamics equation, the population of the excited states given by $|\psi_D\rangle$ can be obtained by

$$P(t) = Tr\{\rho_r(t)|\psi_D\rangle\langle\psi_D|\}$$ (1.68)

Then, the mean electron transfer rate constant can be calculated from [153]

$$\bar{k}_{et} = \left[\int_0^\infty P(t)\,dt\right]^{-1}$$ (1.69)

The matrix density approach offers a suitable tool for the investigation of electron transfer processes, for which the traditional theories based on the formulation of a transfer rate fail to give a complete reproduction of the experimental observed features. For instance, the application of the Redfield theory is especially well-suited for the description of ultrafast photo-induced electron transfer and the prediction of oscillatory behavior of the population of excited states $P(t)$ that are experimentally detected by oscillatory features observed by time-resolved spectroscopic techniques [154].

REFERENCES

1. Metz, B., Davison, O., Bosch, P., Dave R. and L. Meyer. 2007. *Climate Change 2007: Mitigation.* Intergovernmental panel on climate change. Cambridge: Cambridge University Press.
2. McLaughlin Green, C. and M. Lomask. 1970. *Vanguard: A history.* Washington DC. The NASA Historical Series, National Aeronautics and Space Administration.
3. Lutgens, F.K. and E.J. Tarbuck. 2000. *Essentials of geology,* 7th ed. New Jersey: Prentice Hall.
4. Ito, M., Kato, K., Komoto, K., Kichimi, T. and K. Kurokawa. 2008. A comparative study on cost and life-cycle analysis for 100 MW very large-scale PV (VLS-PV) systems in deserts using m-Si, a-Si, CdTe and CIS modules. *Prog. Photovolt: Res. Appl.* 16:17.
5. Contreras, M.A., Egaas, B., Ramanathan, K., Hiltner, J., Swartzlander, A., Hasoon, F. and R. Noufi. 2003. Progress toward 20% efficiency in Cu(In,Ga)Se$_2$ polycrystalline thin-film solar cells. *Prog. Photovolt: Res. Appl.* 11:243.
6. Schock, H.W. 2004. Properties of chalcopyrite-based materials and film deposition for thin-film solar cells. In *Thin-film solar cells, next generation photovoltaics and its applications,* ed. Y. Hamakawa, Ch. 10. Berlin: Springer-Verlag.
7. Regan, B.Ó. and M. Grätzel. 1991. A low-cost, high efficiency solar cell based on dye-sensitized colloidal TiO$_2$ films. *Nature* 353:737.
8. Grätzel, M. 2001. Photoelectrochemical solar cells. *Nature* 414:6861.
9. Ferrere, S., Zaban, A. and B.A. Gregg. 1997. Dye sensitization of nanocrystalline tin oxide by perylene derivatives. *J. Phys. Chem. B* 101:4490.
10. Youngblood, W.J., Lee, S.-H.A., Maeda, K. and T.E. Mallouk. 2009. Visible light water splitting using dye-sensitized oxide semiconductors. *Acc. Chem. Res.* 42:1966.

11. Repins, I., Contreras, M.A., Egaas, B., DeHart, C., Scharf, J., Perkins, C.L., To, B. and R. Noufi. 2008. 19.9%-efficient ZnO/CdS/CuInGaSe$_2$ solar cell with 81.2% fill factor. *Prog. Photovolt.: Res. Appl.* 16:235.

12. Green, M.A., Emery, K., Hishikawa, Y. and W. Warta. 2009. Solar Cell Efficiency Tables (Version 34). *Prog. Photovolt: Res. Appl.* 17:320.

13. Grampp, G. 1993. Der inverse Marcus-Bereich—von der Theorie zum Experiment. *Angew. Chem.* 105:724.

14. Barbir, F. 2005. PEM Electrolysis for production of hydrogen from renewable energy sources. *Solar Energy* 78:661.

15. Khaselev, O., Bansal, A. and J.A. Turner. 2001. High-efficiency integrated multijunction photovoltaic/electrolysis systems for hydrogen production. *Int. J. Hydrogen Energy* 26:127.

16. Lewis, N. 2001. Light work with water. *Nature* 414:589.

17. Schottky, W. 1938. Halbleitertheorie der Sperrschicht. *Die Naturwissensch.* 26:843.

18. Mott, N.F. 1938. Note on the contact between a metal and an insulator or semiconductor. *Proc. Cambr. Philos. Soc.* 34:568.

19. Mönch, W. 1996. Electronic properties of ideal and interface-modified metal-semiconductor interfaces. *J. Vac. Sci. Technol. B* 14:2985.

20. Gerischer, H. 1969. Charge transfer processes at semiconductor-electrolyte interfaces in connection with problems of catalysis. *Surf. Sci.* 18:97.

21. Gerischer, H. 1991. Electron-transfer kinetics of redox reactions at the semiconductor/electrolyte contact. A new approach. *J. Phys. Chem.* 95:1356.

22. Abe, R., Sayama, K. and H. Arakawa. 2002. Efficient hydrogen evolution from aqueous mixture of I$^-$ and acetonitrile using a merocyanine dye-sensitized Pt/TiO$_2$ photocatalyst under visible light radiation. *Chem. Phys. Lett.* 362:441.

23. Kim, Y., Atherton, S.J., Brigham, E.S. and T.E. Mallouk. 1993. Sensitized layered metal oxide semiconductor particles for photochemical hydrogen evolution from nanosacrificial electron donors. *J. Phys. Chem.* 97:11802.

24. Würfel, P. 2000. *Physik der Solarzellen.* Heidelberg-Berlin: Spektrum Akademischer Verlag GmbH.

25. Hamman, C.H. and W. Vielstich. 1998. *Elektrochemie.* Weinheim: Wiley-VCH.

26. Sato, N. 1998. *Electrochemistry at Metal and Semiconductor Electrodes.* Amsterdam: Elsevier.

27. Guillon, C., Langot, P., Del Fatti, N. and F. Vallée. 2003. Nonequilibrium electron-loss kinetics in metal clusters. *New Journal of Physics* 5:13.1.

28. Del Fatti, N., Voisin, C., Achermann, M., Tzortzakis, S., Christofilos, D. and F. Vallée. 2000. Nonequilibrium electron dynamics in noble metals. *Phys. Rev. B* 61:16956.

29. Sun, C.-K., Vallée, F., Acioli, L.H., Ippen, E.P. and J.G. Fujimoto. 1994. Femtosecond-tunable measurement of electron thermalisation in gold. *Phys. Rev. B* 50:15337.

30. Iwasita, T., Schmickler, W. and J.W. Schultze. 1985. The influence of the metal on the kinetics of outer sphere redox reactions. *Ber. Bunsenges. Phys. Chem.* 89:138.

31. Sung, Y.-E. and A.J. Bard. 1998. Enhancement of electrochemical hot electron injection into electrolyte solutions at oxide-covered tantalum electrodes by thin platinum films. *J. Phys. Chem. B* 102:9806.

32. Iwasita, T., Schmickler, W., Hermann, J. and U. Vogel. 1983. The kinetic parameters of the Fe(CN)$_6^{3-/4-}$ redox system. New results obtained with the ring electrode in turbulent pipe flow. *J. Electrochem. Soc.* 130:2026.

33. Marcus, R.A. 1965. On the theory of oxidation-reduction reactions involving electron transfer. I. *J. Chem. Phys.* 24:966.

34. Heller, A., Aharon-Shalom, E., Bonner, W.A. and B. Miller. 1982. Hydrogen-evolving semiconductor photocathodes. Nature of the junction and function of the platinum group metal catalyst. *J. Am. Chem. Soc.* 104:6942.

35. Heller, A. and R. G. Vadimsky. 1981. Efficient solar to chemical conversion: 12% efficient photoassisted electrolysis in the p-type InP(Ru)/HCl-KCl/Pt(Rh) cell. *Phys. Rev. Lett.* 46:1153.

36. Becquerel, A.E. 1839. Memoire sur les effets électriques produits sous l'influence des rayons solaires. *Comptes Rend. Hebdom. Des Séances de l'Academie de Sciences, Paris* 9:561.

37. Williams, R. 1960. Becquerel photovoltaic effect in binary compounds. *J. Chem. Phys.* 32:1505.

38. Smith, W. 1873. Effect of light on selenium during the passage of an electric current. *Nature* 7:303.

39. Adams, W.G. and R.E. Day. 1876. *Proceedings of the Royal Society London* 25:113.

40. Lange, B. 1930. New photovoltaic cell. *Z. Phys.* 31:139.

41. Schottky, W. 1930. Cuprous oxide photoelectric cells. *Z. Phys.* 31:913.

42. Planck, M. 1900. Zur Theorie des Gesetzes der Energieverteilung im Normalspektrum. *Verhandlungen der Deutschen Physikal. Gesellschaft* 2:237.

43. Chapin, D.M., Fuller, C.S. and G.L. Pearson. 1954. A new silicon p-n junction photocell for converting solar radiation into electrical power. *J. Appl. Phys.* 25:676.

44. Reynolds, D.C., Leies, G., Antes, L.L. and R.E. Marburger. 1954. Photovoltaic effect in cadmium sulfide. *Phys. Rev.* 96:533.

45. Gremmelmeier, R. 1955. GaAs-Photoelement. *Z. Naturforsch. A* 10:501.

46. Shockley, W. and H.J. Queisser. 1961. Detailed balance limit of efficiency of p-n junction solar cells. *J. Appl. Phys.* 32:510.

47. The term $\exp[(E_e^*-E_h^*)/kT]$ in the expression for the calculation of emitted photon flux accounts for the deviation from the radiative emission in thermal equilibrium conditions. In equilibrium, the flux of absorbed photons equals that of emitted photons or in an equivalent way, that the generation rate G_γ^0 equals the recombination rate R_γ^0 given by $C\, n_e^0\, n_h^0 = n_i^2$, where C is a proportionality constant. Under illumination the blackbody attains a stationary state and the equality $G_\gamma = R = C\, n_e^0\, n_h^0$ is now applicable. As $C = G_\gamma^0/n_i^2$, then $G_\gamma = G_\gamma^0\, n_e^0\, n_h^0/n_i^2$. Because of the direct dependence of photoemission on the recombination rate, $j_{\gamma,em} = j_{\gamma,em}^0\, n_e^0\, n_h^0/n_i^2$, where $j_{\gamma,em}^0$ is the emission rate in equilibrium without illumination.

48. The notation AM 1.5, which means *air mass* 1.5, is used to refer to the sunlight spectrum obtained at the earth surface at an angle of incidence of 41.8°, for which the light has covered 1.5 times the vertical path through the atmosphere. This represents the yearly average in west developed countries and is used as standard for efficiency comparisons.

49. Zhao, J., Wang, A., Green, M.A. and F. Ferrazza. 1998. Novel 19.8% efficient honeycomb textured multicrystalline and 24.4% monocrystalline silicon solar cells. *Appl. Phys. Lett.* 73:1991.

50. Repins, I., Contreras, M., Romero, Y., Yan, Y., Metzger, W., Li, J., Johnston, S., Egaas, B., DeHart, C., Scharf, J., McCandless, B.E. and R. Noufi. 2008. Characterization of 19.9%-efficient CIGS Absorbers. *IEEE Photovoltaics Specialists Conference Record* 1:33.

51. Yoshimi, M., Sasaki, T., Sawada, T., Suezaki, T., Meguro, T., Matsuda, T., Santo, K., Wadano, K., Ichikawa, M., Nakajima, A. and K. Yamamoto. 2003. High efficiency thin film silicon hybrid solar cell module on 1 m²-class large area substrate. *Conf. Record, 3rd World Conference on Photovoltaic Energy Conversion Osaka, May, 2003* 1:1566.

52. Brown, A.B., Green, M.A. 2002. Detailed balance limit for the series constrained two terminal tandem solar cell. *Phys. E* 14:96.

53. Würfel, P. 1982. The chemical potential of radiation. *J. Phys. C* 15:3967.

54. The concept of chemical potential for an emitted photon was introduced by Würfel in 1982. As already discussed in Ref. 47, the emission rate of a photon in a nonequilibrium systems is given by that in equilibrium, j_γ^0, multiplied by a factor $n_e n_h/n_i^2$. Introducing

the energy dependence of the charge carrier concentration of the semiconductor under illumination, given by the Fermi-distribution function, the following expression (in its differential form) is obtained:

$$dj_\gamma = \frac{\Omega}{4\pi\hbar^3 c^2} \frac{(h\nu)^2}{\exp\left[\dfrac{h\nu - (E_n{}^* - E_p{}^*)}{kT}\right] - 1} dh\nu$$

which was already introduced without demonstration in the integral of Eq. (1.18). If one compares the denominator of this expression, which corresponds to the Bose–Einstein distribution of photons, with the Fermi-distribution function, one can make a correspondence of the term $(E_n{}^* - E_p{}^*)$ with an electrochemical potential, or Fermi level. This expression can be considered as a general formulation of Planck's law of radiation with the special cases $E_n{}^* - E_p{}^* = 0$ for thermal emitted radiation and $E_n{}^* - E_p{}^* \neq 0$ for luminescence radiation.

55. King, R.R., Law, D.C., Edmondson, K.M., Fetzer, C.M., Kinsey, G.S., Yoon, H., Sherif, R.A. and N.H. Karam. 2007. 40% efficient metamorphic GaInP/GaInAs/Ge multijunction solar cells. *Appl. Phys. Lett.* 90:183516.

56. Hanna, M.C. and A.J. Nozik. 2006. Solar conversion efficiency of photovoltaic and photoelectrolysis cells with carrier multiplication absorbers. *J. Appl. Phys.* 100:074510.

57. Schaller, R.D. and V.I. Klimov. 2004. High efficiency carrier multiplication in PbSe nanocrystals: implications for solar energy conversion. *Phys. Rev. Lett.* 92:186601.

58. Schaller, R.D., Petruska, M.A. and V.I. Klimov. 2005. Effect of electronic structure on carrier multiplication efficiency: comparative study of PbSe and CdSe nanocrystals. *Appl. Phys. Lett.* 87:253102.

59. Beard, M.C., Knutsen, K.P., Yu, P., Luther, J.M., Song, Q., Metzger, W.K., Ellingson, R.J. and A.J. Nozik. 2007. Multiple exciton generation in colloidal silicon nanocrystals. *Nano Lett.* 7:2506.

60. Würfel, P. 1997. Solar energy conversion with hot electrons from impact ionization. *Sol. Energy Mater. Sol. Cells* 46:43.

61. Jiang, C.W., Cho, E.C., Conibeer, G. and M.A. Green. 2004. Silicon quantum dots: application for energy selective contacts to hot carrier solar cells. *19th EPSEC, Paris, 2004.*

62. Conibeer, G.J., Jiang, C.W., König, D., Shrestha, S., Walsh, T. and M.A. Green. 2008. Selective energy contacts for hot carrier solar cells. *Thin Solid Films* 516:6968.

63. Wolf, M. 1960. Limitations and possibilities for improvement of photovoltaic solar energy converters. *Proceedings of the IRE* 48:4117.

64. Keevers, M.J. and M.A. Green. 1996. Extended infrared response of silicon solar cells and the impurity photovoltaic effect. *Solar Energy Mater. Solar Cells* 41–42:195.

65. Keevers, M.J. and M.A. Green. 1994. Efficiency improvements of silicon solar cells by the impurity photovoltaic effect. *J. Appl. Phys.* 75:4022.

66. Schmeits, M. and A.A. Mani. 1999. Impurity photovoltaic effect in c-Si solar cells. A numerical study. *J. Appl. Phys.* 85:2207.

67. Karazhanov, S.Zh. 2001. Impurity photovoltaic effect in indium-doped silicon solar cells. *J. Appl. Phys.* 89:4030.

68. Beaucarne, G., Brown, A.S., Keevers, M.J., Corkish, R. and M.A. Green. 2002. The impurity photovoltaic effect (IPV)in wide-bandgap semiconductors: an opportunity for very-high-efficiency solar cells? *Prog. Photovolt: Res. Appl.* 10:345.

69. Barnham, K.W.J. and G. Duggan. 1990. A new approach to high-efficiency multi-bandgap solar cells. *J. Appl. Phys.* 67:3490.

70. Barnham, K., Ballard, I., Barnes, J., Connolly, J., Griffin, P., Kluftinger, B., Nelson, J., Tsui, E. and A. Zachariou. 1997. Quantum well solar cells. *Appl. Surf. Sci.* 113–114:722.
71. Anderson, N.G. 2002. On quantum well solar cell efficiencies. *Physica E* 14:126.
72. Araujo, G.L. and A. Martí. 1994. Absolute limiting efficiencies for photovoltaic energy conversion. *Solar Energy Mater. Solar Cells* 33:213.
73. Corkish, R. and M. Green. 1993. *Proc. 23rd IEEE Photovoltaic Specialists Conf. (WEE, 1993)* 1:675.
74. Bremner, S.P., Corkish, R. and C.B. Honsberg. 1999. *IEEE Transactions on electron Devices* 46:1932.
75. Luque, A., Martí, A. and L. Cuadra. 2001. Thermodynamic consistency of sub-bandgap absorbing solar cell. *IEEE Trans.* 48:2118.
76. Saif, B. and J. Khurgin. 2001. Graded band gap multiple quantum well solar cell. *US Patent WO/2001/047031.*
77. Luque, A., Marti, A., López, N., Antolín, E., Cánovas, E., Stanley, C., Farmer, C., Caballero, L.J., Cuadra, L. and J.L, Balenzategui. 2005. Experimental analysis of the quasi-Fermi level split in quantum dot intermediate-band solar cells. *Appl. Phys. Lett.* 87:083505.
78. Cho, E., Green, M.A., Xia, J., Corkish, R., Reece, P. and M. Gal. 2004. Clear quantum-confined luminescence from crystalline silicon/SiO2 single quantum wells. *Appl. Phys. Lett.* 84:2286.
79. Luque, A. and A. Marti. 1997. Increasing the efficiency of ideal solar cells by photon induced transitions at intermediate levels. *Phys. Rev. Lett.* 78:5014.
80. Brown, A.S. and M.A. Green. 2004. Intermediate band solar cell with many bands: ideal performance. *J. Appl. Phys.* 94:6150.
81. Chamberlain, G.A. 1983. Organic solar cells: a review. *Solar Cells* 8:47.
82. Wöhrle, D. and D. Meissner. 1991. Organic solar cells. *Adv. Mater.* 3:129.
83. Tang, C.W. 1986. Two-layer organic photovoltaic cell. *Appl. Phys. Lett.* 48:183.
84. Zhu, X.Y., Yang, Q. and M. Muntwiler. 2009. Charge-transfer excitons at organic semi-conductor surfaces and interfaces. *Acc. Chem. Res.* 42:1779.
85. Hoppe, H. and N. Serdar Sariciftc. 2004. Organic solar cells, an overview. *J. Mater. Res.* 19:1924.
86. Koch, N. 2007. Organic electronic devices and their functional interfaces. *ChemPhysChem* 8:1438.
87. Yu, G., Gao, J., Hummelen, J.C., Wudl, F. and A.J. Heeger. 1995. Polymer photovoltaic cells: enhanced efficiencies via a network of internal donor-acceptor heterojunctions. *Science* 270:1789.
88. Svensson, M., Zhang, F., Veenstra, S.C., Verhees, W.J.H., Hummelen, J.C., Kroon, J.M., Inganäs, O. and M.R. Andersson. 2003. High performance polymer solar cells of an alternating polyfluorene copolymer and a fullerene derivative. *Adv. Mater.* 15:988.
89. Kroon, J.M., Wienk, M.M., Verhees, W.J.H. and J.C. Hummelen. 2002. Accurate efficiency determination and stability studies of conjugated polymer/fullerene solar cells. *Thin Solid Films* 403–404:223.
90. Tennakone, K., Kumara, G.R.R., Kottegoda, I.R.M. and V.S.P. Perera. 1999. An efficient dye-sensitized photoelectrochemical solar cell made from oxides of tin and zinc. *Chem. Commun.* 15:
91. Tennakone, K., Kumara, G.R.R.A., Kumarasinghe, A.R., Wijayantha, K.G.U. and P.M. Sirimanne. 1995. A dye-sensitized nano-porous solid-state photovoltaic cell. *Semicond. Sci. Technol.* 10:1689.
92. Grätzel, M. 2003. Dye-sensitized solar cells. *J. Photochem. Photobiol. C: Photochem. Rev.* 4:145.
93. Grätzel, M. 2001. Photoelectrochemical cells. *Nature* 414:338.

94. Nazeeruddin, M.K., Kay, I., Rodicio, A., Humphry-Baker, R., Müller, E., Liska, P., Vlachopouloss, N. and M. Grätzel. 1993. Conversion of light to electricity by cis-X2 bis(2,2'-bipyridyl-4,4'-dicarboxylate)ruthenium(II) charge-transfer sensitizers (X = Cl-, Br-, I-, CN-, and SCN-) on nanocrystalline titanium dioxide electrodes. *J. Am. Chem. Soc.* 115:6382.

95. Marcus, R.A. 1970. High-order time-dependent perturbation theory for classical mechanics and for other systems of first-order ordinary differential equations. *J. Chem. Phys.* 52:4803.

96. Onuchic, J.N., Beratan, D.N. and J.J. Hopfield. 1986. Some aspects of electron-transfer reaction dynamics. *J. Phys. Chem.* 90:3707.

97. Rossetti, R., Ellison, J.L., Gibson, J.M. and L. E. Brus. 1984. Size effects in the excited electronic states of small colloidal CdS crystallites. *J. Chem. Phys.* 80:4464.

98. Gerischer, H. and M. Lobke. 1986. A particle size effect in the sensitization of TiO_2 electrodes by a CdS deposit. *J. Electroanal. Chem.* 204:225.

99. Robel, I., Subramanian, V., Kuno, M. and P.V. Kamat. 2006. Quantum dot solar cells. Harvesting light energy with CdSe nanocrystals molecularly linked to mesoscopic TiO_2 films. *J. Am. Chem. Soc.* 128:2385.

100. Leschkies, K.S., Beatty, T.J., Kang, M.S., Norris, D.J. and E.S. Aydil. 2009. Solar cells based on junctions between colloidal PbSe nanocrystals and thin ZnO films. *ACS Nano* 3:3638.

101. Chang, C.-H. and Y.-L. Lee. 2007. Chemical bath deposition of CdS quantum dots onto mesoscopic TiO_2 films for application in quantum-dot-sensitized solar cells. *Appl. Phys. Lett.* 91:053503.

102. Yu, P., Zhu, K., Norman, A.G., Ferrere, S., Frank, A.J. and A.J. Nozik. 2006. Nanocrystalline TiO_2 solar cells sensitized with InAs quantum dots. *J. Phys. Chem. B* 110:25451.

103. Luque, A., Martí, A. and A.J. Nozik. 2007. Solar cells based on quantum dots: multiple exciton generation and intermediate bands. *MRS Bull.* 32:236.

104. Hanna, M.C. and A.J. Nozik. 2006. Solar conversion efficiency of photovoltaic and photoelectrolysis cells with carrier multiplication absorbers. *J. Appl. Phys.* 100:074510.

105. Murphy, J.E., Beard, M.C., Norman, A.G., Ahrenkiel, S.P., Johnson, J.C., Yu, P., Micic, O.I., Ellingson, R.J. and A.J. Nozik. 2006. PbTe colloidal nanocrystals: synthesis, characterization and multiple excitation generation. *J. Am. Chem. Soc.* 128:3241.

106. Schaller, R.D., Sykora, M., Pietryga, J.M. and V.I. Klimov. 2006. Seven excitons at a cost of one: redefining the limits for conversion efficiency of photons into charge carriers. *Nano Lett.* 6:424.

107. Singh, S., Jones, W.J., Siebrand, W., Stoicheff, B.P. and W.G. Schneider. 1965. Laser generation of excitons and fluorescence in anthracene crystals. *J. Chem. Phys.* 42:330.

108. Geacintov, N., Pope, M. and F. Vogel. 1969. Effect of magnetic field on the fluorescence of tethracene crystals: exciton fission. *Phys. Rev. Lett.* 22:593.

109. Katoh, R. and M. Kotani. 1992. Fission of a higher excited state generated by singlet exciton fusion in an anthracene crystal. *Chem. Phys. Lett.* 196:108.

110. Austin, R.H., Baker, G.L., Etemad, S. and R. Thompson. 1989. Magnetic field effects on triplet exciton fission and fusion in a polydiacetylene. *J. Chem. Phys.* 90:6642.

111. Österbacka, R., Wohlgenannt, M., Shkunov, M., Chinn, D. and Z.V. Vardeny. 2003. Excitons, polarons, and laser action in poly-(d-phenylene vinylene) films. *J. Chem. Phys.* 118:8905.

112. Papagiannakis, E., Dar, S.K., Gall, A., van Stokkum, I.H.M., Robert, B., van Grondelle, R., Frank, H.A. and J.T.M. Kennis. 2003. Light harvesting by charotenoids incorporated into the B850 light-harvesting complex from Rhodobacter sphaeroides R-26. I Excited-state relaxation, ultrafast triplet formation, and energy transfer to bacteriochlorophyll. *J. Phys. Chem. B* 107:5642.

113. Müller, A.M., Avlasevich, Y.S., Müllen, K. and C.J. Bardeen. 2006. Evidence for exciton fission and fusion in a covalently linked tetracene dimmer. *Chem. Phys. Lett.* 421:518.
114. Greyson, E.C., Stepp, B.R., Chen, X., Schwerin, A.F., Paci, I., Smith, M.B., Akdag, A., Johnson, J.C., Nozik, A.J., Michl, J. and M.A. Ratner. 2009. Singlet exciton fission for solar cell applications: energy aspects of interchromophore coupling. *J. Phys. Chem.* DOI 10.1021/jp909002d.
115. Reichman, J. 1980. The current-voltage characteristics of semiconductor-electrolyte junction photovoltaic cells. *Appl. Phys. Lett.* 36:574.
116. Sah, C.T., Noyce, R.N. and W. Shockley. 1957. Proc. IRE 45:1228; Henry, C.H., Logan, R.A. and F.R. Merritt 1978. *J. Appl. Phys.* 44:3530.
117. Trasatti, S. 1984. Electrocatalysis in the anodic evolution of oxygen and chlorine. *Electrochim. Acta* 29:1503.
118. Harriman, A., Pickering, I.J., Thomas, J.M. and P.A. Christensen. 1988. Metal oxides as heterogeneous catalysts for oxygen evolution under photochemical conditions. *J. Chem. Soc. Faraday Trans. I* 84:2795.
119. Kleiman-Shwarsctein, A., Hu, Y.S., Stucky, G.D. and E.W. McFarland 2009. NiFe-oxide electrocatalysts for the oxygen evolution reaction on Ti doped hematite photoelectrodes. *Electrochem. Commun.* 11:1150.
120. Rüttinger, W. and G.C. Dismukes. 1997. Synthetic water-oxidation catalysts for artificial photosynthetic water oxidation. *Chem. Rev.* 97:1.
121. Najafpour, M.M. 2011. Mixed-valence manganese calcium oxides as efficient catalysts for water oxidation. *Dalton Trans.* 40:3793.
122. Kok, B., Forbush, B. and M. McGloin. 1970. Cooperation of charges in photosynthetic O_2 evolution I: a linear four step mechanism. *Photochem. Photobiol.* 11:457.
123. Jursinic, P.A. and R.J. Dennenberg. 1990. Oxygen release time in leaf discs and thylakoids of peas and Photosystem II membrane fragments of spinach. *Biochim. Biophys. Acta* 1020:195.
124. Razeghifard, M.R. and R.J. Pace. 1999. EPR Kinetic Studies of Oxygen Release in Thylakoids and PSII Membranes: A Kinetic Intermediate in the S3 and S0 Transition. *Biochemistry* 38:1252.
125. Kanan, M.W. and D.G. Nocera. 2008. In situ formation of an oxygen-evolving catalyst in neutral water containing phosphate and Co^{2+}. *Science* 321:1072.
126. Glunz, S.W., Sproul, A.B., Warta, W. and W. Wetting. 1994. Injection-level-dependent recombination velocities at the Si-SiO2 interface for various dopant concentrations. *J. Appl. Phys.* 75:1611.
127. Grant, N.E. and K.R. McIntosh. 2009. Passivation of a (100) silicon surface by silicon dioxide growth in nitric acid. *Elec. Dev. Lett.* 30:922.
128. Kobayashi Asuha, H., Maida, O., Takahashi, M. and H. Iwasa. 2003. Nitric acid oxidation of Si to form ultrathin silicon dioxide layers with a low leakage current density. *J. Appl. Phys.* 94:7328.
129. Maldonado, S. and N.S. Lewis. 2009. Behavior of electrodeposited Cd and Pb Schottky junctions on CH3-terminated n-Si(111) surfaces. *J. Electrochem. Soc.* 156:H123.
130. Henry de Villeneuve, C., Pinson, J., Bernard, M.C. and P. Allongue. 1997. Electrochemical formation of close-packed phenyl layers on Si(111). *J. Phys. Chem. B* 101:2415.
131. Buriak, J.M. 2002. Organometallic chemistry on silicon and germanium surfaces. *Chem. Rev.* 102:1271.
132. Würfel, P. 2000. Physik der Solarzellen, Spektrum Akademischer Verlag GmbH, Heidelberg-Berlin.
133. Barter, L.M.C., Klug, D.R. and R. van Grondelle. 2005. Energy trapping and equilibration: a balance of regulation and efficiency in Photosystem II. The light-driven water: plastoquinone Oxidoreductase, T. Wydrzynski, K. Satoh (Ed.), *Advances in photosynthesis and respiration*, Vol. 22, Springer Verlag, Dordrecht.

134. Zhu, X.-Y., Wang, Q. and M. Muntwiller. 2009. Charge-transfer excitons at organic semiconductor surfaces and interfaces. *Acc. Chem. Res.* 42:1779.
135. Wannier, G.H. 1937. The structure of electronic excitation levels in insulating crystals. *Phys. Rev.* 52:191.
136. Frenkel, J. 1931. On the transformation of the light into heat in solids, I. *Phys. Rev.* 37:17.
137. M.L. Ghirardi, L. Zhang, J.W. Lee, T. Flynn, M. Seibert, E. Greenbaum, A. Melis. 2000. Microalgae: a green source of renewable H2, *TIBTECH* 18: 506.
138. Gaffron, H. and J. Rubin. 1942. *J. Gen. Physiol.* 26:219.
139. Alstrum-Acevedo, J.H., Brennaman, M.K. and T.J. Meyer. 2005. Chemical approaches to artificial photosynthesis, 2. *Inorg. Chem.* 44:6802.
140. Banwell, C.N. and E.M. McCash. 1999. Molekülspektroskopie, Ein Grundkurs, Oldenburg Verlag, München.
141. Herek, J.L., Wohlleben W., Cogdell R.J., Zeidler D. and M, Motzkus 2002. Quantum control of energy flow in light harvesting. *Nature* 417:533.
142. May, V. and O. Kühn 2000. *Charge and energy transfer dynamics in molecular systems*, Wiley-VCH Verlag, Berlin.
143. Ulstrup, J. and J. Jortner. 1975. The effect of intramolecular quantum modes on free energy relationships for electron transfer reactions. *J. Chem. Phys.* 63:4358.
144. Förster, Th. 1948. Zwischenmolekulare Energiewanderung und Fluoreszenz. *Analen der Physik* 2:55.
145. Dexter, D.L. 1953. A Theory of Sensitized Luminescence in Solids. *J. Chem. Phys.* 21:836.
146. Naqvi, K.R. and C. Stee. 1993. Inverted region in intermolecular electronic energy transfer. *Spectrosc. Lett.* 26:1761.
147. Gerischer, H. 1960. Über den Ablauf von Redoxreaktionen an Metallen und an Halbleitern. I. Allgemeines zum Elektronenübergang zwischen einem Festkörper und einem Redoxelektrolyten. *Z. Phys. Chem. n. F.* 26:223.
148. Landau, L. 1932. Zur Theorie der Energieubertragung. II. *Physics of the Soviet Union* 2: 46.
149. Zener, C. 1932. Non-adiabatic crossing of energy levels. *Proc. Royal Soc. London A* 137: 696.
150. Redfield, A.G. 1965. The theory of relaxation processes. *Adv. Magn. Reson.* 1:1.
151. Jean, J.M., Friesner, R.A. and G.R. Fleming. 1992. Applications of a multilevel Redfield theory to electron transfer in condensed phases. *J. Chem. Phys.* 96:5827.
152. Egorova, D., Thoss, M., Domcke, W. and H. Wang. 2003. Modeling of ultrafast electron-transfer processes: validity of multilevel Redfield theory. *J. Chem. Phys.* 119:2761.
153. Jean, J.M., Friesner, R.A. and G.R. Fleming 1992. Applications of a multilevel Redfield theory to electron transfer in condensed phases. *J. Chem. Phys.* 96:5827.
154. Egorova, D., Thoss, M., Domcke, W. and H. Wang 2003. Modeling of ultrafast electron-transfer processes: validity of multilevel Redfield theory. *J. Chem. Phys.* 119:2761.

2 Nanodimensioned Systems Based on Semiconductor–Oxide–Metal–Electrolyte Contacts

The conversion of solar energy into chemical or electrical energy in inorganic photo-electrocatalytic systems is based on an efficient collection of photogenerated charges under nanodimensioned semiconductor–metal Schottky type junctions in contact with the reacting electrolyte. The metal phase is separated from the semiconductor by a stabilizing film consisting of a thin dielectric or semiconducting oxide, which protects the underlying light absorber from (photo)-chemical attack. A method to achieve this type of interfaces is by using an electrochemical treatment, sometimes referred to as surface conditioning including chemical or electrochemical etching, anodic or cathodic passivation, and metal deposition steps. In this chapter some fundamental aspects of these different steps will be described and illustrated with practical examples.

2.1 FUNDAMENTALS OF ELECTROCHEMICAL PHASE FORMATION

The formation of metal particles of reduced dimensions by electrochemical reduction of metal ions onto a foreign substrate takes place as the electrode potential is driven externally to a cathodic value over the equilibrium potential for the system M/M^{z+}. The overpotential, defined as $\eta = V_{ap} - V^0_{M/M^{z+}}$, represents the work necessary to overcome the birth and growth energy [1] of the new phase. The equilibrium potential we refer to is in reality the potential drop established under equilibrium of the metal with its cations in solution at the metal–electrolyte interface. Thermodynamically, the equilibrium potential of the system $M/M^{z+}_{(aq)}$ is described by the condition

$$\Delta G_r = \sum_i \nu_i \tilde{\mu}_i = 0 \tag{2.1}$$

where ν_i are the stoichiometric coefficients of the species i in the reduction reaction. Thus, according to the following generic dissolution reaction,

$$M + n\,H_2O \rightarrow M(H_2O)_n^{z+} + z\,e^- \tag{2.2}$$

we have

$$\tilde{\mu}_M + n\tilde{\mu}_{H_2O} = \tilde{\mu}_{M(H_2O)_n^{z+}} + z\tilde{\mu}_e \tag{2.3}$$

which after its expansion in chemical and electrical terms gives

$$\mu_M + n\mu_{H_2O} = \mu_{M(H_2O)_n^{z+}} + ze\varphi_s + z\mu_e - ze\varphi_M \tag{2.4}$$

where the subscripts s and M refer to the liquid and metal phase, respectively. Solving Eq. (2.4) for $(\varphi_s - \varphi_M)$, we can see that the potential drop at the metal–solution interface is a linear function of the chemical potential of electrons in the metal phase:

$$(\varphi_s - \varphi_M) = \frac{\mu_M + n\mu_{H_2O} - \mu_{M(H_2O)_n^{z+}} - \mu_e}{ze} \tag{2.5}$$

Equation (2.5) is the well-known Nernst equation, which gives the concentration dependence of the equilibrium potential. This equation can be viewed as the thermodynamic description of the electrified interface. Figure 2.1 shows a schematic representation of a metal cluster in equilibrium with its cations in solutions. The cation to be reduced is placed at the outer Helmholtz plane, defined by the minimal approach distance of hydrated ions.

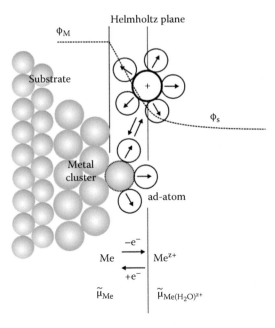

FIGURE 2.1 Schematic showing a growing metal cluster in contact with the solution containing its cations.

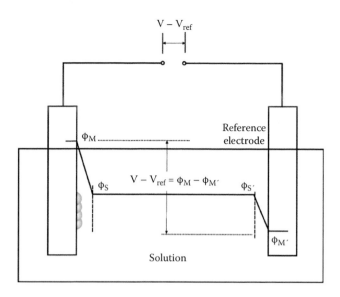

FIGURE 2.2 Measurement of electrode potentials.

The absolute value of the interfacial potential drop can only be determined by means of an additional reference interface. It can be demonstrated that the voltage measured at the terminals of a two-electrode constructed cell is given by (see Figure 2.2)

$$V = (\varphi_s - \varphi_M) - \mu_e^M - (\varphi_s^{ref} - \varphi_M^{ref}) + \mu_e^{Mref} = V - V_{ref} \tag{2.6}$$

According to this expression, the following definition of the absolute potential can be derived [1]:

$$V_{abs} = (\varphi_M - \varphi_s) - \mu_e^M / e \tag{2.7}$$

The absolute potential is in principle an abstract term, which is not experimentally accessible. However, it was estimated for the case of the hydrogen evolution reaction (her) by means of a thermodynamic cycle using quantities extracted from vacuum physics and theoretically calculated terms. A value of $eV_{abs\,(H+/H2)} = -4.6 \pm 0.1$ eV was obtained [2,3]. All of these fundamental definitions concerning the thermodynamics of the metal–electrolyte interface for nonpolarizable electrodes aim to find a relation between the adopted interfacial potential drop and the physical properties of the metal. The outer or Volta-potential φ consists of an electrical contribution, the inner or Galvani potential ϕ, and a dipole term χ. Thus, the absolute potential can be expressed as

$$V_{abs} = (\phi_M - \phi_s) + \chi_m - \chi_s - \mu_e^M / e \tag{2.8}$$

On the other hand, the electron work function of the metal will be defined as

$$\Phi_m = -\mu_e^M + e\chi_m \tag{2.9}$$

Introducing (2.9) in (2.8), we have

$$V_{abs} = (\phi_M - \phi_s) - \chi_s + \Phi_m / e \tag{2.10}$$

Equation (2.10) predicts a direct proportionality of the electrode potential of the metal in equilibrium with its ions in solution and the corresponding metal work function. Figure 2.3 shows the reduction potentials for various metals plotted as a function of their work functions. In principle, it can be observed that this tendency is followed by a large amount of elements, which are placed basically around two lines with a slope of +1. The scatter of point is related with the different values of the surface dipole at the solution side χ_s and the electrical potential drop arising at the interface $\phi_M - \phi_s$. Trasatti [4,5] has investigated in details this relationship for the case of polarizable electrodes (i.e., without the occurrence of electrochemical reactions) of different metals at the potential of zero charge (discharged interface). Thus, $\phi_M - \phi_s = 0$. It was found that transition and sp-metals can be placed on two defined lines with some deviations. This splitting was ascribed to different orientations adopted by adsorbed water molecules at the interface and the consequent different created dipoles. The relation between the work function and the equilibrium potential for the dissolution-deposition reaction can also be analyzed by regarding the energy dependence of the electron distribution at the metal surface. The electron density distribution at the interface can be described by the Jellium-model [6], according to which electrons spread out from the lattice of positive metal cores. Although this model was developed for the metal surface in a vacuum, it is in principle acceptable to extrapolate it to the metal–solution interface. The Jellium model predicts the formation of a stabilizing surface electrical dipole, as the Fermi level, given by the electrochemical potential, is equal to the work function. In contact with the solution, the energy of a stable interface will be modified to balance the dipole redistribution at the interface.

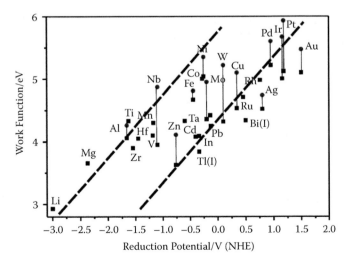

FIGURE 2.3 Dependence of the equilibrium potential for the metal reduction dissolution reaction on the electron work function of the metal.

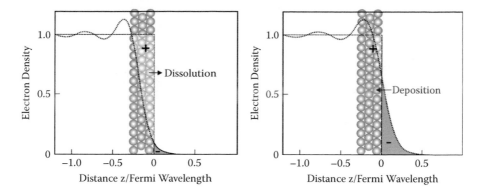

FIGURE 2.4   Electron density distribution at the metal–solution interface as modeled by the Jellium model at, below, and above the Fermi level.

Figure 2.4 shows a schematic representation of the electron density distribution expected above and below the Fermi level. As the Fermi level is driven downward, that is, toward lower (more negative) energies, an interfacial charge imbalance with an excess of positive charge at the metal surface weakens the metal bonds and the dissolution process is thus favored. On the other hand, an excess of electrons at the surface by an upwards shift of the Fermi level accelerates the charge transfer rate in the reduction of metal cations and the growth of the metal phase takes place. These concepts refer to a metal-solution interface with a surface area large enough, so that size effects can be disregarded. Electrochemical deposited particles of Rh have for instance a radius between 100 nm to 10 nm, which corresponds to an amount between 3.041×10^8 and 3.042×10^5 atoms. This amount is still beyond that necessary for the appearance of quantum effects. On the other hand, the metal-solution electrified double layer (Helmholtz-layer) extends 0.1 to 0.2 nm from the metal surface. Thus, the electrochemistry of the particles can be regarded as the same of planar surface if we compare the double layer thickness with a radius of 10 nm. Size effects however become important during the first stages of growth of nucleated stable clusters, as we will see later.

The formation of a new metal phase on a foreign substrate starts with the clustering of few atoms to a stable nucleus. Figure 2.5 shows in a simplified way the atomic process involved in the formation of a new phase. The first step consists of the electroreduction of a metal cation situated at the outer Helmholtz-plane (see Chapter 4). The reduced atoms remain absorbed at the surface for a determined time. Within their lifetimes, the ad-atoms can join to another group of ad-atoms to build a stable nucleus or be dissolved again. As the reader can envisage, we are dealing with a dynamic process and therefore, we have to speak about the probabilities when referring to the rate of nucleus formation, as will be later explained. The new metal phase adopts different morphologies depending on the relation of the binding energies of atoms of the substrate, $E(S\text{-}S)$, that of atoms of the metal, $E(M\text{-}M)$, and that of atoms of the substrate and the metal, $E(M\text{-}S)$ (see Figure 2.5). When $E(M\text{-}M) \gg E(S\text{-}S)$ there is a tendency of the depositing metal to form isolated islands. This growth mode is also referred to as the Volmer–Weber mechanism [7], and is the most frequent mode seen

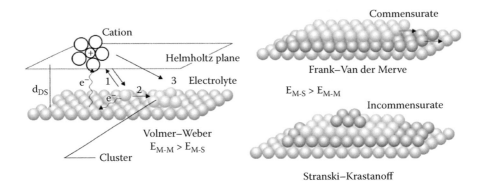

FIGURE 2.5 Atomistic representation of the electrodeposition of a metal cluster on a metal substrate and different types of growth modes: (1) reduction of cations and formation of ad-atoms on the substrate; (2) surface diffusion of ad-atoms; (3) reduction of cations and formation of ad-atoms on the growing cluster.

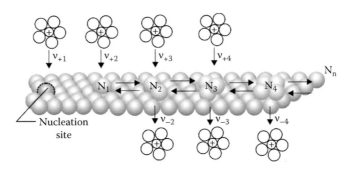

FIGURE 2.6 Atomistic representation of the electrochemical nucleation and growth of new metal phases on foreign substrates.

when electrodepositing metals onto a foreign substrate in the absence of additives or complexing agents. In the case where $E(M\text{-}S) \gg E(M\text{-}M)$, the metal nucleus tends to grow in a two-dimensional mode. Here, two cases can be distinguished depending on to what extent the structures of the metal and substrate match. If the metal and substrate have very similar lattice structures and comparable atomic distances, the formed nucleus extends up to cover the whole or a large part of the surface. The metal phase grows in a layer-by-layer mode, also known as the Frank–van der Merve mechanism. In the case that the metal and substrate have very different lattice parameters, large stress forces arise in the first deposited layers. The stressed interface structure relaxes by a 3D growth of islands on the first epitaxial layer; this is known as the Stranski–Krastanoff mechanism.

One of the first thermodynamic descriptions of the electrochemical metal electrodeposition, often referred to as electrocrystallization, was made in 1928 by Gibbs [8]. Following this were the contributions of Volmer [9,10], Stranski [11], Kaischew [12], and Böcker and Döring [13], who initiated a theoretical treatment of the kinetic and statistical aspects of the process. Kaischew [14], particularly, presented the first molecular-kinetic analysis of the electrocrystallization process by introduction of the attachment–detachment frequency concept, which was further completed by Conway and Bockris [15] and Gerischer [16] by considering the mechanisms of incorporation of the deposited ad-atoms in the crystal structure.

Continuing with the analysis of the nucleation rate in terms the attachment–detachment frequency, as introduced by Kaischew, Stoyanov developed a theory to quantify the nucleation process [17]. After this theory, the probability of formation of a stable growing nucleus is directly proportional to the difference of the frequency of attachment attempts and that of redissolution of the ad-atoms [18,19]. In reality, this concept can be regarded as an extension of the probabilistic analysis of the cluster formation formulated by Becker and Döring [20] for electrochemical systems. According to this analysis, the formation of a stable nucleus is preceded by a steady state system constituted by a series of nuclei with increasing number of atoms which exchange, with each other by attachment or detachment of one atom, as represented schematically in Figure 2.6. Thus, the rate of formation of cluster with n atoms (i.e., the number of cluster births as a function of time) is given by

$$\frac{dN_n}{dt} = \frac{v_{+1}N_1}{1 + \sum\limits_{j=2}^{n} \prod\limits_{i=2}^{j} \frac{v_{-i}}{v_{+i}}} \tag{2.11}$$

where v_{+i} is the attachment frequency of one atom to form a cluster with i atoms and v_{-i} is the detachment frequency of one atom from the cluster with i atoms. N_i is the number of clusters with i atoms. In particular, N_1 corresponds to the number of nucleation sites occupied by an only ad-atom. Let us assume that the cluster becomes stable as the number of atoms forming it is $\geq N_{nc}$. This cluster will designed as *critical nucleus*. The product corresponding to n_c in the sum of the denominator of Eq. (2.11) is much larger than all other terms. Thus, Eq. (2.11) can be reduced to

$$\frac{dN_{n_c}}{dt} = \frac{v_{+1}N_1}{\dfrac{v_{-2}}{v_{+2}} \dfrac{v_{-3}}{v_{+3}} \dfrac{v_{-4}}{v_{+4}} \cdots \dfrac{v_{-n_c}}{v_{+n_c}}} = \frac{v_{+n_c}N_1}{\dfrac{v_{-2}}{v_{+1}} \dfrac{v_{-3}}{v_{+2}} \dfrac{v_{-4}}{v_{+3}} \cdots \dfrac{v_{-n_c}}{v_{+n_c-1}}} \tag{2.12}$$

Now, we consider the attachment of one atom to the cluster with $(i-1)$ atoms and the detachment of one atom from the formed cluster with i atoms. Under supposition that the reduction of cations and the oxidation of ad-atoms can be described by an activation mechanism, the frequency of attachment and detachment will be given by

$$v_{+i} = k_{+i} \times c \times \exp\left[-\frac{\Delta G^*_{+i}}{kT}\right] = k_{+i} \times c \times \exp\left[-\frac{\mu_i - \alpha_c ze(V_0 + \eta)}{kT}\right] \tag{2.13}$$

$$v_{-i} = k_{-i} \times \exp\left[-\frac{\Delta G^*_{-i}}{kT}\right] = k_{-i} \times \exp\left[-\frac{\mu'_i + (1-\alpha_c)ze(V_0 + \eta)}{kT}\right] \tag{2.14}$$

where μ_i and μ'_i are the Gibbs free energy of a cluster of $(i-1)$ and i atoms, respectively. As explained in more detail in Chapter 4, the terms $\alpha_c z e(V_0 + \eta)/kT$ and $(1-\alpha_c)z e(V_0 + \eta)/kT$ represent the contribution of the potential drop at the metal–solution interface to the activation energy. α_c is the charge transfer coefficient, V_0 is the equilibrium potential of the reaction (2.2), and η is the applied overpotential $(V-V_0)$. c is the concentration of metal cations in the solution and k_{+i} and k_{-i} are specific rate constants. The relation v_{-i}/v_{+i} is equivalent to the relation v_{-i}/v_{+i-1} of Eq. (2.12). Thus, from Eq. (2.13) and Eq. (2.14), we have

$$\frac{v_{-i}}{v_{+i}} = \frac{k_{+i}}{k_{-i}} \times \frac{1}{c} \times \exp\left(-\frac{W_i}{kT}\right) \exp\left[-\frac{ze(V_0 + \eta)}{kT}\right] \tag{2.15}$$

W_i represents the work required to remove one atom from the cluster with i atoms. V_0 is defined as the equilibrium potential, which can be related with the cation concentration by the Nernst relation:

$$\exp\left(-\frac{W_{1/2}}{kT}\right) \exp\left(-\frac{ze V_0}{kT}\right) = \left(\frac{k_{+1/2}}{k_{-1/2}}\right) c \tag{2.16}$$

Relation (2.16) was derived considering the dissolution–deposition of an atom at a kink site in an infinitely large crystal surface. In the particular case of a kink site, denoted with the subscript ½, the number of their concentration does not change after removing an atom from this site. On the other hand, N_1 refers to the surface concentration of ad-atoms occupying nucleation sites on the surface. This quantity can be calculated considering the equilibrium equation for the adsorption process as follows:

$$N_1\left(\frac{k_{-1}}{k_{+1}}\right) = N_0 c \exp\left(\frac{W_1 - ze(V_0 + \eta)}{kT}\right) \tag{2.17}$$

where N_0 is the concentration of active sites. Introducing Eqs. (2.14–17) into Eq. (2.13) and after some rearrangement, we obtain

$$\frac{dN}{dt} = \frac{\left(\frac{k_{-1/2}}{k_{+1/2}}\right)^n}{\prod_1^n \left(\frac{k_{-j}}{k_{+j}}\right)} k_{+1} \times c \times N_0 \times \exp\left(\frac{W_n - \alpha z e V_0}{kT}\right) \tag{2.18}$$

$$\times \exp\left(-\frac{\Phi(n)}{kT}\right) \times \exp\left(\frac{(n+\alpha)}{kT} z e \eta\right)$$

In this expression,

$$\Phi(n) = n W_{1/2} - \sum_1^n \mu'_i$$

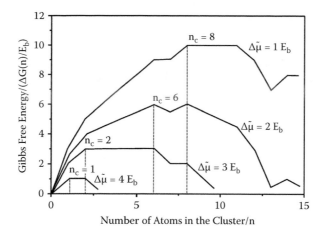

FIGURE 2.7 Theoretical calculation of the change of the Gibbs free energy of a growing cluster after the atomistic theory of nucleation.

This term can be regarded as the energy difference between a group of n-atoms inside an infinitely large crystal and that of the same number of atoms in the cluster. $W_{1/2}$ accounts for the work necessary to remove an atom from a kink site. Since the removal of a kink site leaves another kink site, the total energy to sublimate the cluster can be regarded as $n\,W_{1/2}$. According to the thermodynamic criterion, the precondition for stable growth of a cluster formed by n-atoms is the negative change of the Gibbs free energy $\Delta G(n)$ with the addition of more atoms to the cluster. This change can be expressed as

$$\Delta G(n) = -nze(\tilde{\mu}_K - \tilde{\mu}_s) + G_s(n) = -nze\eta + G_s(n) \qquad (2.19)$$

The first term on the right-hand side of Eq. (2.19) represents the driving force for the crystal growth, where $\tilde{\mu}_K$ and $\tilde{\mu}_s$ are the electrochemical potential of an atom in an infinitely large crystal and that of the metal cation in the solution, respectively. The difference between these two values, $\Delta\tilde{\mu} = (\tilde{\mu}_K - \tilde{\mu}_s)$, is denoted as electrochemical supersaturation and is a function of the cation concentration. The second term in Eq. (2.19) represents the work required for the creation of the metal–electrolyte interface, which is given by

$$\Phi(n) = nW_{1/2} - \sum_{1}^{n} \mu_i'$$

The maximum shown by the energy balance (2.19) represents the inflexion point at which further growth is thermodynamically favored. The atomic cluster at this point corresponds to the critical nucleus. It is important to observe that the number of atoms in the critical nucleus decreases with the supersaturation. Values of $\Delta G(n)$ for an increasing number of atoms in the cluster were calculated by Stoyanov et al. [21].

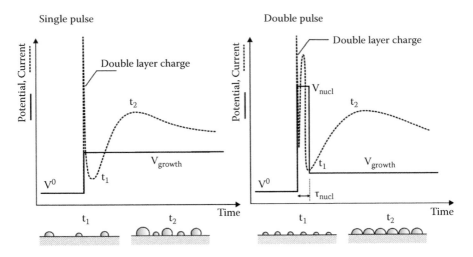

FIGURE 2.8 Pulse modalities and general form of current transients.

Their results are shown in Figure 2.8 for various level of supersaturation indicated as a multiple of the atom binding energy E_b.

At sufficiently low supersaturation values, the critical nucleus may reach a large amount of atoms, so that its properties can be regarded as macroscopic. In that case, the surface energy will be given by $G_s = \gamma \times a_g \times n^{2/3}$, where γ is the specific surface energy, a_g is a geometric factor, and n is the number of atoms per unit of volume. Introducing the late expression in Eq. (2.19), the number of atoms in the critical nucleus is found for the condition $(\partial \Delta G(n)/\partial n) = 0$. Thus, we obtain

$$n_c = \left(\frac{2\gamma a_g}{3ze|\eta|} \right)^3 \tag{2.20}$$

Although this situation is seldom found in conventional electrochemical deposition experiments, this analysis offers a qualitative description of the nucleation behavior. Thus, the nucleation rate, for instance, can be described by the first-order expression

$$\frac{dN}{dt} = k_{0n}c \times \exp\left(-\frac{\Delta G(n_c)}{kT} \right) = k_{0n}c \times \exp\left(\frac{n_c ze\eta - \gamma \times a_g \times n_c^{2/3}}{kT} \right) = \tag{2.21}$$

$$= k_{0n}c \times \exp\left(\frac{1}{27} \frac{4}{kT} \frac{(\gamma a_g)^3}{(ze)^2} \frac{1}{\eta^2} \right)$$

This expression gives a quantitative relationship between the nucleation rate and the overpotential η, which was found to apply for the nucleation of some electrochemical systems at low overpotentials.

2.2 EXPERIMENTAL TECHNIQUES AND EXPERIMENT DESIGNS

Investigations of the electrochemical nucleation and growth processes follow in general two modalities: one indirect, by analyzing the time-dependence of the current on applying a potential pulse and one straightforward method by counting the number of particles using a microscopy technique such as scanning electron microscopy (SEM) or atomic force microscopy (AFM). Due to the relative simple experimental implementation of the method, the pulse technique is the most used. The critical point of this technique, however, is the selection of an adequate theory that can describe the mechanism of incorporation of atoms to the growing crystal and the mass transfer of the cations in the solution. The former includes the knowledge of the electron transfer kinetics and the surface diffusion of ad-atoms. In its primary form, the technique consists in applying a potential pulse to the investigating system from a potential $V_0 \geq$ equilibrium potential for the reduction reaction to a potential V. Initially, the measured current shows a spike due to the charge of the interface. To this follows an increasing course reflecting the enlargement of the crystal area or the diffusion zone in the case of a mass-transport controlled process (see left scheme in Figure 2.8). The second mode consists in the application of a first short (τ_{nucl}) and large pulse for the nucleation of clusters, which grow in a second, long, much lower pulse, so that further nucleation is hindered (see right scheme in Figure 2.8).

The type of control depends on the concentration of cations and the stirring conditions. If the transport of the depositing metal is high enough (high cation concentration or strong solution stirring), the deposition rate is controlled by the electron transfer reaction or by the surface diffusion of ad-atoms to the growing centers. Surface films may arise, however, as a consequence of chemical changes at the very proximity of the surface in the presence of side reactions during the deposition process. For instance, the precipitation of hydroxides due to a strong alkalization produced by the hydrogen evolution [22] can considerably hinder the reduction current.

One of the first models for the interpretation of the current-time curves (current transients) by application of potential pulses was formulated by Fleischmann and Thirsk [23] and later completed by Armstrong and Harrison [24,25]. The model is based on the formation of many nuclei, the growth of which is kinetically controlled by the electron transfer process [26]. The model considers the growth of particles with a defined geometrical form as of right cones or spheres. For the derivation of an expression for the current transient, an adequate formulation of the nucleation rate is required. This is given by the phenomenological equation:

$$N(t) = N_0\left[1 - \exp(-k_n t)\right] \tag{2.22}$$

where $N(t)$ is the time-dependent number of formed nuclei, N_0 [cm^{-2}] is the total number of nucleation sites at the surface, and k_n is the specific nucleation rate [s^{-1}]. Usually, two limit cases are considered: If $k_n t \gg 1$, all nucleation sites are rapidly saturated on applying the potential pulse, that is, $N(t) = N_0$. This case receives the name *instantaneous nucleation*. On the contrary, if $k_n t \ll 1$, the expression (2.22) reduces to $N(t) = N_0 k_n t$. This case is known as *progressive nucleation* and the term $J_n = N_0 k_n t$ is the nucleation rate [cm^{-2}s^{-1}].

The derivation of the current-time equation in electrocrystallization processes also requires the calculation of the fractional area of growing centers, which are expected to overlap after a defined growth time. Here, the Avrami [27] theorem will be applied, according to which the fractional area of the growing centers $\theta = A_n/A_0$ is related to the extended area without overlap $\theta_{ext} = A_{ext}/A_0$ as

$$\theta = 1 - \exp(-\theta_{ext}) \tag{2.23}$$

On the basis of Eqs. (2.22) and (2.23), the following current density–time relation was derived for the growth of right cones [28]:

$$j = zFk_{out}\left\{1 - \exp\left[-\frac{\pi M^2 k_{lat}^2 N_0}{\rho^2}\left(t^2 - \frac{2}{k_n}t + 2\frac{1}{k_n^2} - 2\frac{1}{k_n^2}\exp(-k_n t)\right)\right]\right\} \tag{2.24}$$

For hemispheres, the current is given by the more complex expression [29,30]

$$j = 2zFk_{out}\left(\frac{\pi M^2 k_{lat}^2 N_0}{\rho^2}\right)\left\{1 - \exp\left[-\frac{\pi M^2 k_{lat}^2 N_0}{\rho^2}\left(t^2 + 2\frac{1}{k_n^2}(k_n t - 1)\right)\right]\right\} \times \tag{2.25}$$

$$\times \int_0^t \left[t - \frac{1}{k_n} - \left(\frac{k_n u - 1}{k_n}\right)\exp(-k_n(t-u))\right] \times$$

$$\times \exp\left[\frac{\pi M^2 k_{lat}^2 N_0}{\rho^2}\left(u^2 - \frac{2}{k_n^2}(k_n u - 1)\exp(-k_n(t-u))\right)\right]du$$

where M and ρ are the molar mass and the density of the deposited metal, respectively. These expressions include a rate constant for the lateral, k_{lat}, and outward growth k_{out}, of the crystal. The general course of the current transients given by Eqs. (2.24) and (2.25) are shown in Figure 2.9. These expressions reproduce the main features of the current transients observed experimentally in kinetically controlled systems: an initial increase up to a maximum, which indicates the overlap of the growing centers. At the beginning of the transient, Eq. (2.24) and (2.25) reduce to the expressions:

$$j(t) = \gamma z Fk_{out}\left(\frac{\pi M^2 k_{lat}^2 N_0}{\rho^2}\right)t^2 \tag{2.26}$$

and

$$j(t) = \gamma z Fk_{out}\left(\frac{\pi M^2 k_{lat}^2 J_n}{3\rho^2}\right)t^3 \tag{2.27}$$

for the limits cases of an instantaneous and a progressive nucleation respectively, with the coefficient $\gamma = 2$ for hemispheres and $\gamma = 1$ for right cones. These equations

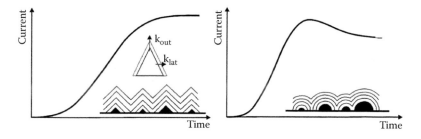

FIGURE 2.9 Form of the current transients given by the theoretical expressions derived by assuming the nucleation and growth of right cones (left) and hemispheres (right).

are useful for determining the type of nucleation and the calculation of its parameters by fitting the initial rising part of the transients to one of them.

Scharifker and Hills developed a model for the description of systems whereby the crystal growth is controlled by mass transport [31,32]. The theory is based essentially on the calculation of the evolution of hemispherical diffusion zones around growing crystallites. Further improvements of this first formulation were reported by Scharifker and Mostany [32], Sluyters-Rehbach et al. [33], Mirkin and Nilov [34], and Heerman and Tarallo [35]. These upgrades of the original theory aim to resolve some oversimplifications as the treatment of limit cases of nucleation rate and the susceptibility of the nucleation probability on nucleation sites to the depletion of cations caused by outward expansion of the diffusion zones.

The model of Scharifker and Hill considers that a hemispherical diffusion zone arises around the freshly nucleated crystallites [31,32,36]. The current for each individual crystallite increases as a consequence of the radial expansion of the mass transport field until the growing hemispheres overlap (see scheme in Figure 2.10). Hence, a transition from a hemispherical to a planar diffusion arises. The problem was resolved by assuming that the material flux to the electrode is equivalent to a planar diffusion over the projected area of the overlapping diffusion zones, the radius of which increase at a rate given by

$$r_{diff}(t) = \sqrt{C \times Dt} \tag{2.28}$$

where C is a constant determined by the experimental conditions and equal to $(8\,\pi\,c_i M/\rho)^{1/2}$, and D is the diffusion coefficient of the reducing species. The projected area is calculated by the Avrami equation (2.23). Hereby, the fraction area covered by diffusion zones without overlapping is given by

$$\theta_{ext} = N_0 \pi CDt \tag{2.29}$$

for an instantaneous nucleation and by

$$\theta_{ext} = \frac{4}{3}\pi CD \int_0^t J_n t\, dt = \frac{4}{6}\pi CDt^2 \tag{2.30}$$

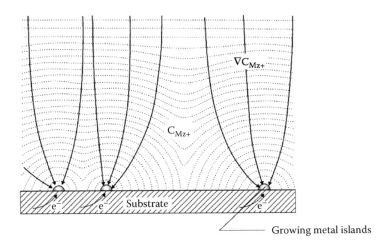

FIGURE 2.10 Schematic resuming the formulation of the Scharifker–Hills model.

for a progressive nucleation. Introducing Eqs. (2.29) and (2.30) into Eq. (2.23) yields the following expressions for the limit cases of an instantaneous (Eq. 2.31) and a progressive (Eq. 2.32) nucleation:

$$j(t) = \frac{zF\sqrt{D_i}\,c_i}{\sqrt{\pi}\sqrt{t}}\left\{1 - \exp\left[-N_0\pi D_i\left(\frac{8\pi c_i M}{\rho}\right)^{1/2} t\right]\right\} \tag{2.31}$$

$$j(t) = \frac{zF\sqrt{D_i}\,c_i}{\sqrt{\pi}\sqrt{t}}\left\{1 - \exp\left[-\frac{2}{3}J_n\pi D_i\left(\frac{8\pi c_i M}{\rho}\right)^{1/2} t^2\right]\right\} \tag{2.32}$$

It was shown that these expressions are able to accurately describe the potentiostatic deposition of many metals. This theory, however, has been frequently criticized because it fails in the prediction of the number of nuclei when corroborated by direct counting on microscopy examinations. Numerical simulations of nucleation and 3D-diffusional controlled growth of crystallites reported by Cao et al. [37] have shown that models in Refs. [32–34] lead to relatively small errors in the calculation of the density of nuclei for the instantaneous type of nucleation. For finite nucleation rates, the model of Mirkin and Nilov gives a better representation of the real situation than the models of Scharifker and Hills and Sluyters-Rehbach et al. One of the major handicaps in the formulation of nucleation and diffusion-controlled growth models is the lack of an estimation of the time-dependent concentration at the non-nucleated sites. This can lead to a considerable reduction of the nucleation probability near growing crystallites and the growth rate of newly born nuclei. Hence, zones with reduced or even zero nucleation rates expand around the growing nuclei. These are referred to as exclusion zones. Markov, Boynov, and Toschev reported experimental evidence for the formation of exclusion zones [38,39]. They deposited silver and mercury by a three-pulse train, where the length of the first pulse was adjusted to the formation of a few nuclei, which continued growing in the second pulse. The height of the third pulse is selected to cause a massive nucleation. In this way, the formation of a zone free of deposits around the crystallites formed during the first pulse was observed. One of the simplifications made in the model of Scharifker and Hills is to suppose that the birth of new nuclei will take place only on the fraction of the surface still uncovered by the diffusion zone. Hence, the time dependence of the formed nuclei can be determined by integrating the probability to find a site still uncovered by a diffusion zone, given by the Avrami equation, where the birth of a nucleus may take place. Thus, we have

$$N(t) = \int_0^t J_n(1-\theta) = J_n \int_0^t \exp\left[-\frac{4}{6} J_n \pi CD\tau^2\right] d\tau \qquad (2.33)$$

$$= N_s erf\left[\frac{\pi^2 \sqrt{J_n \pi \frac{4}{6} CD}}{2} t\right]$$

The nearest-neighbor distribution obtained by numerical simulations of the nucleation process assuming a complete inhibition of nucleation inside the diffusion zones was reported by Scharifker and Hills in Ref. [36]. These results are in good agreement with the experimentally observed distribution. This provides a theoretical basis for the existence of exclusion zones and supports the validity of Eq. (2.33), at least as a good approach.

As in the case of a kinetically controlled deposition, Eqs. (2.31) and (2.32) shows two limit behaviors. For t → 0, these expressions reduce to

$$j(t) = \frac{zFM^{1/2}c_i^{3/2}}{\rho^{1/2}} N_0 \pi D^{3/2} 2^{3/2} \sqrt{t} \qquad (2.34)$$

$$j(t) = \frac{zFc_i^{3/2}M}{\rho^{1/2}\sqrt{\pi}} \frac{2^{5/2}}{3} J_n \pi D_i^{3/2} t^{3/2} \qquad (2.35)$$

Thus, the type of nucleation and the parameters can be determined by analyzing the rising part of the current transients. This is a common practice that circumvents uncertainties related with the mathematical description of the overlap of the diffusion zones. Experimental data are plotted as $j(t)^2$ or as $j(t)^{2/3}$ as a function of time. The alignment of the points in a straight line in one of the plotting forms allows identification of the type of nucleation and its parameters can be calculated from the slope. This type of plotting also takes the induction time, t_0, into account, indicating the delay for the onset of nucleation:

$$j(t)^2 = C(t - t_0) \qquad (2.36)$$

$$j(t)^{2/3} = C'(t - t_0) \tag{2.37}$$

The diffusion constant, necessary for the calculation of the constants C and C', can be calculated from the decreasing part of the transient, which is represented by the Cottrell equation for planar diffusion as

$$j(t) = \frac{zFc_i\sqrt{D_i}}{\sqrt{\pi}\sqrt{t}} \tag{2.38}$$

The quantification of current transients by using Eqs. (2.36) and (2.37) requires the absence of side reactions such as hydrogen evolution. Thus, in this case, this method provides only a qualitative analysis.

The double-pulse technique is thought to allow disregarding an uncertain description of the nucleation kinetics under the overlapping diffusion hemispheres [40]. This technique consists of applying a very short potential pulse after conditioning of the electrode at the equilibrium potential of the reduction reaction (see Figure 2.8). Thus, the birth of crystallites during the pulse is followed by their growth at an overpotential lower than a critical value to avoid further nucleation. The nucleation rate can thus be determined by direct counting of the formed crystallites at different nucleation pulse lengths. This method, however, may imply a redistribution of the particle size during the second pulse due to a dissolution–redeposition process, which is caused by the dependence of the supersaturation on the particle size. The electrochemical potential of small metal particles in equilibrium with its cations in solution differs from that adopted by a large surface as predicted by the Gibbs–Thomson equation:

$$\tilde{\mu}_r = \tilde{\mu}_{r \to \infty} + \frac{2\gamma_{s-l}v_m}{r} \tag{2.39}$$

where γ_{s-l} is the surface energy of the solid–liquid interface, v_m is the molar volume of the metal, and r is the radius of the particle. According to this expression, the equilibrium potential for the reaction (2.2) varies with the particle radius as

$$V_r = V_{r \to \infty} - \frac{1}{ze}\frac{2\gamma_{s-l}v_m}{r} \tag{2.40}$$

This means that particles that have reached their critical radius during the first potential pulse will be undersaturated during the second pulse and hence will redissolve in favor of the growth of larger particles. The validity of the Gibbs–Thomson equation was corroborated experimentally (see Figure 2.11) for silver [41,42], cooper, and mercury particles [43]. Equation (2.40) loses its physical meaning for particles constituted by a small group of atoms, since surface tension is a macroscopic property. The atomistic theory of nucleation, on the other hand, predicts a stepwise dependence of the critical nucleus on the supersaturation.

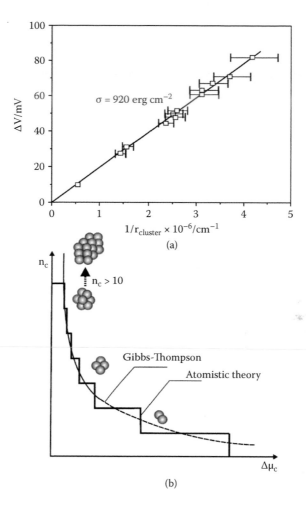

FIGURE 2.11 (a) Change of the equilibrium potential with the particle size; (b) comparison of the relation between the critical nucleus size and the supersaturation predicted by the Gibbs–Thomson equation with that from the atomistic theory (after Ref. 41).

2.3 THE ELECTRODEPOSITION OF SELECTED SYSTEMS

The methods and crystallization theories discussed in Section 2.2 will be exemplified with some particular cases of practical interest. The first example we will discuss is the electrodeposition of In onto vitreous carbon. Vitreous carbon has found large acceptance in electrochemical science due to its inertness, high electrical conductivity, and high overpotential for the hydrogen evolution reaction [44]. Hence, a large potential window without the onset of side reaction makes this material ideal for the study of electrocrystallization processes. The deposition–dissolution mechanism of In is technologically relevant in the fabrication of sacrificial anodes based on aluminum-zinc alloys. The presence of this third alloy element leads to a considerable improvement in the performance of these anodes: an even dissolution and the maintenance of a very negative activation potential (–1.1 V vs. SCE) [45]. The particular decrease of the anodic dissolution potentials in the presence of In can be explained in terms of the threshold potential for the adsorption of chloride on In, which yields a surface concentration of chloride high enough to inhibit the formation of alumina and hence to avoid the surface repassivation [46]. The adsorption of Cl⁻ on In surfaces also has important implications for the photoelectrochemical behavior of a photoelectrochemically conditioned InP surface. The adsorption of Cl⁻ on In is not merely specific [47] but involves the transfer of charge and the formation of a surface dipole, which increases the electron affinity of the metal as we will discuss later.

2.3.1 ELECTRODEPOSITION OF INDIUM

The electrodeposition of In onto vitreous carbon from chloride solutions containing In^{3+} solution involves a three-electron process. The reaction can be written as [48]

$$InCl_n(H_2O)_m^{3-n} + 3e^- \rightarrow In + n\ Cl^- + m\ H_2O \qquad V^0 = -0.656 \text{ V vs. SCE} \qquad (2.41)$$

The process was analyzed by means of potentiostatic pulses of increasing heights. The general course of the current-time curves is characteristic for a nucleation- and diffusion-controlled 3D-growth (see Figure 2.12). The initial increase follows a $j(t) \propto t^{3/2}$ function, thus indicating a progressive type of nucleation. The current density reaches a maximum and then it drops according to $j(t) \propto 1/\sqrt{t}$, as expected after the onset of a planar diffusion regime. Figure 2.13 shows SEM pictures of In particles obtained after electrodeposition from chloride solutions onto vitreous carbon (a) and p-Si(111) (b). In both cases a wide size distribution is observed, in agreement with

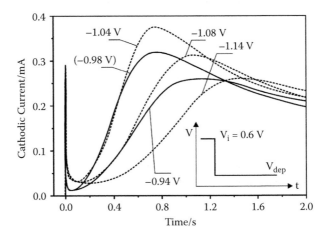

FIGURE 2.12 Current-time curve recorded on applying a potentiostatic pulse of different lengths on a glassy carbon electrode in 0.01 M In^{3+} + 0.5 M NaCl solution of pH 3. V_i = initial potential. V_{dep} = deposition potential.

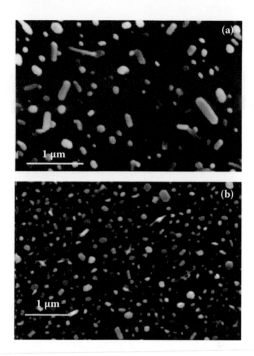

FIGURE 2.13 (a) SEM picture of a glassy carbon electrode after 10 s In deposition from 10 mM In^{+3} + 0.5 M Cl^-, pH 3 at –0.85 V; (b) SEM picture of a p-type Si(111) electrode after deposition of In at –0.6 V from 1mM $InCl_3$ + 0.5 M KCl, pH 4 under illumination.

the progressive type of nucleation indicated by the course of current transients. Here, it is also interesting to note that some particles adopt a baton form, indicating a larger growth rate in a preferential crystallographic direction. This particularity allows the formation of long wires by electrodeposition of In in hole arrays made in an insulating silicon oxide film on the conductive substrate [49] (see Figure 2.14).

The nucleation rate J_n as a function of potential is calculated by fitting Eq. (2.35) to the raising part of the experimental transients, plotted as $j^{2/3}$ vs. t. The log J_n vs. V plot (Figure 2.15) indicates a linear increase between –0.75 V and –1.0 V with a change of the slope of about –0.8 V. According to the atomistic theory of nucleation, the slope of the log J_n – V lines is given by the term $(n + \alpha)$ $ze/2.303$ kT. From –0.75 V to –0.8 V a slope of 55.1 mV dec^{-1} is measured, that is, $(n + \alpha)z = 1.08$. From –0.8 V to –1.0 V, the slope is 113.8 mV dec^{-1}, that is, $(n + \alpha)z = 0.53$. Taking into account that α is about 0.5, it can be inferred that the change of slope is caused by a change of the rate-determining step in the multi-electron reduction process. The electron exchange takes place in three consecutive steps: $In(III) \xrightarrow{1} In(II) \xrightarrow{2} In(I) \xrightarrow{3} In$. Investigations of the reduction kinetics of In(III) have shown that the second step is much faster than the third and first ones. Therefore, the reduction reaction can be written as

$$InCl_n^{3-n} + 2\ e^- \xrightarrow{k_1} InCl + (n-1)\ Cl^- \tag{2.42}$$

$$InCl + e^- \xrightarrow{k_3} In + Cl^- \tag{2.43}$$

FIGURE 2.14 SEM picture of In nanowires grown by passing –0.6 C cm⁻² (referred to the exposed area) at –1.2 V on hole arrays with hole diameter of 150 nm and pitch of 300 nm. G. Hautier et al. [49] are acknowledged. Reproduced by permission of The Electrochemical Society (2008).

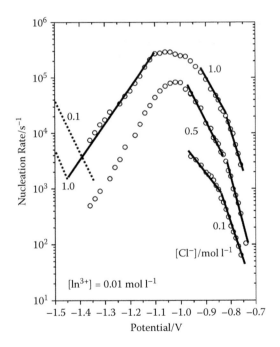

FIGURE 2.15 Nucleation rate of In as a function of the pulse potential from solutions containing 10 mM In³⁺ and different chloride concentrations.

If step (2.42) controls the whole reduction process, we have $z = 2$. Otherwise, reaction (2.43) is the controlling one and hence $z = 1$. In both cases, the slope value indicates $n = 0$. The physical interpretation of this particular result is that the formation of a stable growing nucleus appears just after the adsorption of an ad-atom on the nucleation site [50]. This case is fundamentally observed in heterogeneous nucleation, whereby attachment forces between substrate and ad-atoms play an essential role. The nucleation sites on vitreous carbon can in principle be ascribed to the presence of terminal acetate and aldehyde groups [44]. These sites are especially reactive as a consequence of the van der Waals forces and electrostatic and hydrogen-bridge bonds involved in the electron transfer process. This fact explains the observed strong dependence of the preparation, as well as the prepolarization of the electrode on the electrochemical response upon applying a potential pulse. Another particularity of the nucleation of In particles is the accelerating effect caused by chloride ions. Figure 2.15 shows that the rate of nucleation is practically proportional to the chloride concentration. On the other hand, it drops on increasing the cathodic potential beyond -1.05 V vs. SCE. This potential is close to the threshold potential for the adsorption of chloride on In, indicating that the adsorption of Cl^- is involved in the nucleation mechanism, probably due to an acceleration of the charge transfer by formation of electron bridges. The reduction of crystallization overvoltages in the presence of halogenides is a long-known experimental fact that is commonly attributed to the high electron density and the large polarization constant of anions [51]. Nevertheless, the mechanism by which this occurs is still unknown.

The adsorption step in the formation of In clusters is supposed to occur together with the ad-atom formation, thus contributing to its stabilization. In the absence of chloride, higher energies are required for nucleation, $G_s(n)$, and hence larger overpotentials, η, must be applied. As a consequence, the linear relation log J_n vs. V shifts toward more negative potentials. The drop of the nucleation rate beyond the adsorption potential threshold (i.e., $V < -1.05$ V vs. SCE) is also reflected by an increase of the induction time, t_0. This time is experimentally determined by the point at which the abrupt current increase at the transients occurs. Figure 2.16 shows two current transients recorded close to and beyond the adsorption threshold potential. This quantity can be regarded as inverse of the statistic mean value of the frequency

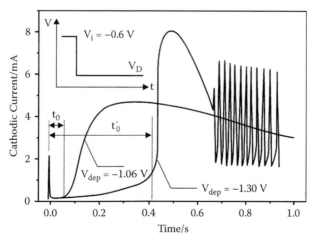

FIGURE 2.16 Current-time curves recorded on a glassy carbon electrode by applying a deposition pulse within the desorption-potential region in a solution containing 10 mM In^{3+} + 0.5 M NaCl at pH 3.

of formation of stable ad-atoms: $t_0 = 1/v_{+1}$. Here, it is also interesting to note that the current transients show the onset of oscillations after the current maximum in the desorption-potential region: -1.05 V $< V < -1.4$ V vs. SCE. Current oscillations in electrochemical systems appear in the case of negative reaction resistances [52], $R_k = (\partial j/\partial V)^{-1}$, as is common in the case of surface passivation by formation of an oxide or in adsorption-desorption processes, whereby an overvoltage increase leads to a retarding of the electron transfer process.

2.3.2   ELECTRODEPOSITION OF CO ONTO SILICON

Semiconductors introduce a new factor in the analysis of the electrodeposition of metals concerning the rate of metal reduction. Here, the presence of chemical moieties on the surface gives rise to particular spatial electronic distributions, which generates preferential sites for nucleation, as for instance oxidized surface atoms, as later discussed. In the following, we will discuss the electrodeposition of Co onto p-Si(111) as a model system. This analysis is intended to form a basis for a further discussion of the deposition of noble metals used for the photo-assisted hydrogen evolution.

The electrodeposition of cobalt and Co-based alloys onto Si-substrates finds application in the fabrication of magnetic thin films, Schottky-type diodes [53,54], and magnetic sensors based on the giant magneto resistance phenomenon [55,56]. The electrochemical deposition offers the possibility of obtaining control of the morphology and size of the metal islands. This allows tailoring the magnetic properties of the conditioned surface, such as the formation of magnetic domains [57,58] or the fabrication of nanomagnets as storage units [59]. The reactivity of Si against the electrolyte introduces an unavoidable modification of the chemical and electronic structure of the metal-semiconductor interface with important consequences for the performance of these systems when working as photoelectrodes in conversion systems.

The reduction of Co^{2+} involves an outer charge transfer, which can be written as

$$Co(H_2O)_6^{2+} + 2\,e^- \rightarrow Co + 6\,H_2O \qquad V^0 = -0.52 \text{ V} \qquad (2.44)$$

Co, like other transition metals such as Pt, Ni, Rh, and Cu, is coordinated with six water molecules, forming an octahedron. They can be replaced by ligands such as SCN^-, Cl^-, and NH_3, which leads to significant changes in the reduction rate. Figure 2.17 shows a current-potential curve recorded on n-type Si(111) and p-type Si(111) in the dark and under illumination, respectively, in a sulfate solution containing Co^{2+}. A previous surface treatment consisting of 5 s immersion in a 40% HF solution followed by a longer immersion (10 min) in a 40% NH_4F solution was carried out to eliminate the native oxide and to obtain an H-termination [60]. The cyclovoltammogram on the p-type Si shows two cathodic peaks at -0.65 V and -1.12 V in the forward scan (indicated as I and II). Each one reflects the onset of different electrochemical processes: the reduction of cobalt ions and the onset of hydrogen evolution on the freshly deposited metal islands, respectively. Deposited metal particles are dissolved in the anodic reverse scan at potentials more positive than the equilibrium potential for the reaction (2.44). The reader will note that the deposition onto n-type Si starts at -1.1 V, that is, at

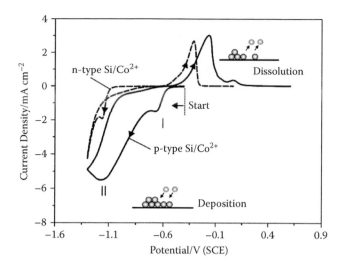

FIGURE 2.17 Current-potential curves obtained on n-type and p-type Si(111) in 0.01 M Co^{2+} +0.5 M Na_2SO_4 + 0.5 M H_3BO_3 at pH 4.5. $V_i = -0.4$ V. Scan rate: 0.010 V s^{-1}. The experiment with p-type silicon was carried out under illumination with a W-I lamp at an intensity of 80 mW cm^{-2}.

a cathodic overpotential of 0.58 V. This fact cannot be attributed to a rectifying effect, since the flat band potential is around -0.5 V. The cause, as discussed in Chapter 5, is related to the surface chemistry and electronics. Oxidized atoms on the p-type Si(111) introduce localized energy states in the semiconductor band gap and act as preferential sites for the metal nucleation. The interaction of the terminal $-OH$ groups and the coordination water molecules, on the other hand, decreases the activation energy for the adsorption of ad-atoms. The large hole concentration in p-type Si(111) leads to a rapid surface oxidation at the start potential of -0.4 V (see subsection 2.4) and the deposition starts as the electron quasi-Fermi level reaches the redox level of reaction (2.44). The H-terminated surface, on the contrary, presents a relatively large stability in aqueous solutions, so increasing the nucleation overpotential, probably as a consequence of its hydrophobicity. The H-termination is a metastable state at potentials more positive than the standard potential for oxide formation: $V^0_{ox} = -0.98$ V (SCE, pH 4.5). The higher stability of n-type H:Si is provided by the low concentration of holes, which are necessary for breaking the stable Si-H bonds.

The presence of localized energy states or *surface states* generated by oxidation of surface atoms is manifested by the appearance of a peak in the differential-capacitance vs. voltage curves, measured at a frequency of 300 Hz (Figure 2.18). This peak reflects the dynamics of the charge-discharge of surface states by applying an additional alternating voltage on the polarization voltage.* It must be noted that this peak increases under illumination because of the trapping of photoelectrons from the conduction band and the consequent increase of the filled surface states. The course of the

---

* A detailed electrochemical and spectroscopic analysis of surface and interface states on Si surfaces can be read in Chapter 5.

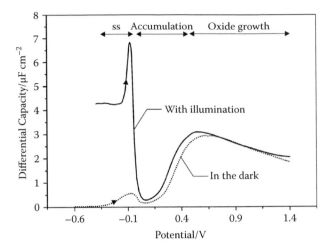

FIGURE 2.18 Potential dependence of the differential surface capacitance recorded at 300 Hz on p-type Si(111) in 0.5 M Na_2SO_4 + 0.5 M H_3BO_3 at pH 4.5. The electrode was illuminated with a W-I lamp at an intensity of 80 mW cm^{-2}.

capacitance-voltage curve for p-type Si shows a series of features that reflects different electronic processes: (i) the initial peak due to the charge of surface states, (ii) an exponential increase of the capacitance in accumulation followed by (iii) a capacitance drop due to the formation of an oxide film. This latter is given by the oxide capacity, which is proportional to $(V-V^0_{ox})^{-1}$ with V^0_{ox} as the oxide formation potential [61].

Because of the concomitant hydrogen evolution, it is difficult to analyze the nucleation and growth of metal islands by the current-time response only. The growth progress of metal particles can be followed by AFM-pictures taken at successive deposition times as shown in Figure 2.19 for –1.3 V. The heterogeneity of the particle size after 5 s deposition suggests a progressive type of nucleation. The nucleation and growth process continue up to a complete surface coverage and coalescence of metal islands. It is to be noted that the protrusions generated after 15 s of deposition are smoothed with deposition time. This effect is probably caused by boric acid, usually used as an additive in plating baths. Boric acid is supposed to adsorb at the metal surface, thus reducing the mobility of ad-atoms [62]. This in turn is expected to contribute to an increase of the deposition rate onto the uncovered surface.

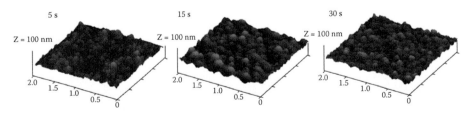

FIGURE 2.19 2 μm × 2 μm AFM pictures of a p-type Si(111) surface taken after deposition of Co at –0.9 V. The image sequence shows the surface development with the deposition time.

The formation of a thin oxide film on Si has important consequences due to its dielectric and insulating properties. Terminal Si-O bonds introduce reactive nucleation sites—that is, the deposition overpotential is reduced. The electron transfer occurs by discharge of photogenerated electrons from the quasi-Fermi level upon states arising at the Si-SiO$_x$ interface and tunneling through the oxide film. Hence, an increase of the thickness of the oxide film limits the growth of metal islands due to the tunneling barrier. As a consequence of the dielectric behavior of the oxide, the total potential drop (φ_{sc}-φ_s) will be distributed between the oxide film and the semiconductor in a relation which depends on the capacitance of the semiconductor and that of the oxide. Thus, the semiconductor band bending is reduced, and thus the nucleation overpotential [63]. These facts are exemplified in Figure 2.20, which shows the current-time curves obtained by applying a potentiostatic pulse after the formation of a thin oxide film of increasing thickness by a preceding anodic pulse. After an anodic pulse of 0.9 V, the current density shows values twice those observed without a previous anodic pulse. On applying the deposition pulse, the current rises abruptly and drops following a course $j \propto t^{-1/2}$. The current response is dominated by the diffusion-controlled hydrogen evolution on the surface of growing Co particles (peak II in the current-voltage characterization in Figure 2.17). Further increase of the oxide thickness leads to a considerable decrease of the nucleation rate. The current increases now from zero up to reach a constant value after 60 s of deposition (curve for an anodic potential of 1.9 V). Further increase of the anodic pulse potential reduces the current value of the deposition pulse; the nucleation rate is slower. The deposition on a Si surface covered by a thin oxide film hints at the formation of coordinated localized states in the oxide band gap that forms some kind of conduction paths, which end in hot spots at the oxide–electrolyte surface where nucleation may take place. A direct tunneling mechanism seems to be improbable. The oxide thickness after an anodic pulse of 1.9 V for 5 s can be estimated at 2.70 nm (dd_{ox}/ $dV \times (V_a$-$V_{ox}^0) = 0.94$ nm V$^{-1} \times 2.88$ V).

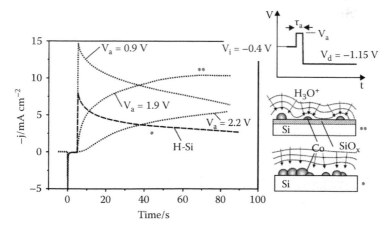

FIGURE 2.20 Current-time curves obtained on p-type Si(111) by applying the potential program shown in the upper right corner of the figure in 0.01 M Co^{2+} + 0.5 M Na$_2$SO$_4$ + 0.5 M H$_3$BO$_3$ at pH 4.5. $\sigma_a = 5$ s. Light intensity: 80 mW cm^{-2}.

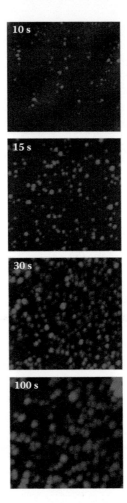

FIGURE 2.21   Sequence of 6 μm × 6 μm AFM pictures taken at different states of deposition at −1.15 V onto p-type Si(111) in 0.01 M Co^{2+} + 0.5 M Na$_2$SO$_4$ + 0.5 M H$_3$BO$_3$ at pH 4.5. The surface was previously anodized for 5s at 1.4 V. Light intensity: 80 mW cm^{-2}.

Figure 2.21 shows AFM pictures of the successive studies of the deposition of Co onto Si after the formation of an anodic film at 1.4 V. It should be noted that the formation of fewer particles than on the surface without oxidation (see Figure 2.19) does not lead to a complete surface coverage and the formation of a continuous film. With the density and the mean size of particles measured by means of AFM after defined potential and time deposition for increasing anodic oxide thicknesses, the deposition efficiency was calculated according to

$$\eta_{dep} = \frac{NV_p \rho_{Co}}{\int_0^{t_{dep}} j\, dt} \frac{zF}{M_{Co}} \tag{2.45}$$

where z is the ion valence, N is the density of particles [cm^{-2}], V_p is the mean particle volume [cm^3], ρ_{Co} is the density of Co, t_{dep} is the deposition time, and is the atomic mass of Co.

Figure 2.22 shows that density of deposited particles. The deposition efficiency decreases exponentially with the circulated anodic charge. The inspection of AFM pictures allows identification of a charge threshold at 0.8 mC cm^{-2}, which separates continuously from island-like films. The right side of Figure 2.22 shows the relation of the density of grown particles with the inverse of the oxide thickness calculated by taking a factor of 0.93 nm C^{-1} cm^2. A linear relationship can be seen. In principle, this result can be interpreted in terms of a direct relation between the nucleation rate and the electric field in the oxide film, given by $\overline{E} = \Delta V_{ox} / d_{ox}$ [64]. This means that the density of nucleation sites should be related with the density of hot spots generated in the oxide. These conduction paths, as discussed in the next subsection, arise as a consequence of the structural changes generated by the transport of oxygen anions during the field-induced oxide growth. The existence of a relation between the nucleation rate and the electric field is further supported by the linear change of the density of particles with the deposition potential for a constant oxide thickness as shown in Figure 2.23. The interpretation presented above is based on the assumption that the change of the potential drop silicon-electrolyte on changing the applied deposition potential, V_{dep}, is held to a large extent by the oxide film. As explained in more detail in Chapter 5, this implies a Fermi-level pinning at the energy states arising at the silicon-silicon oxide interface. In the deposition potential range, the semiconductor is in depletion and its capacitance is much lower than that of the oxide: $C_{sc} \ll C_{ox}$. This means that, in the absence of Fermi level pinning, the potential drop would occur in the semiconductor only.

The decrease of the deposition efficiency with the oxide thickness adheres to an enhancement of the hydrogen evolution rate, which can be ascribed to a higher diffusional hemispherical flux of protons to the dispersed particles than the planar

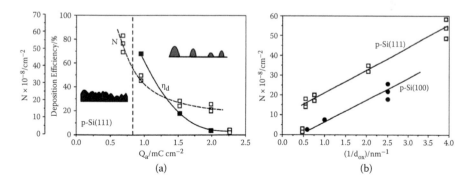

(a)   (b)

FIGURE 2.22 (a) Density of deposited Co-particles on p-type Si(111) and deposition efficiency as a function of the anodic charge circulated during the previous surface anodization. The anodization was carried out by applying an anodic pulse of 5 s duration at potentials between 0.9 V and 2.2 V. The metal deposition succeeded after applying a cathodic pulse of 80 s of duration at −1.15 V. (b) Density of deposited particles as a function of the inverse of the oxide thickness. Deposition potential: −1.15 V, p-type Si(111); −1.1 V, p-type Si(100).

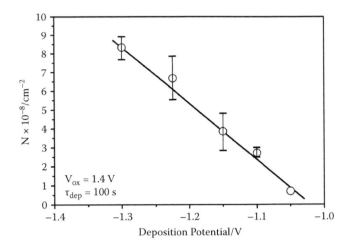

FIGURE 2.23 Density of Co-particles deposited onto p-type Si(111) as a function of the deposition potential after 5 s surface anodization at 1.4 V.

diffusion of a dense distribution. The Cottrell equation for a spherical electrode is given by [65]

$$j_p = \frac{F}{\sqrt{\pi D_{H^+}}} \frac{C_{H^+}}{\sqrt{t}} + FD_{H^+}C_{H^+}\frac{1}{r_p} \tag{2.46}$$

where r_p is the radius of the sphere. For $t \to \infty$, $J_p = 2FD_{H^+}C_{H^+}r_p$, and hence the total measured current is given by $J_T = N \times 2FD_{H^+}C_{H^+}\bar{r}_p$. On the other hand, the adsorption of boric acid is expected to slow the island growth and only the hydrogen evolution current is recorded. The increase of hydrogen evolution on a dispersed amount of Co particles is also manifested in the current-voltage curves performed with and without a previous anodic pulse (Figure 2.24). Here, a shift of the onset of the cathodic current of $\Delta V \sim 0.42$ V and a doubling of the current peak are observed.

In the following, the electron transfer in the deposition process will be discussed with the help of energy band diagrams. Figure 2.25 represents the p-type semiconductor–electrolyte interface at a deposition potential of –0.5 V (SCE) without previous anodization. As explained in more detail in Chapter 4, Fermi-level pinning occurs as it reaches the neutrality level interface state energy distribution, which for silicon is placed about the middle of the band gap. Assuming a flat band potential for p-Si of 0.1 V, the middle of the band gap should be reached at about –0.25 V. Therefore, it is expected that further change of the electrode potential will be taken by the potential drop at the semiconductor–electrolyte (double layer) or at the oxide film (in the case of applying an anodic pulse). This means that the band bending remains practically constant.

Under illumination, photogenerated electrons in the bulk reach the interface by diffusion and drift. At the semiconductor surface, the energy of electrons is given by the quasi-Fermi level. Hence, the overpotential for deposition is $\eta_{dep} = ({}_n E^*_{F(s)} - E_{redox(Co/Co^{2+})})/e$.

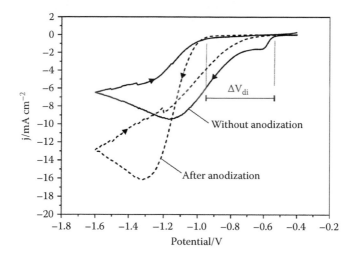

FIGURE 2.24 Current-voltage curves of p-type Si(111) without anodization and after 5 s surface anodization at 1.4 V in 0.01 M Co^{2+} + 0.5 M Na_2SO_4 + 0.5 M H_3BO_3 at pH 4.5. Scan rate: 0.010 V s^{-1}. The anodization was carried out in the same deposition solution.

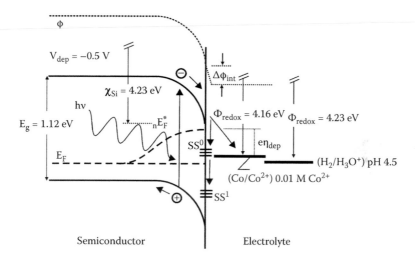

FIGURE 2.25 Energy band diagram of the p-type Si-electrolyte interface at -0.5 V.

The overpotential for the hydrogen evolution reaction (her) $\eta_{her} = (_nE^*_{F(s)} - E_{redox(H_2/H^+)})/e$ is at this potential not large enough. Co, with an exchange current density $j_0 = 10^{-5.3}$ A cm^{-2}, is not a good catalyst for the *her* in comparison with Pt, with $j_0 = 10^{-3.0}$ A cm^{-2}. Part of the photoelectrons arriving to the interface will recombine at the interface states. The energy distribution of interface states at the Si-SiO_x interface, which receive the denomination *Pb-centers*, present two peaks, one above and other below the middle of the band gap named SS^0 and SS^1, respectively. SS^0 states show a conduction band character (donors). SS^1 states show a valence band character (acceptors).

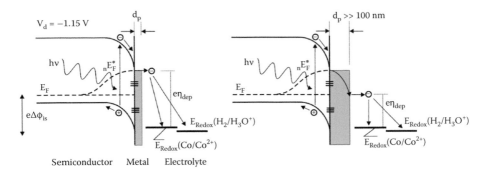

FIGURE 2.26 Energy band diagram of the p-type Si–metal–electrolyte interface at –1.15 V (SCE) for an early and later stadium of the growth of metal islands.

At the deposition potential of –1.15 V (see Figure 2.26), the reduction of Co^{2+} occurs together with the hydrogen evolution reaction at the surface of growing particles. The growth of metal particles modifies the electron dynamics due to the transfer of electrons through the metal phase. Electrons are emitted at the quasi-Fermi level at the semiconductor–metal interface. In principle, it can be assumed that electrons reach the metal–electrolyte interface by ballistic emission. The inelastic mean free path for electrons with a kinetic energy of 1 to 2 eV is 200–300 nm. For longer distances, electrons thermalize up to the Fermi level of the metal, equal to that of bulk semiconductor. Electrons are transferred at the metal–solution interface at a lower overpotential, and hence the deposition rate onto metal particles larger than the inelastic mean free path decreases.

The thin oxide film formed by a preceding anodic pulse takes part of the interfacial potential drop (see Figure 2.27). The charging of interface states generates an electric field inside the oxide, which behaves as a dielectric capacitor. As a consequence, photoelectrons at the semiconductor–oxide interface have to be transferred by tunneling to the oxide–electrolyte or oxide–metal interface. As already mentioned, it seems to be improbable that the current experimentally observed involves a direct tunneling mechanism through oxide thicknesses larger than 1 nm. Electron transfer has to occur by a resonance tunneling upon localized states in the oxide band gap. Based on spectroscopic measurements, Jungblut and Lewerenz have assumed the existence of a defect band near the Si conduction band [66]. The energy of photogenerated electrons drops in the decay process by $\Delta E = \overline{E} \times (d_{ox} - d_{tunnel})$, where d_{tunnel} is the direct tunneling distance from one point inside the oxide.

2.3.3 Electrocrystallization of Noble Metals onto Silicon

The electroplating of semiconductor surfaces with noble metals is customarily carried out using solutions of their chloride complexes. In the particular case of platinum deposition, one can start from Pt(IV) or Pt(II) salt solutions. The transfer of four electrons for the reduction of $PtCl_6^{2-}$ to Pt metal occurs in two consecutive steps [67,68]:

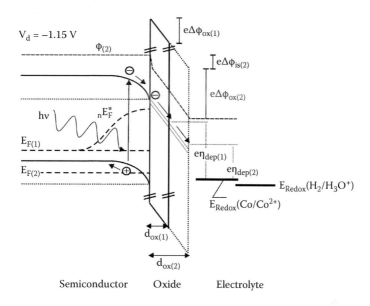

FIGURE 2.27 Energy band diagram of the p-type Si–oxide–electrolyte interface at –1.15 V (SCE).

$$PtCl_6{}^{2-} + 2e^- \rightarrow PtCl_4{}^{2-} + 2Cl^- \qquad V^0 = 0.708 \text{ V(SCE)} \qquad (2.47)$$

$$PtCl_4{}^{2-} + 2e^- \rightarrow Pt + 4Cl^- \qquad V^0 = 0.535 \text{ V(SCE)} \qquad (2.48)$$

Figure 2.28a and b shows cyclic voltammetries performed on n-type Si(111) surface is in solutions containing Pt(IV) in sulfate and chloride solutions as ground electrolytes. In the former case, the forward scan is characterized by the presence of two voltammetric peaks at –0.3 V and –0.55 V. The steep current increase at potentials about –1.0 V corresponds to the hydrogen evolution by reduction of water. It was demonstrated that the first peak, appearing within the depletion potential region of the semiconductor, corresponds to the formation of metal clusters. The latter corresponds to the hydrogen evolution by reduction of protons at the surface of freshly formed Pt metal islands [69]. The reduction of Pt-complexes onto the semiconductor takes place by injection of holes from the empty states of the metal complex into the valence band or into surface states of the semiconductor arising from oxidized surface atoms as explained in detail in Chapter 5. It should be noted that the onset of electroreduction current at about –0.2 V is observed precisely as the Fermi level of the semiconductor is located above the electroneutrality level of surface states, near the mid gap. Regarding a flat band potential for n-type Si(111) of –0.55 V in sulfate solution and a bulk Fermi level 0.25 eV below the conduction band edge, the electroneutrality level must be 0.6 eV below the conduction band (see diagram in Figure 2.29). Injection of holes leads to the formation of an oxide film as discussed in detail in Chapter 5. This induces a potential drop in the oxide by charging of surface

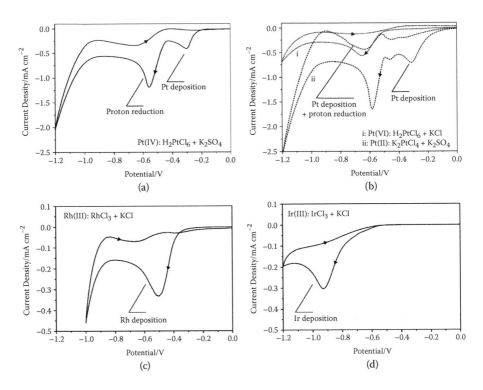

FIGURE 2.28 Current density-voltage curves (cyclo-voltammetries) performed with n-type Si(111) surface in different electrolytes. (a) 1 mM H_2PtCl_6 + 0.1 M K_2SO_4; (b) 1 mM K_2PtCl_4 + 0.1 M K_2SO_4 (pH 2.8) and 1 mM H_2PtCl_6 + 0.5 M KCl; (c) 2 mM $RhCl_3$ + 0.5 M NaCl; (d) 5 mM $IrCl_3$ + 0.5 M KCl. Scan rate: 10 mV s^{-1}.

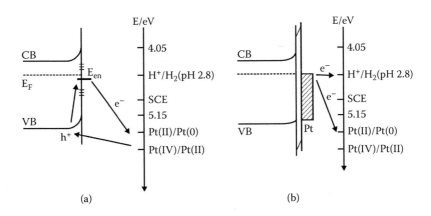

FIGURE 2.29 Energy band diagrams corresponding to the n-type Si/electrolyte interface in semiconductor depletion showing possible charge transfer pathways. After formation of Pt islands, the reduction of protons and further reduction of metal complexes occurs by direct injection of electrons from the conduction band over the metal to the empty states of metal complexes in the solution.

states and reduces the semiconductor band bending. Thus, further charge transfer is allowed to occur by injection of electrons from the conduction band.

If one compares the potential scans performed in solutions containing Pt(IV) and Pt(II) with potassium sulfate as a supporting electrolyte, one can note that practically the same features appear in both cases. A larger current is, however, observed for the first reduction peak of K_2PtCl_4 solutions. This is unexpected, considering that the reduction of $PtCl_4^{2-}$ involves only two electrons. In aqueous solutions, Pt(IV), Ir(III), and Rh(III) ions form octahedral coordination structures by means of an hybrid molecular orbital of type sp^3d^2. Chloride ligands can be replaced by water to form chloride-aquo-complexes of the type $PtCl_n(H_2O)_{6-n}^{[4-n]}$, $IrCl_n(H_2O)_{6-n}^{[3-n]}$, and $RhCl_n(H_2O)_{6-n}^{[3-n]}$.

The number of coordinated water molecules is given by the following equilibrium:

$$Pt(H_2O)_nCl_{6-n}^{n-2} + m\ Cl^- \rightarrow Pt(H_2O)_{n-m}Cl_{6-n+m}^{n-m-2} + m\ H_2O \qquad (2.49)$$

The increase of chloride concentration in the solution brings about a retardation of the reduction rate of platinum chloride complexes as reported by Lau and Hubbard by using Pt as a substrate [70]. Similarly, the inhibition effect appears by the reduction of $PtCl_6^{2-}$ by using KCl as the supporting electrolyte for the deposition of Pt onto Si (see Figure 2.28b). Here, the inhibition is pointed out by a shift of the onset of cathodic current toward more negative potentials. A similar argument, as stated in Ref. 71, based on a blockage of the adsorption of the reactive species by a preferential adsorption of Cl^-, cannot be extrapolated for silicon. A reduction control given by a previous replacement of coordinated chloride by water is in line with results reported by Gregory et al. [71], who attributed the inhibition effect exerted by chloride to the decrease of the concentration of the electroactive species $Pt(H_2O)_4^{+4}$ by preferential coordination with chloride. An equilibrium constant of 4.1×10^{-5} was reported for the equilibrium [72]:

$$PtCl_4^{-2} + 4\ H_2O \rightarrow Pt(H_2O)_4^{+2} + 4\ Cl^- \qquad (2.50)$$

Accordingly, a shift of the equilibrium potential of reaction (2.48) by 0.38 V can be estimated by an addition of 0.5 M of Cl^- in the deposition solution (taking the rough approximation concentration = activity).

This explains the experimentally observed fact that the deposition of Rh and Ir from their chloride salts in chloride-supporting electrolytes starts at potentials more negative than −0.35 V and −0.6 V, respectively (Figure 2.28c and d). In general, the different position and width of the cathodic peaks appearing in a voltammogram reflects the different inertness of the complexes against reduction. This case was discussed by Pletcher and Urbana [73], particularly for the electroduction of Rh(III). This cation exists in an aqueous chloride solution as a mixture of $[RhCl_n(H_2O)_{6-n}]^{3-n}$ species. There is also evidence confirming that the interconversion of these complexes is slower than the electron transfer processes. In spite of this, only one cathodic peak appears.

The electrocrystallization of metals onto n-type Si(111) surfaces can be analyzed by current-time curves measured upon applying potential pulses at increasingly negative polarizations. Figure 2.30a and b exemplifies the deposition of Pt and Ir from solutions containing 1 mM H_2PtCl_6 in sulfate-supporting electrolytes and 10 mM

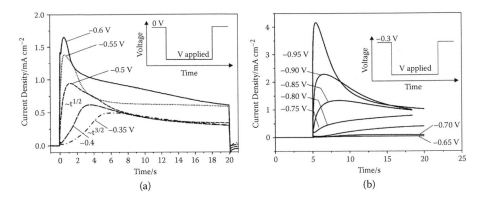

FIGURE 2.30 Current-time curves registered upon applying a potential pulse on n-type Si(111) in Pt (a) and Ir (b) solutions. (a) 1 mM H_2PtCl_6 + 0.1 M K_2SO_4; (b) 10 mM $IrCl_3$ + 0.5 M KCl + 5% i-propanol. Cyclo-voltammograms of n-type Si(111) in different plating solutions. (a) 1 mM H_2PtCl_6 + 0.1 M K_2SO_4; (b) 1mM H_2PtCl_6 + 0.5 M KCl, 1mM K_2PtCl_4 + 0.1 M K_2SO_4; (c) 10 mM $IrCl_3$ + 0.5 M KCl + 5% i-propanol; (d) 2 mM $RhCl_3$ + 0.5 M NaCl. Scan rate: 10 mV s^{-1}.

$IrCl_3$ in chloride-supporting electrolytes, respectively. The initial part of transients for Pt deposition reveals a change of the nucleation type from progressive ($j \sim t^{3/2}$) to instantaneous ($j \sim t^{1/2}$) as the applied potential is driven from semiconductor depletion to accumulation. Only instantaneous deposition is observed, however, for Ir deposition. In the case of Pt deposition, an increase of the falling current density after the transient maximum caused by the hydrogen evolution accompanying the growth of metal islands is observed.

Under depletion, the nucleation mechanism is linked to the charge transfer by surface states, the concentration of which increases by the simultaneous surface oxidation triggered by the injected holes during the electroreduction of metal complexes. As further discussed in Chapter 5, nucleation occurs on oxidized surface atoms that introduce the local energy levels (surface states). Near the flat band potential, the electroreduction occurs predominantly by injection of electrons from the conduction band. The charge transfer process is not bound to the oxidation process and, hence, the electrodeposition is controlled by some of the electrochemical steps of the reduction process. The oxidation process accompanying the electrodeposition of noble metals is indicated by the noted transformation of the substrate topology along with the growth of the nanodimensioned metal islands (see Figure 2.31). Note how the initial zigzag shaped terraces of atomic dimensions transforms rapidly to a steeped surface with steps of a height ranging various atomic layers. It is evident, that a step-bunch type of dissolution takes place as discussed in more detail in Chapter 5. Structure features are not discerned after longer deposition time because of leveling brought about by the formation of an oxide film beneath the growing particles. The progressive type of nucleation is reflected by the size distribution of particles. This contrasts with the more even particle size distribution observed in the case of Ir deposition, for which the current-time curves indicate an instantaneous type of nucleation.

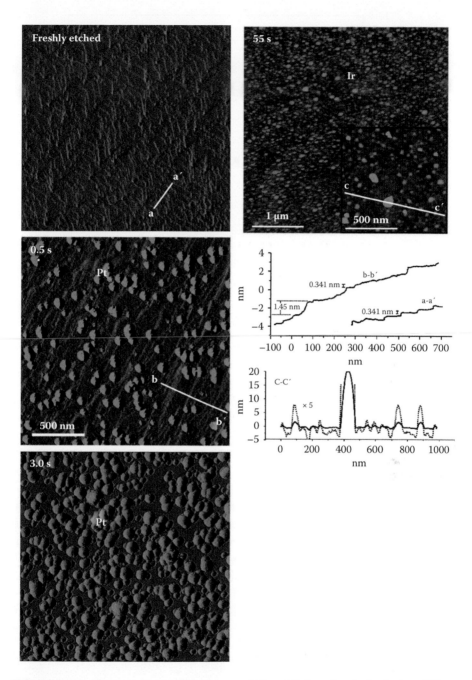

FIGURE 2.31 Tapping-mode AFM pictures of Pt and It deposits obtained after different plating times at −0.35 V(SCE) for Pt and −0.9 V(SCE) for Ir from solutions indicated in Figure 2.31. Inset: cross sections performed as indicated in the pictures.

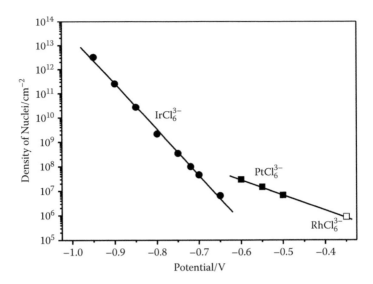

FIGURE 2.32 Density of nuclei formed upon potentiostatic Pt and Ir plating of n-type Si(111) as a function of potential. For the sake of comparison, the density of Rh nuclei deposited on p-type Si(111) under illumination is also plotted.

The density of nuclei was calculated from the raising part of transients by applying Eq. (2.34). Diffusion coefficients were calculated from the falling part of transients by applying the Cottrell equation (2.38); a value of 6.4×10^{-6} cm^2 s^{-1} and 4.27×10^{-6} cm^{2}s^{-1} were calculated for $PtCl_6^{2-}$ and $IrCl_6^{3-}$, respectively (in chloride solutions, Ir(III) forms sixfold coordinated chloride complexes [68]).

The calculated density of nuclei plotted as a function of applied potential is shown in Figure 2.32. It should be noted, that the density of nuclei increases exponentially with the applied potential. Slopes of 51 mV dec^{-1} and 160 mV dec^{-1} are observed for the Ir and Pt nuclei, respectively. In principle, the potential dependence of the nucleation density can be related to the electron transfer kinetic involved in the formation of the critical nucleus. In the case of semiconductors, it is difficult to ascertain to what extent the change of applied potential is transduced in a change of potential drop at the double layer, which is the driving force for the reduction of the ion in the solution. It is, in principle, feasible, because the Fermi level is pinned at the surface states present at the surface fresh etched in 40% NH$_4$F, as customary.

Let us suppose that the formation of stable nuclei is controlled by the adsorption of a metal atom on an oxidized site as described by

$$Ir(H_2O)_nCl_{6-n}^{n-3} + Si\text{-}OH_3 + 2e^- \rightarrow Si_s\text{-}O\text{-}Ir + n\,H_2O + (6-n)Cl^- + H^+ \qquad (2.51)$$

Further assumptions are (i) that the reaction follows a Langmuir type of site occupancy and (ii) that only one of the three electron transfer steps controls the whole formation of ad-atoms. Thus we arrive at

$$N_{ad} = N_s \frac{K \exp\left(-\alpha_+ \dfrac{e}{kT} \Delta\varphi_H\right)}{K \exp\left(-\alpha_+ \dfrac{e}{kT} \Delta\varphi_H\right) + \exp\left(\alpha_- \dfrac{e}{kT} \Delta\varphi_H\right)} \tag{2.52}$$

where K is a constant including the activity of the Ir complex, $\Delta\varphi_H$ is the potential drop at the Helmholtz layer, and N_{ad} and N_s are the surface concentration of ad-atoms and active sites, respectively. If $N_s \gg N_{ad}$, this situation is justified by noting that N_s (~ 10^{14}–10^{15} cm^{-2}: density of kink and edge sites on Si(111)) and $N_{ad} < 10^{13}$ (cf. Figure 2.31). Thus, Eq. (2.52) reduces to

$$N_{ad} = N_s K \exp\left[-(\alpha_+ + \alpha_-)\frac{e}{kT}\Delta\varphi_H\right] \tag{2.53}$$

This electrosorption reaction is thought to immediately reach an equilibrium state in comparison with extended growth time.

2.3.4 ELECRODEPOSITION ONTO NIOBIUM OXIDE

In some photoelectrode designs, oxides with semiconducting properties are selected as absorber materials. The interest in these materials is sometimes justified by their low production cost by using, for instance, simple anodization procedures. Another attractive feature of these materials is the possibility of generating self-organized nanostructures by choosing adequate operation conditions: electrolyte, anodization voltage, temperature, and so on [74–76]. One of the best-known examples is the formation of honeycomb porous aluminum oxide by anodization at high voltages (50 to 100 V) in phosphoric acid and sulfuric acid. Because of the insulating properties of alumina, this type of structure has rapidly found application as masks for the electrodeposition of metal nanowires [77,78] and also as optical waveguides [79,80]. The formation of nanostructures by anodization has been extended to other materials such as Ti, W, and Nb [81], as we will refer to in the next subsection.

The knowledge of the electronic properties of the semiconducting oxide–electrolyte interface and their influence on the electrodeposition process is essential for the design of novel photoelectrochemical conversion devices. The electrodeposition of Co onto oxide covered Nb-surfaces will be presented as a model system.

As likely in the electrochemical science, a first overview of the electrodeposition process is obtained by a current-potential curve in one Co^{2+}-containing sodium sulfate solution with the addition of boric acid. Figure 2.33 shows the results obtained with a Nb surface covered with a native oxide and other after potentiodynamic anodization to 4.4 V. The abrupt cathodic current increase at –1.1 V indicates the onset of deposition at an overpotential of 0.52 V. A similar behavior is observed with the anodized surface, where the overtension was reduced by 0.050 V.

Niobium oxide behaves as an n-type semiconductor with a donor concentration 3×10^{19} eV^{-1} cm^{-3} and a flat band potential $V_{fb} = -0.5 \pm 0.08$ V (SCE) under the

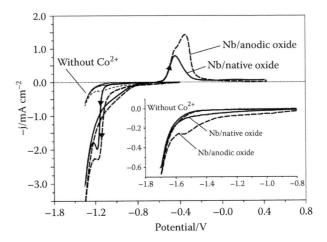

FIGURE 2.33 Current-voltage curves of Nb covered with a native oxide (full line) and after surface anodization at 4.4 V (dashed line) in 0.01 M Co^{2+} + 0.5 M Na_2SO_4 + 0.5 M H_3BO_3 at pH 4.5. $V_i = 0.4$ V. Anodization was carried out potentiodynamically by a cycling scan between –0.4 V and 4.4 V with a scan rate of 0.1 V s^{-1}.

anodizing conditions of the above-described experiment [82]. This means that the electrodeposition takes place in semiconductor accumulation. The anodic course in the reverse potential scan indicates the dissolution of deposited Co. The double peak observed in the anodic course for the anodized surface can be ascribed to the appearance of a second anodic process: the oxidation of H intercalated in the oxide structure during the forward potential scan. The intercalation process is reflected by the cathodic peak observed in the forward scan in a solution without Co^{2+} (see detail in Figure 2.33).

Three regions can be distinguished in the current transients upon applying pulses in the deposition potential region: (i) the initial interface charging, (ii) a transition region with a linear current increase, and (iii) the beginning of crystal growth with the characteristic form for a nucleation and mass transport-controlled growth (Figure 2.34). The transition zone represents the induction time for the onset of nucleation, quantified by extrapolation of the rising part of the transient to $j = 0$. It must be noted that this time decreases with the overpotential and is shorter after the surface anodization. Induction times of 1 s to 20 s in the potential region of -1.1 V $> V > -1.25$ V are related to the modification of the electronic structure of the surface by intercalation of H in the amorphous oxide structure. H diffuses to the oxide bulk with diffusion coefficients $D_H = 10^{-8}$–10^{-9} $cm^2 s^{-1}$ [83,84] and leads to the formation of bronzes of the type $H_x Nb_x^{(+4)} Nb_{2-x}^{(+5)} O_5$ at the surface:

$$x\ H^+ + x\ e^- + Nb_2^{(+5)}O_5 \rightarrow H_x Nb_x^{(+4)} Nb_{2-x}^{(+5)}O_5 \tag{2.54}$$

The presence of a bronze thin layer modifies the conductivity [85] and the photoelectrochemical behavior of the surface [86,87]. The formation of terminal $Nb_2O_{5-x}(OH)_x$ bonds, on the other hand, lowers the activation energy for nucleation, because of the stronger coupling of them with solvated $Co(H_2O)^{2+}$ [88]. Hence, the charge transfer

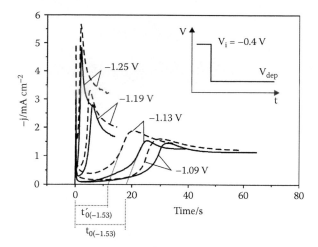

FIGURE 2.34 Current-time response upon applying potential pulses within the deposition potential region on Nb covered with a native oxide and after the anodization at 4.4 V. Deposition solution: 0.01 M Co^{2+} + 0.5 M Na_2SO_4 + 0.5 M H_3BO_3 of pH 4.5. $V_i = -0.4$ V. Anodization was carried out potentiodynamically by a cycling scan between -0.4 V and 4.4 V with a scan rate of 0.1 V s^{-1}.

process and thus the nucleation rate are accelerated. In any case, the nucleation process is limited by the formation of these nucleation sites formed by H intercalation. At pH 4.5 the intercalation of H takes place at supersaturation larger than 0.5 V. As a result, a high density of nuclei is formed and the surface is rapidly covered with a thin metal film by coalescence of growing particles (see Figure 2.35).

Current-voltage curves performed after anodizing the Nb thin film at increasing potentials have shown a shift of the onset of deposition with the oxide thickness for Ag and Cu (Figure 2.36). These metals have a standard reduction potential more positive than the flat band potential. The larger the difference between the standard reduction potential and the flat band potential, the stronger is the dependence of the overpotential, η_d, on the anodization potential, V_a (Figure 2.37). The overpotential increases linearly with the oxide thickness, given by $d_{ox} = k_{ox} (V_a - V^0_{ox})$, where k_{ox} is the oxide growth rate. The change of the deposition overpotential with the oxide thickness seems to be related to the change of donor concentration with the oxide thickness as reported by Torresi and Nart [90] and Heusler and Schultze [91]:

$$N_D \sim \frac{1}{d^2_{ox}} \tag{2.55}$$

A similar behavior was also observed upon the anodic growth of Ta-oxide [92] and Ti-oxide [94]. If one compares relation (2.55) with the expression for the length of the space charge layer:

$$W = \sqrt{\frac{2\varepsilon_{ox}\Delta\varphi_{sc}}{eN_D}} \tag{2.56}$$

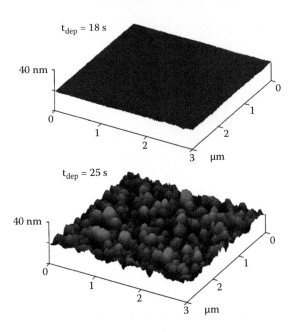

FIGURE 2.35 3 μm × 3 μm AFM images showing a Nb surface covered with native oxide taken after increasing deposition time at −1.13 V in M Co^{2+} + 0.5 M Na_2SO_4 + 0.5 M H_3BO_3 at pH 4.5. $V_i = -0.4$ V.

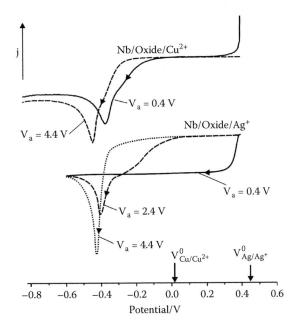

FIGURE 2.36 Current-voltage curves of Nb anodized at different potentials: in 0.01 M Cu^{2+} + 0.5 M Na_2SO_4 + 0.5 M H_3BO_3, pH 4.5 and 0.01 M Ag^+ + 0.5 M Na_2SO_4 + 0.5 M H_3BO_3, pH 4.5V. V_a: anodizing potential. $V_i = 0.4$ V. Scan rate: 0.01 V s^{-1}.

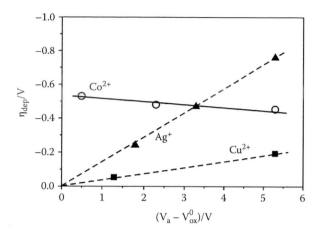

FIGURE 2.37 Deposition overpotential as a function of the anodizing potential for different metals. V^0_{ox}: oxide formation standard potential $= -0.89$ V(SCE) [82].

where $\Delta\varphi_{sc}$ is the semiconductor band bending, one can see that thickness of the space charge layer increases linearly with the oxide thickness. This means that the deposition of metals with a standard reduction potential more positive than the flat band potential is limited by the potential barrier given by the space charge layer. The variation of the donor concentration with the oxide thickness arises as a consequence of the reduction of O^{2-} vacancies with the anodization potential. Possible causes are (i) the acceleration of the surface reaction for the formation of oxygen vacancies due to the increases of the potential drop at the oxide–electrolyte interface and (ii) the formation of a nonstoichiometric oxide film (NbO_x) at the metal–oxide interface that influences the semiconductor behavior [94,95]. Figure 2.38 summarizes the electronic processes in the electrodeposition of metal particles of niobium oxide. The first energy band diagram represents the Nb-Nb_2O_5-electrolyte interface at the flat band potential. According to the values of work function of Nb and the affinity of the Nb oxide, the metal–oxide interface presents an ohmic contact [96,97]. Hence, any change of the Fermi level leads to a band bending at the oxide–electrolyte interface. The mobility band edges, characteristic of amorphous oxides, are represented by dashed lines below and above the conduction and valence band edge, respectively (see next subsection). The band diagram in the middle of Figure 2.38 represents the Nb-Nb oxide–electrolyte interface at the standard potential for the couple Ag/Ag^+ for an anodic oxide grown at 4.4 V. Assuming a dielectric constant for the oxide $\varepsilon_{ox} = 45$ and a donor concentration $N_D = 3 \times 10^{19}$ cm^{-3}, a space layer thickness $W \sim 13$ nm can be calculated by Eq. (2.56). On the other hand, an oxide thickness of ~ 16 nm can be calculated by taking an oxide growth rate of ~ 3 nm V^{-1}. For an oxide grown at 2.4 V, a thickness of ~ 10 nm can be calculated. This means that $W > d_{ox}$ and the oxide behaves as a dielectric material at 0.56 V. The electron transfer through the oxide takes place by a resonance tunneling mechanism upon electron traps. For this transfer mechanism, the following dependence of the current on the oxide thickness applies [98,99]:

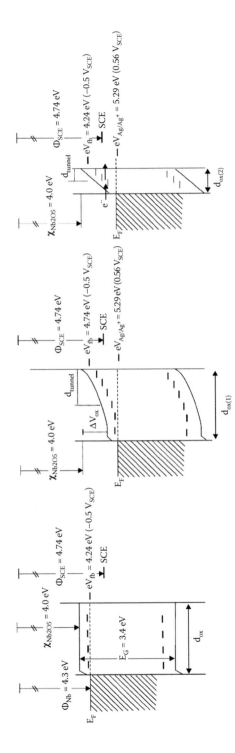

FIGURE 2.38 Left: Energy band diagrams of the Nb-NbO$_x$–electrolyte interface at the flat band potential. Oxide thickness corresponds to a 4.4 V–oxide. Energy band diagram of the Nb-NbO$_x$–electrolyte interface in equilibrium with the redox pair Ag/Ag$^+$ for a Nb-surface anodized at 4.4 V (middle) and 2.4 V (right).

$$\log j = \log j_0 - \frac{d_{tunnel}}{d_0} + \frac{e}{kT}\Delta V_{ox}\frac{d_{tunnel}}{d_{ox}} \qquad (2.57)$$

where j_0 represents a kind of exchange current density, d_{tunnel} is the tunnel distance (i.e., the mean separation of electron traps), d_0 is a constant, and ΔV_{ox} is the potential drop in the oxide.

2.4 ANODIC OXIDE GROWTH: MECHANISMS, METHODS, AND PROPERTIES

The anodic growth of thin oxide films on so-called valve metals takes place by migration and diffusion by hoping between interstitial, lattice, or defect sites. The hoping process is thermally activated and assisted by an electric field. A simplified model considers a symmetrically ordered array of potential barriers along which ions of a charge ze move. This simplified picture of the process does not consider the electronic and atomic processes related to the ion hopping. Figure 2.39 shows the periodic barrier before and after applying an electric field. From the elementary theory of transport kinetics, the ion flux in stationary state is given by

$$J_{ion} = a\nu\exp\left(-\frac{E_a}{kT}\right)\left[n_{(x-a)}\exp\left((1-\alpha)ze\frac{(\varphi_{x-a}-\varphi_x)}{kT}\right) - n_{(x+a)}\exp\left(\alpha ze\frac{(\varphi_{x+a}-\varphi_x)}{kT}\right)\right] \qquad (2.58)$$

where φ is the potential, E_a is the activation energy, a is the lattice constant, n is the concentration of mobile ions, α is the potential fraction from the equilibrium position to the energy maximum, and ν is the attempt frequency. According to the formalism of the absolute reaction rate, ν can be expressed as

$$\nu \cong \frac{kT}{h}\exp\left(\frac{S_a}{k}\right) \qquad (2.59)$$

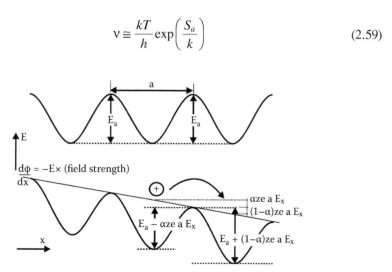

FIGURE 2.39 Energy course of a mobile ionic defect in a crystalline material before and after applying an electric field.

with S_a as the activation entropy. Thus, the activation energy corresponds to the activation energy barrier H_a. The expansion of Eq. (2.56) by a Taylor series at the point x yields (see derivation in Ref. 100):

$$J_{ion} = -\frac{Dn}{kT}\frac{\partial \tilde{\mu}}{\partial x}\frac{\sinh\left(\frac{\alpha ze}{kT}aE_x\right)}{\frac{\alpha ze}{kT}aE_x} \tag{2.60}$$

where D and $\tilde{\mu}$ are the diffusion coefficient and the electrochemical potential of the mobile ion, respectively. For low field strengths, Eq. (2.60) reduces to

$$J_{ion} = -\frac{Dn}{kT}\frac{\partial \tilde{\mu}}{\partial x} = -D\frac{\partial n}{\partial x} + n\bar{v}E_x \tag{2.61}$$

where $\bar{v} = \alpha zeD/kT$ represents the ion mobility. Equation (2.61) indicates a linear dependence of the ion flux on the electric field E_x. For $\partial n/\partial x = 0$, Eq. (2.61) yields the known Mott–Cabrera relation [101]:

$$J_{ion} = 4a\bar{v}n\exp\left(-\frac{E_a}{kT}\right)\sinh\left(\frac{\alpha ze}{kT}aE_x\right) \tag{2.62}$$

which for high field strengths adopts the form

$$J_{ion} = 4a\bar{v}n\exp\left(-\frac{E_a}{kT}\right)\exp\left(\frac{\alpha ze}{kT}aE_x\right) \tag{2.63}$$

According to this expression, the observed anodic current will be given by

$$j_a = zeJ_{ion} = j_0\exp\left(\beta E_x\right) \tag{2.64}$$

where j_0 represents a kind of exchange current and $\beta = \alpha zea/kT$. Hence, the oxide growth rate can be expressed, under the supposition of 100% efficiency [102,103],

$$\frac{d\,d_{0x}}{dt} = \frac{\overline{M}_{ox}}{\rho_{ox}zF}j_0\exp\left(\beta E_x\right) = \frac{\overline{M}_{ox}}{\rho_{ox}zF}j_0\exp\left[\beta\frac{(V-V_{ox}^0)}{d_{ox}}\right] \tag{2.65}$$

where $\overline{M}_{ox}$ and ρ_{ox} are the molecular weight and the density of the oxide, respectively. Equation (2.65) must be solved numerically. One approximation is given by logarithmic inverse law [104,105]:

$$\frac{1}{d_{ox}} = A - \frac{1}{\beta(V-V_{ox}^0)}\log t \tag{2.66}$$

where A is a constant. For $t \to \infty$ the oxide thickness reaches a stationary value, given by

$$d_{ox} = k_{ox}(V - V_{ox}^0) \qquad (2.67)$$

where k_{ox} is the so-called oxide growth factor, a empiric parameter that depends on the experimental conditions such as polarization time (potentiostatic mode) or potential scan rate (potentiodynamic mode).

Characteristic of the field assisted oxide growth are (i) a linear relation between the inverse of the oxide capacity and the anodizing potential and (ii) a constant anodizing current density in the potentiodynamic mode of growth. The former can be seen from (2.60) supposing a dielectric behavior of the oxide, that is,

$$\frac{1}{C_{ox}} = \frac{d_{ox}}{\varepsilon_0 \varepsilon_{ox}} = \frac{k_{ox}}{\varepsilon_0 \varepsilon_{ox}}(V - V_{ox}^0) \qquad (2.68)$$

with ε_0 as the dielectric permittivity in vacuum and ε_{ox} as the dielectric constant of the oxide. The experimental fact (ii) arises as a consequence of the high-field mechanism, where

$$E_x = \left(\frac{\partial V}{\partial d_{ox}}\right) = \left(\frac{\partial V}{\partial t}\right) \times \left(\frac{\partial t}{\partial d_{ox}}\right) = v \times \frac{1}{\left(\frac{\partial d_{ox}}{\partial t}\right)} = v \times \frac{\rho_{ox} z F}{\overline{M}_{ox} j_0 \exp(\beta E_x)}, \qquad (2.69)$$

where v is the potential scan rate. Equation (2.69) predicts a constant E_x because all other factors are constant. Hence, j_a remains also constant.

In the following, the electrochemical oxide growth on three types of substrates will be analyzed: silicon, niobium, and titanium indium phosphide.

The high current densities involved in the first instants of the potentiostatic anodization limit the study of the oxide growth because of electrolyte resistance when using usual electroactive areas of 0.1 to 1 cm². This drawback is eliminated by using especially designed micro-electrochemical cells such as that schematized in Figure 2.40 [106]. This cell consists of an acrylic body with a glass microcapillary driven by an x-y-x-positioner. The end of the microcapillary has a diameter of 10 μm to 100 μm. The electrochemically active surface is defined by a sealing rubber ring and the working solution is dropped into the capillary by a syringe. A micro-auxiliary electrode embedded in the acrylic body and a gold wire as counter electrode complete the three-electrode cell system.

Figure 2.41 exemplifies a typical experiment for the investigation of the anodic oxide growth: a cyclic current-voltage curve and the simultaneous recording of the interfacial capacitance [107,108]. For p-type silicon, the curve shows two characteristic anodic peaks between 0 V and 1 V. The first peak is ascribed to the oxidation of the H-terminated surface. The second peak arises reflects the relaxation of accumulated ions generated at the interface by an abrupt change of the electric field [109–111]. The delayed oxide formation brings about a current overcompensation

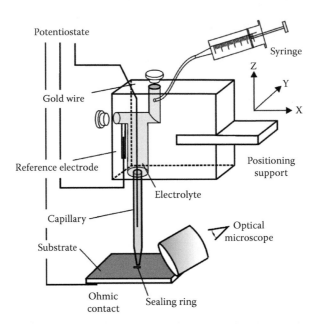

FIGURE 2.40   Schematic of a microcapillary cell.

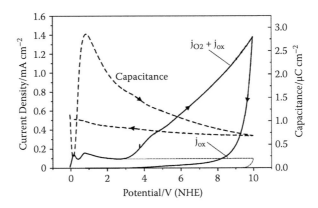

FIGURE 2.41   Current-potential and capacitance-potential curves recorded during the anodic potentiodynamic oxide growth on p-type Si(100) in acetate buffer of pH 5.6. Frequency: 1 kHz. Scan rate: 0.1 V s^{-1}.

[111], the so-called *overshoot*. Thereafter, the oxide continues its growth according to the high-field mechanism, and hence a constant anodic current is observed upon further potential scanning. The anodic oxidation of p-type silicon can be written

$$Si + 2\ H_2O + 4\ h_{VB}^+ \rightarrow SiO_2 + 4\ H^+ \tag{2.70}$$

The oxide growth mechanism involves the formation of O^{2-} at the oxide–electrolyte interface and its subsequent transport to the Si-SiO$_x$ interface. It was found that the movement of the O^{2-} takes place by breaking and formation of Si-O bonds on adjacent Si atoms. This process leads to the formation of positive charged oxygen vacancies which move in the opposite direction [112,113]. The movement of Si atoms toward the oxide–electrolyte interface is just only apparent and the transport number $t_+ = 0$.

The increase of the anodic current at $V > 3$ V(NHE) is ascribed to the onset of oxygen evolution reaction, which occurs simultaneously with the oxide growth. The dashed line in Figure 2.41 indicates the expected current density according to the high-field model, j_{ox}, without the concomitant oxygen evolution reaction. The course of the potential dependence of the differential capacitance can be interpreted in terms of the potential dependence of the individual capacity elements of the semi-conductor-oxide–electrolyte system, given by

$$\frac{1}{C_T(V,f)} = \frac{1}{C_{sc}(V,f) + C_{SS}(V,f)} + \frac{1}{C_{ox}(V,f)} + \frac{1}{C_H(f)} \tag{2.71}$$

where C_{sc}, C_{SS}, C_{ox}, and C_H are the semiconductor, the surface state, the oxide, and the Helmholtz-layer capacitances, respectively. Equation (2.71) corresponds to an electrical equivalent circuit where the semiconductor capacity connected in parallel with the surface state capacity are connected in series with the capacitance of the oxide and that of the Helmholtz layer. At $V < 0.2$ V, $C_{SS} > C_{sc}$ and $C_{SS} \ll C_{ox}$, C_H, so that $C_T \sim C_{SS}$. In the potential region 0.2 V $< V <$ 1 V the p-type semiconductor is under accumulation and the capacitance increases exponentially with the anodic potential. The oxide grows and the oxide capacity decreases inversely with d_{ox}. Hence, as $C_{sc} > C_{ox}$, C_H, the measured capacity corresponds to C_{ox}. As shown in Figure 2.42,

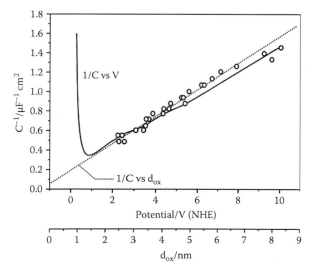

FIGURE 2.42 Inverse of the interfacial capacitance as a function of applied voltage (-----) and of the oxide thickness (acitance aspotentiodynamic oxide growth on p-type Si(100) in acetate buffer solution. The capacity was measured by a frequency of 1 kHz. Scan rate: 0.1 V s^{-1}. The oxide thickness was determined by a micro-ellipsometer.

the inverse of the oxide capacitance increases linearly with the anodic potential. The linearity is also observed when plotting $1/C_{ox}$ as a function of the oxide thickness determined by ex-situ micro-ellipsometric experiments. These experiments confirm the oxide growth according to a high-field mechanism, as described by Eq. (2.68). On the other hand, the oxygen evolution does not affect the oxide growth in an acetate buffer of pH 5.9. From Eq. (2.68), a dielectric constant of the oxide $\varepsilon_{ox} = 6.5$ can be calculated. This value differs from 3.9 found for a thermally grown oxide. This reveals the high content of coordinated water, broken Si-Si bonds, and Si-OH groups in anodic oxides [114,115].

The anodic oxide growth involves several electronic and ionic processes character-ized by time constants, which differ in some orders of magnitude. Therefore, these processes can only be distinguished by recording the current-time response to poten-tiostatic pulses by a nonlinear data collection system. Such a system was developed at the University of Düsseldorf in the 1990s. This recording system is constituted by a fast-ramp potentiostat, an auto-range amplifier, and a nonlinear generator, which allow the measurement of the current in a time scale between 10^{-6} s and 10^3 s (see Ref. 105 for more details). Figure 2.43 shows current-time curves obtained by apply-ing the potential program shown in the upper right corner of the figure on p-type silicon. The current was recorded during the second pulse, after the formation of an initial oxide film of a thickness d_0 during the first pulse. The course of the current-time curves in a double logarithmic plot shows essentially three time regions. The

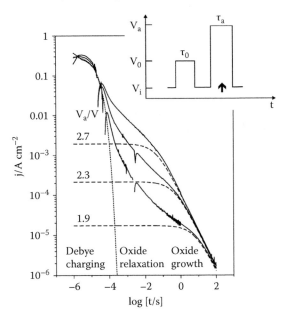

FIGURE 2.43 Current-time curves recorded on p-type Si(100) by applying potentiostatic pulses in a borate buffer solution of pH 8.4. An initial oxide film was formed potentiostatically at $V_0 = 1.5$ V for $\tau_0 = 100$ s: $d_{ox}^0 \sim 1.5$ nm. Dotted line: Debye charging. Dashed line: curve obtained by integration of Eq. (2.65) taking: $V_{0x}^0 = -0.86$ V (NHE), log $(j_0/\text{A cm}^{-2}) = -13.4 \pm 1$, $\beta = 10.6 \pm 1$ nm V^{-1}.

initial region up to 0.1 ms is characterized by the dielectric Debye charging. The dashed line in Figure 2.43 is a calculated dielectric answer of a series RC-system with $R = 8\ \Omega$ cm^2 and $C = 2.5\ \mu$F cm^{-2}. The following region between 10 ms and 0.1 s reflects the dielectric oxide relaxation arising from correlated dipole transitions between two preferential positions upon a sudden change of the electric field. The formation of electrical dipoles in the oxide is closely related with oxide defects [116]. Hence, relaxation effects involve thermally activated ionic transitions between two lattice sites, as well as collective displacements around the lattice site: the so-called domino reactions. The course of the current-time response in this time-region reflects the time-dependence of the dielectric constant, which can be described by the universal exponential law [117,118]. Accordingly, the depolarization current will be given by

$$j_{relax} = \frac{d\overline{P}}{dt} = \varepsilon_0 \overline{E} \times const \times t^{-n} \qquad (2.72)$$

where $\overline{P}$ is the polarization vector.

The physical meaning of the exponent n is related to the extent of correlation of the dipole transitions. Thus, $n = 0$ corresponds to completely uncorrelated while $n = 1$ to completely correlated transitions. Accordingly, this parameter can also be related to the order of the lattice structure [111]. The third time region after 0.1 s reflects the oxide growth. The experimental curves were fitted to the numerical integration of Eq. (2.65), whose parameters j_0 and β were previously calculated by applying the limit expression:

$$\lim_{t \to 0} \log j = \log j_0 + \beta \frac{(V - V_{ox}^0)}{d_{ox}^0} \qquad (2.73)$$

Figure 2.44 shows the limit current density values as a function of anodization potential for various electrolytes. Values of $j_0 = 3.98 \times 10^{-14}$ A cm^{-2} and $\beta = 10.6 \pm 1$ nm V^{-1} are obtained by the extrapolation to $V = V_{ox}{}^0$ and from the slope of the linear part of the plots, respectively.

The horizontal shift of the linear relation obtained in sulfuric acid is probably related to the high oxide dissolution rate in this medium. Deviations of the linear relation at high current densities reflect the effects of ohmic potential drops. From the expression of the β-coefficient, $\beta = (\alpha z e/kT)a$, a lattice constant $a = 0.27$ nm can be calculated by assuming $\alpha = 0.5$. This value is in agreement with the distance between two adjacent O-atoms in the SiO$_2$-structure [119], as assumed in a hopping mechanism upon oxygen vacancies.

The particularly high oxygen evolution current during anodization of p-type silicon in sulfuric acid solution at $V > 4.5$ V (HESS) (see Figure 2.45) may be related to the enhanced oxide dissolution in this electrolyte. This fact is pointed out by a distinct drop of the inverse of the oxide capacitance ($1/C_{ox}$) upon the onset of oxygen evolution and its subsequent slower increase with the potential than in the buffer solutions. This latter indicates an increase of the dielectric constant according to Eq. (2.68). An implicit high-field-assisted mechanism involved in the oxygen evolution

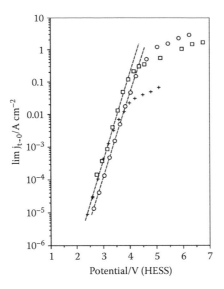

FIGURE 2.44   Logarithm of the limit initial oxidation current density as a function of the anodization potential in the HESS-scale (hydrogen electrode in the same solution) measured in various electrolytes: (+) borate buffer of pH 8.4, $d_{ox}^0 \sim 1.5$ nm; (o) acetate buffer of pH 5.9, $d_{ox}^0 \sim 1.7$ nm; (□) acetat$_2$SO$_4$, $d_{ox}^0 \sim 2.2$ nm.

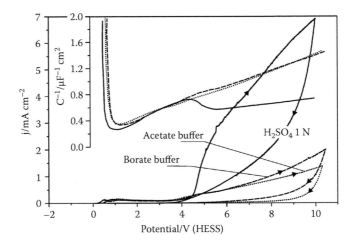

FIGURE 2.45   Current-voltage curves with simultaneous measurement of the interfacial capacity of p-type Si(100) in different solutions. Scan rate: 0.1 V s^{-1}; Frequency: 1 kHz.

reaction can be inferred from these experiments: a higher oxygen evolution current by a thinner oxide film.

The oxygen evolution current in current-time experiments was separated from the oxide formation current by calculating the oxide formation current by integrating Eq. (2.65) (Figure 2.46). j_0 and β parameters were taken from Figure 2.44 at $V < 4$V. The right part of Figure 2.46 shows the values of log j_{O2} plotted against $(V_a - V_{ox}^0)^{-1}$ for

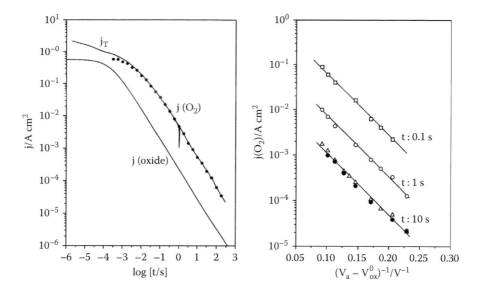

FIGURE 2.46 Left: current-time curve recorded by applying a pulse of 8 V (j_T) and oxide growth current density calculated by taking: $V_{ox}^0 = -0.86$ V(NHE), log (j_0/A cm^{-2}) = -13.4 und $\beta = 1.06 \times 10^{-6}$ cm V^{-1}. $d_{ox}^0 \sim 3$ nm. Right: logarithm of the oxygen evolution current density as a function of the inverse of the oxide potential drop for different constant times.

different times, where linear relations can be easily seen. This result is a clear hint of a control of the oxygen evolution reaction by the electric field in the growing oxide. The experimental results can mathematically be expressed as

$$\log j_{O_2} = A' - \frac{B}{(V_a - V_{ox}^0)} \log t \tag{2.74}$$

where A' and B are constants. The introduction of the logarithmic inverse law for the oxide growth (Eq. 2.66) into Eq. (2.74) yields

$$\log j_{O_2} = (A' + BA\beta) - \beta \frac{B}{d_{ox}} \tag{2.75}$$

It should be noted that this expression shows an interesting similarity with that formulated for the electron transfer by resonance tunneling in oxides (Eq. 2.57). Lewerenz and Jungblut have proposed the existence of a shallow defect band near below the silicon conduction band for interpreting the results of their photoelectron-spectroscopic studies [66]. These defects may be ascribed to positive-charged under-coordinated Si atoms in the SiO$_2$ structure. Other types of defects can also be identified in silicon oxide. Positive charged over-coordinated O-atom forms localized states 1.4 eV below the oxide conduction band. The so-called E' centers are a type of charging defects situated between 4.78 eV and 5.92 eV below the oxide conduction band [116]. The coupled charging and discharging of E' centers is a possible electron transfer pathway through the dielectric oxide.

Figure 2.47 shows the energy band diagram of a p-type Si-SiO$_x$-electrolyte interface, where different electron-transfer routes are illustrated. The band structure was constructed according to the reported data in Ref. 66. The diagram represents the situation at 2.8 V, just above the anodic potential for the onset of the oxygen evolution reaction in an acetate buffer solution. Here, it is supposed that the electric field during the potentiodynamic scan is given by $\overline{E} = (V_a - V_{ox}^0)/d_{ox}$, where d_{ox} is the oxide thickness at the anodic potential V_a, and that the whole potential difference between the semiconductor and the electrolyte falls practically entirely inside the oxide. Under these conditions, the electron transfer can occur by hole transfer upon E′ centers or by resonance tunneling of electrons from the redox pair in the solution via the shallow band. The fraction of potential drop at the oxide–electrolyte interface can be roughly estimated from the current-potential curve, log j vs. V_a by 0.2 $\Delta\phi_{ox}$ [108]. This potential fraction determines the rate of the electron transfer at the oxide–electrolyte interface by a relation of the Butler–Volmer type (see Chapter 4).

The anodic oxide growth on silicon takes place exclusively by transport of O-atoms from the oxide–electrolyte to the silicon–silicon oxide interface: that is, the transport number $t_O = 1$. The accumulation of holes from the semiconductor valence band at the oxide–electrolyte interface leads to the formation of $^+SiO_3$ species. The translocation of Si-O bonds between fully coordinated SiO_4 entities and $^+SiO_3$ was

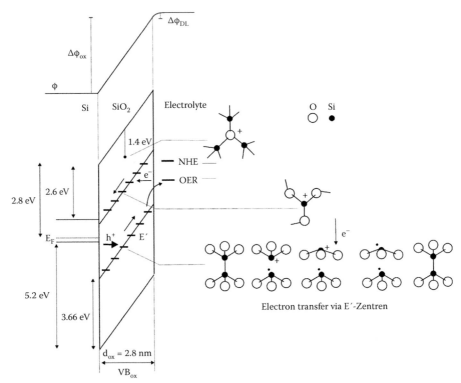

FIGURE 2.47 Energy band diagram of the p-type Si–silicon oxide–electrolyte interface at a potential of 2.8 V.

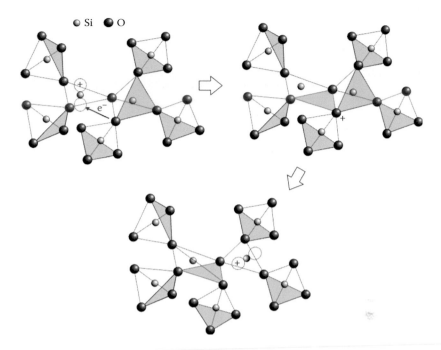

Si ⊙   O ●

FIGURE 2.48 Three-dimensional representation of the SiO_2 structure as tetrahedrons showing the elemental steps for the ionic mobility.

proposed as a possible transport mechanism for O-atoms [113] (a 3D representation is given in Figure 2.48). As a result of the thermally activated process, positive charges move toward the oxide–electrolyte interface, whereas O-atoms move in the opposite direction. New Si-O bonds are generated at the oxide–electrolyte interface by water splitting under the electric field arising at this interface.

The formation of oxygen vacancies in the oxide influences the electron transport processes inside the oxide, because the formation of under-coordinated species gives rise to localized energy states in the oxide band gap which promote a resonance transfer process. On the other hand, an excessive accumulation of positive charged $^+SiO_3$ species at the oxide–electrolyte interface leads to an enhancement of the oxide dissolution rate as a result of an increase of the potential drop $\Delta\varphi_{ox}$ and a reduction of the electric field in the oxide.

Unlike silicon oxide, niobium oxide behaves as an n-type semiconductor with a band gap of 5.3 eV. Electron donors arise as a consequence of oxygen vacancies [89]. Anodization of Nb generates an amorphous oxide, the donor concentration of which changes with the oxide thickness [88,90]. This effect is a consequence of the change of oxide stoichiometry with the progress of growth [94]. The series of cyclic current-voltage experiments with increasing anodic limits performed with Nb in a sulfate solution shows no crossing between the reverse and the forward scan of two successive cycles. The overshoot at the beginning of the first cycle and the anodic current plateau are also characteristic of a field-assisted oxide growth (Figure 2.49a) [120]. The potential dependence of the capacitance shows a more complex behavior.

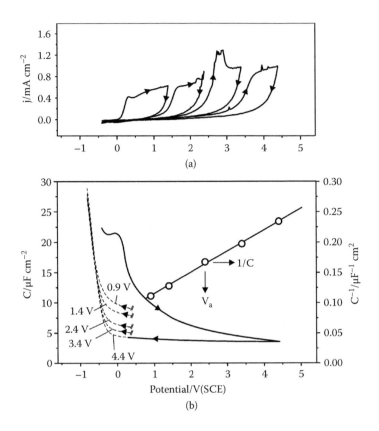

FIGURE 2.49 Potentiodynamic anodization of sputtered thin Nb-film in 0.5 M Na_2SO_4 + 0.5 M H_3BO_3. (a) Cyclic current-voltage curves with increasing anodic limit. Scan rate: 0.1 V s^{-1}; (b) (——) quasi stationary capacitance-voltage curves obtained at 300 Hz: scan rate: 0.005 V s^{-1}; (— —) reverse capacitance-voltage curves at 300 Hz after potentiodynamic (scan rate: 0.1 V s^{-1}) anodization up to indicated anodic limit.

Figure 2.49b shows the interfacial capacitance measured at 300 Hz for the potentiodynamic anodization up to 4.5 V. The oxide capacitance decreases with potential during the forward scan as a consequence of the oxide growth during the scan. It remains practically constant in the reverse scan (constant oxide thickness) and increases abruptly at $V < 0V$. The dashed lines show the reverse scan for $V < 0.5$ V for different reverse anodic potentials. The inverse of the capacitance measured at this potential shows a linear relationship when plotted as a function of anodic limit (V_a, anodization potential), which in turn is proportional to the oxide thickness. On the one hand, these experiments indicate that the growth of niobium oxide occurs according to a high-field-assisted mechanism. On the other hand, the oxide presents a dielectric behavior, at least at potentials higher than 0.5 V. The exponential capacitance increase at $V < 0$ V indicates that the semiconductor is in accumulation.

The semiconductor behavior of amorphous oxides can be quantified by using the Cohen–Fritzsche–Ovishinsky model [121]. This model considers a continuous distribution of the density of localized states inside the mobility band as a consequence of the potential fluctuations in the disordered lattice structure, which arise from translational and compositional disorder. This model differs from that presented by Mott and Davis [122] in that the later makes a distinction between localized states which originate from lack of long-range order and others which are due to defects in the structure. Long-range disorder leads to the formation of tails in the distribution of localized states extending from the mobility edges toward the gap center [123] but they not reach the gap center, so that they are unable to pin the Fermi level.

The capacitive response of amorphous semiconductors depends on the capability of electronic states to react to the rapid changes of potentials. This is represented by the relaxation time for the electron emission from the localized states, which is related to their energy position by

$$\tau = \tau_0 \exp\left(\frac{E - E_{CB}}{kT}\right) \tag{2.76}$$

where τ_0 is a constant characteristic for each material. Thus, only those states for which the condition $\omega\tau \ll 1$ holds will follow the alternating potential, while those for which $\omega\tau \gg 1$, will not response at all. According to this criterion, a frequency dependent threshold in the band bending, z_c, is defined as follows (see also Figure 2.50):

$$e\varphi(z_c) = e\varphi_c = -kT \ln(\omega\tau_0) - (E_{F(bulk)} - E_{CB}). \tag{2.77}$$

For a low band bending, the semiconductor capacitance can be regarded as a series connection of a capacity element for $z > z_c$ with a semiconducting behavior and a capacity element for $z < z_c$ with a dielectric behavior. Hence, the total capacitance can be written as [124]

$$\frac{1}{C(\varphi_{sc},\omega)} = \frac{1}{C(\varphi_c,0)} + \frac{z_c}{\varepsilon_0\varepsilon_{ox}} = \frac{1}{\sqrt{\varepsilon_0\varepsilon_{ox}e^2 N_D}} + \frac{1}{\sqrt{\varepsilon_0\varepsilon_{ox}e^2 N_D}} \ln\left(\frac{\varphi_{sc}}{\varphi_c}\right) \tag{2.78}$$

where φ_{sc} represents the semiconductor band bending, given by $\varphi_{sc} = V - V_{fb}$, and φ_c is the potential difference between das potential in the bulk and at the point z_c. Cohen and Lang [125] have found that for a high semiconductor band bending the occupation of states is pinned at the middle of the gap under electronic equilibrium between the states with the conduction and valence band. Like for low band bending, the semiconductor capacitance for high band bending can be expressed as a series connection of two capacity elements [126,127]:

$$\frac{1}{C(\varphi_{sc},\omega)} = \frac{1}{C(\varphi_g,\omega)} + \frac{z_g}{\varepsilon_0\varepsilon_{ox}}, \tag{2.79}$$

where φ_g is the potential drop between the conduction band edge and the crossing point of the conduction band edge with the pinned Fermi level. The first term

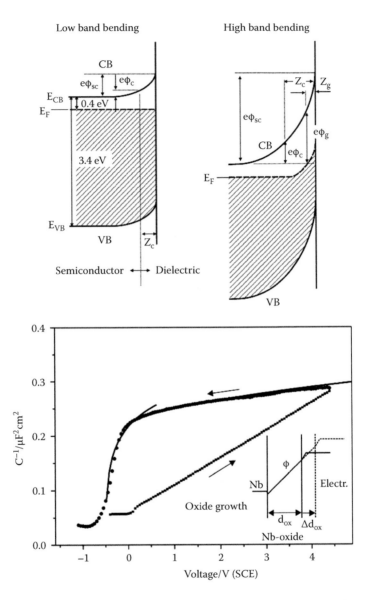

FIGURE 2.50 Top: energy band diagram for an amorphous semiconductor at low and high band bending with indication of the parameters defined in the Cohen–Fritzsche–Ovishinsky model; bottom: quasi-stationary capacitance-voltage curve at 300 Hz in 0.5 M Na_2SO_4 + 0.5 M H_3BO_3; (——) theoretical capacitance-voltage curve predicted by the Cohen–Fritzsche–Ovishinsky model. Detail: potential distribution at the metal–oxide–electrolyte interface during the oxide growth.

of Eq. (2.79) represents the low band bending capacity for $z > z_g$ (see Figure 2.50 and Eq. 2.78). Under the supposition of a constant energy distribution, an analytical expression for z_c can be derived and hence the following expression for the total capacity results [128]:

$$\frac{1}{C(\varphi_{sc},\omega)} = \frac{1}{\sqrt{\varepsilon_0\varepsilon_{ox}e^2 N_D}} \left\{ \ln\frac{\varphi_g}{\varphi_c} + \left[1 + \frac{2(\varphi_s - \varphi_g)}{\varphi_g} \right]^{1/2} \right\} \qquad (2.80)$$

The reverse potential course of the inverse of the capacitance in a potentiodynamic oxide growth experiment can be reproduced by using Eq. (2.78) and Eq. (2.80) with appropriate parameters. The forward course shows a linear increase, as expected for a dielectric behavior during the oxide growth and a high-field assisted growth mechanism. Taking a dielectric constant of $\varepsilon_{ox} = 41–46$ for niobium oxide [129,130], values for the donor concentration of $N_D = 6.5 \times 10^{20}$ eV^{-1}cm^{-3} and $N_D = 1.75 \times 10^{21}$ eV^{-1}cm^{-3} for the low and high band bending regions respectively can be calculated. According to these calculations, the flat band potential is $V_{fb} = -0.5$ V (SCE). The dependence of the capacitance on frequency and anodization potential (oxide thickness) was analyzed in the low band bending potential region [131]. The experimental capacitance values were fitted to Eq. (2.78) by adjusting the parameter φ_c. Figure 2.51 shows that the fraction of surface potential drop with a semiconductor behavior decreases exponentially with frequency as predicted by Eq. (2.77). For a constant frequency, the parameter φ_c drops to values lower than 10 mV for anodizing

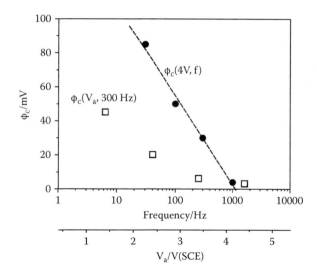

FIGURE 2.51 Dependence of the φ_c parameter on the frequency and anodization potential. Frequency dependence was measured for an oxide grown potentiodynamically up to 4 V. The influence of the anodization potential was determined at 300 Hz. Scan rate for anodic growth: 0.1 V s^{-1}. A donor concentration $N_D = 15 \times 10^{20}$ eV^{-1} cm^{-3} was used for the fitting of experimental curves.

potentials higher than 3 V. Looking at Figure 2.51, a predominant dielectric behavior for potentials higher than 1 V can be seen.

Another anodic oxide with semiconducting properties and potential photo(electro) chemical applications is titanium dioxide. This material, an n-type semiconductor with a band gap of 3.6 eV was extensively investigated with regard to UV photocatalysts for the decomposition of polluting organic substances [132,133] and more recently as support for an electron collector in dye-sensitized cells (see Chapter 1). Ti belongs, like Si, Nb, Al, Ta, and Zr, to the group of so-called *valve metals*, for which the oxide grows by a high-field-assisted mechanism [134,135]. To increase the relation efficiency-mass of the photo(electro)chemical processes, two routes for enlarging the surface-volume ratio are mainly followed: the deposition of films of fine particulated oxide or compact arrangements of nanodimensioned tubes grown perpendicularly on a metal substrate. The former option is usually carried out by synthesis of nanodimensioned particles using a sol-gel procedure [136–138] followed by a sintering step and a series of conditioning treatments such as thermal annealing [139,140] and/or nitrogen [141,142] and carbon doping [143]. The latter concept is generally preferred for photoelectrochemical applications in view of the much better electronic conductivity and larger oxide–electrolyte contact area in this type of structure in comparison with sintered colloidal oxide particles [144–148]. High-ordered porous oxide structures form by high anodization voltages (10 V–100 V) of valve metals in oxide-dissolving electrolytes. The anodic formation of high-ordered porous structures was extensively applied for the formation of alumina masks used in the fabrication of, for instance, ordered arrangements of nanodots. Figure 2.52 shows an upper view of the honeycomb-like porous oxide formed by anodizing in sulfuric and oxalic acid [149]. High-ordered arrays of pores arise as a consequence of differentiated local concentrations, field strengths, and mechanical stresses [149] initiated by inhomogeneous dissolution-oxide growth processes at the oxide–electrolyte interface. The details of the growth mechanism, however, are beyond the scope of this book. In the case of Ti, oxide dissolving fluoride solutions are used for the anodization. Figure 2.53 shows an upper and lateral view of obtained porous oxide under different pHs and anodization times. It should be noted, that in contrast with porous alumina, the walls of adjacent pores in titanium dioxide are separated, forming oxide tubes. In the following analysis of the electrochemical and photoelectrochemical properties, these types of self-organized porous oxides will be referred to as short and long tubes.

General aspects of the electron transfer at the oxide–electrolyte interface of self-organized porous oxides can be analyzed by the investigation of current-voltage experiments performed in solutions containing a redox pair like $Fe(CN)_6^{3-}/Fe(CN)_6^{9-}$ (see Figure 2.54) [150,151]. In the absence of transport limitations, the Fe(II)-Fe(III) redox-reaction is controlled by an outer-sphere charge transfer mechanism. Therefore, it can be modeled by the Marcus–Gerischer theory (see Chapter 1) and the work function for this reaction can be well determined. The current-voltage curves in Figure 2.54 show a characteristic Schottky behavior with a logarithmic linear current increase at potentials more negative than the flat band potential ($V < -0.4$ V) and a lightly decreasing saturation current at potentials more positive than that. The current passes through zero at the equilibrium redox potential at -0.18 V. The slope change near the flat band points out the electron exchange between the redox system and

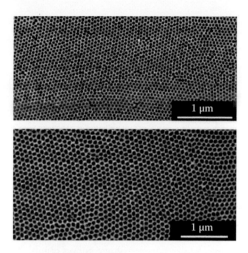

FIGURE 2.52   SEM micrographs of the upper view of anodic alumina layers formed in 0.3 M sulfuric acid at 10°C at 25 V (a) and 0.3 M oxalic acid at 1°C at 40 V. Courtesy of F. Müller and coworkers. Reprinted with permission from Li, A. P., Müller, F., Birner, A., Nielsch, K. and U. Gösele 1998. *J. Appl. Phys.* 84, 6023 (1998). Copyright 1998, American Institute of Physics.

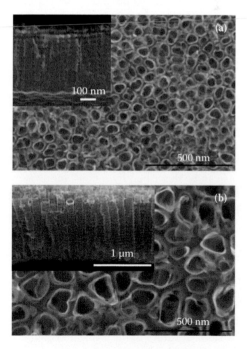

FIGURE 2.53   Upper and lateral view of high-ordered arrays of titanium nanotubes obtained by anodization in fluoride containing solutions: (a) 2 h at 10 V in 1M H3PO4 + 0.3% HF and (b) 2 h at 20 V in 1 M H2NaPO4 + 0.5% HF. Porous oxide was annealed at 450°C for 3 h in the air and a ramp of 20°C s^{-1} was applied for the heating and cooling steps. The transformation to anatase was confirmed by an XRD-test.

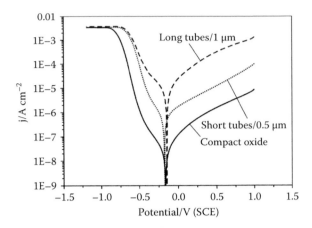

FIGURE 2.54 Quasi-potentiostatic current-voltage curves obtained in 0.1 M Fe(CN)$_6^{3-}$ + 0.1 M Fe(CN)$_6^{4-}$ by using different porous TiO$_2$-electrodes (anatase). Scan rate: 0.001 V s^{-1}.

surface states of the semiconductor. The onset of constant current at the cathodic limit indicates that the reaction is diffusion controlled. The current-voltage curves reflect clearly the effects of the increase of the oxide–electrolyte contact area. However, the increase of the reaction area is not in linear relation with the current increase: the $j_{porous}/j_{compact}$ ratio increases from 9 to 194 for short and long tubes respectively. This implies the influence of additional electronic factors in the porous structure. Impedance measurements performed in depletion potential region and close to the flat band potential can be interpreted in terms of a double oxide structure formed by a bottom compact oxide and the upper arrangement of oxide nanotubes. As shown in Figure 2.55, the impedance spectra (Bode plots) can be fitted by an equivalent circuit consisting of two parallel-RC units connected in series with the electrolyte resistance. One unit represents the impedance behavior of the oxide and the other that of the double layer. The RC-units can be described by a constant-phase-element (CPE), which essentially takes into account the frequency dispersion of the dielectric properties of the material [152]. Thus, the complex impedance is given by

$$Z = \frac{R}{[1+(i\omega RC)^{-n}]} \tag{2.81}$$

where n is the dispersion factor and ω is the angular frequency. The factor n adopts a value of 1 for an ideal behavior, 0 for a pure resistance and 0.5 for a mass-transport controlled system. If we look at values of the oxide and double layer capacitances, we can see that in depletion the double layer capacitance increases with the growth of nanotubes whiles the semiconductor capacitance remains constant (Table 2.1). The effects of the area increase are observed on both semiconductor and double layer capacitance near the flat band potential. The capacity of the double layer results, as expected, independent of potential.

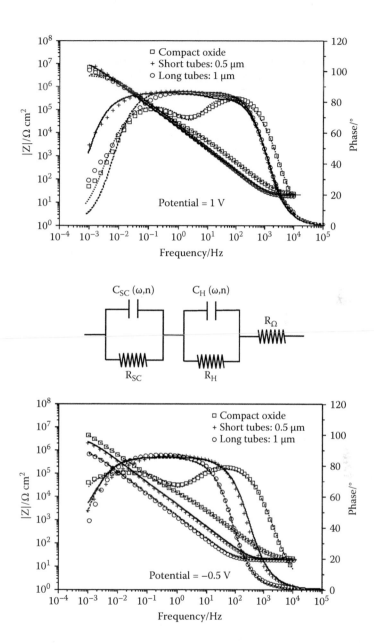

FIGURE 2.55 Impedance spectra, presented as Bode diagrams, recorded in an acetate buffer solution of pH 5.9 with compact and porous titanium electrodes (anatase) close (−0.5 V) and anodic from the flat band potential (1.0 V). Points: experimental values. Lines: result of fitting experimental results with the equivalent circuit shown in the middle of the figure. Amplitude: 10 mV.

TABLE 2.1

Values of Capacity and Resistances Obtained after Fitting the Impedance Response of Compact and Porous Titanium Oxides (Anatase) with the Equivalent Circuit Shown in Figure 2.55

	C_{sc} [μFcm^{-2}]	R_{sc} [MΩ cm^2]	C_H [μFcm^{-2}]	R_H [kΩ cm^2]	n_{sc}
			1 V		
Compact	8.0	4.0	12.0	5.0	0.91
Short tubes	10.0	9.0	90.0	0.2	0.97
Long tubes	10.0	3.0	150.0	0.05	0.95
			−0.5 V		
Compact	20.0	25.0	12.0	5.0	0.90
Short tubes	80.0	6.0	90.0	800.0	0.97
Long tubes	350.0	1.0	150.0	500.0	0.95

The fact that the influence of the nanotubes is decoupled in the depletion potential region is, in principle, ambiguous. It can be interpreted as an increase of the space charge layer length up to the thickness of the walls of the oxide tubes and the appearance of a dielectric behavior. On the other hand, one can arrive at an explanation by assuming a high donor concentration of the tubes, so that they behave as a metal.

One of the interesting features of the impedance behavior of titanium dioxide is the appearance of a bend of the Mott–Schottky plots (see Figure 2.56) [153]. This effect, which was interpreted by some researchers by the attainment of wall thickness by the space charge layer length [153], does not seem to apply in this case where the bending appears at the same potential for two different oxide thicknesses. Other researchers have attributed this effect to the presence of multiple level donors that are progressively

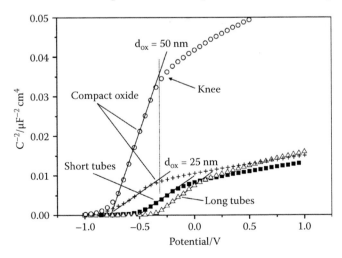

FIGURE 2.56 Mott–Schottky plots of compact and porous titanium oxide (anatase) measured in an acetate buffer of pH 5.9.

ionized with potential increase [154–156]. In a first approach, the first linear part of the Mott–Schottky plots can be quantified by the Mott–Schottky equation:

$$\frac{1}{C_{sc}^2} = \frac{2}{e\varepsilon_{sc}\varepsilon_0 N_D}\left(V - V_{fb} - \frac{kT}{e}\right)$$ (2.82)

where V_{fb} is the flat band potential. From the slope and the extrapolation to potential to $C^{-2} = 0$ the values of N_D and V_{fb} can be calculated. The growth of oxide nanotubes shifts the flat band potential from −0.725 V (compact oxide) to −0.575 V for the short tubes and to −0.33 V for the long tubes. This shift clearly reflects the increase of the potential drop at double layer in comparison with that of the space charge layer as a result of the enlargement of area of the oxide–electrolyte contact. The potential drop can be calculated as

$$\Delta\varphi_H = \frac{q_{surf} \times A_{tubes}}{C_H \times A_{geom}}$$ (2.83)

where q_{surf} [C cm^{-2}] is the interfacial charge, A_{tubes} is the surface area of the tubes, and A_{geom} is the area of the electrode. On the other hand, the length of the space charge layer is given by

$$L_{sc} = 2\left(\frac{\varepsilon_{sc}\varepsilon_0}{2eN_D}\right)^{1/2}\left(V - V_{fb} - \frac{kT}{e}\right)^{1/2}$$ (2.84)

Taking a dielectric constant of 42, Eq. (2.84) yields a length of 1.73 nm for the potential where the bending of the Mott–Schottky plot occurs. This value results much shorter than the oxide thickness of 50 nm and 25 nm. The origin of the bending can be attributed to the presence of multiple ionization levels so that the Cohen–Fritzsche–Ovishinsky model can be applied. As shown in Figure 2.57, the experimental data in depletion can be well represented by this model by using the parameters shown in Table 2.2. It can be seen that no appreciable differences appear in the donor concentration in spite of the large differences in the oxide–electrolyte contact areas. Therefore, it can be argued that the impedance response originates in the underlying compact oxide of the double oxide structure. No impedance response appears, however, that could be attributed to oxide nanotubes. As already mentioned, this fact can be interpreted as (i) a complete depletion of the walls of tubes or (ii) a quasi-metallic behavior as a consequence of a high concentration of defects (donor concentration). There is, in principle, no concrete evidence about the dielectric behavior of the layer of tubes. The wall thickness of tubes oscillates between 10 and 30 nm. On the other hand, the thickness of the space charge layer could reach a value of 3 nm for a band bending of 1.5 V adopting $\varepsilon = 42$ and $N_D = 6.5 \times 10^{20}$ cm^{-3} (Eq. 2.84). Therefore, option (i) seems to be improbable.

In spite of the crystalline nature of the material (anatase), the correspondence of the semiconductor behavior with model of amorphous semiconductors derives probably from the intercalation of anions such as SO_4^{2-} [157,158] and PO_4^{3-} [159] in the

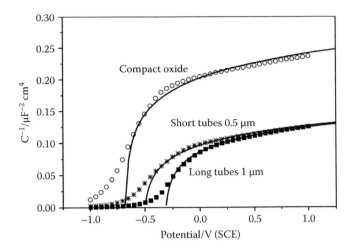

FIGURE 2.57 Potential dependence of the inverse of surface capacity of compact and porous titanium oxide in an acetate buffer of pH 5.9. Symbols: experimental data. Lines: fitting curves (Eq. 2.78).

TABLE 2.2
Fitting Parameters (Eq. 2.78)

	$N_D \times 10^{-20}$ [eV^{-1}cm^{-3}]	V_{fb} [V]	φ_c [V]
Compact	3.1	−0.85	0.15
Short tubes	6.5	−0.72	0.30
Long tubes	6.5	−0.50	0.27

oxide structure. The formation of localized intraband states also arises a consequence of the transport of mobile defects at $V > V_{fb}$. For instance, the presence of a defect band 0.5 eV below the conduction band was proposed to explain the appearance of a photocurrent maximum in titanium electrodes sensitized with Zn-porphyrins [160].

Due to the large surface/mass ratio, one can expect that a large concentration of defects is contained in the walls of oxide tubes. Thus, an n-n$^+$-type junction appears at the bottom of the tubes. On the other hand, surface states pin the Fermi level of the tubes.

Acid–base reactions at the oxide surface (Eq. 2.85 and 86) lead to changes of the double layer charging and thus to a shift of the Fermi level relative to the reference potential in the solution [162].

$$\text{Ti-O}^{2-}_{(surf)} + \text{H}^+ \leftrightarrow \text{Ti-OH}^-_{(surf)} \tag{2.85}$$

$$\text{Ti-OH}^-_{(surf)} + \text{H}^+ \leftrightarrow \text{Ti-(OH}_2)_{(surf)} \tag{2.86}$$

According to this reaction scheme, the potential of the double layer changes as

$$V_{fb} = V_{fb,iep} - 2.3\frac{kT}{e}(\text{pH} - \text{pH}_{iep}) \tag{2.87}$$

where $V_{fb,iep}$ and pH_{iep} are the flat band potential and the pH of the isoelectric point. The latter was found to be 5.8 for anatase [158]. The slope of V_{fp} vs. pH of 59 mV pH^{-1} predicted by Eq. (2.87) is found for the compact oxide (see Figure 2.58). The increase of the surface/mass ratio causes an increase of the slope and a shift of the curve toward more positive potentials. The latter effect is a direct consequence of the increasing negative charge induced at the bottom of the tubes as a compensation for the positive charge accumulated at the oxide surface (pH < pH_{iso}: positive charge).

The larger slope of the V_{fb} vs. pH plots shown by the porous oxides (105 mV pH^{-1} for the short tubes and 166 mV pH^{-1} for the long tubes) at pH < pH_{iso} may be attributed to the intercalation of protons in the oxide structure. It has been suggested that this process takes place by discharge of electrons from the conduction band via electron traps located close to the trap [163,164] according to the reaction

$$\text{Ti}^{(+4)}\text{O}_2 + \text{e}^-_{CB} + \text{H}^+ \rightarrow \text{Ti}^{(+3)}\text{O(OH)} \tag{2.88}$$

The intercalation of protons is a reversible process that largely modifies the optoelectronic properties and wettability of the oxide [165,166]. Wang et al. [167] have found an empiric relation between the trapped charge and the proton concentration: q [C cm^{-2}] = 3.0×10^{-6} [H$^+$]$^{-0.23}$. Li$^+$ can penetrate into the oxide structure as well, modifying its optical properties [168]. Since the effect is a surface one, the effects are enlarged in the porous structures. The intercalation of protons appears as a cathodic peak in the cyclic current-voltage curves shown in Figure 2.59. The corresponding anodic peak reflects the reverse process. The uptake of H$^+$ modifies the reflectivity of the surface as a consequence of the formation of localized intraband energy states

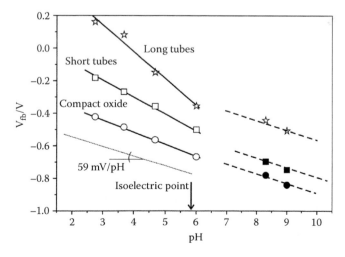

FIGURE 2.58 pH-dependence of the flat band potential for compact and porous oxide (anatase).

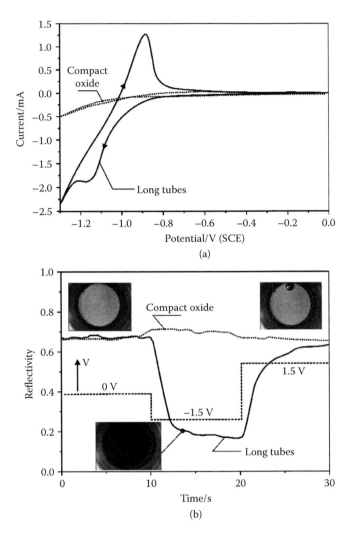

FIGURE 2.59 (a) Cyclic current-voltage curves of compact and porous titanium oxide (anatase) in $HClO_4$ 0.1 M. Scan rate: 0.01 V s^{-1}. (b) Change of reflectance of porous titanium oxide under monochromatic illumination at $\lambda = 480$ nm (2.58 eV) after changing the electrode polarization.

(traps) 0.75 eV and 1.18 eV below the conduction band [169–171] which allow light to be absorbed in the visible range.

Surface Ti atoms partially coordinated with water create surface states which act as electron traps [172]. The charging of these states is indicated by an increase of a capacitance peak appearing superimposed on the semiconductor response by illumination (see Figure 2.60).

The bell form of the capacitance-voltage curve of the surface states arises as a consequence of the charging–discharging kinetics, as discussed in Chapter 5.

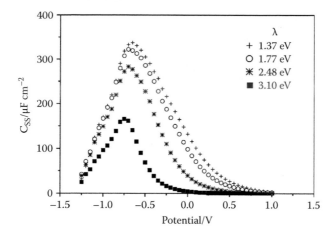

FIGURE 2.60 Spectral behavior of surface states in porous titanium oxide (anatase) in an acetate buffer solution of pH 5.9. Frequency: 3 Hz.

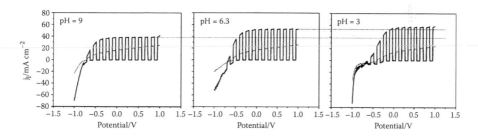

FIGURE 2.61 Photocurrent-voltage behavior of compact (———) and porous (----) titanium oxide (anatase) in solution of different pH's under monochromatic illumination at 3.54 eV.

The photo-effect of proton intercalation is much more pronounced in the oxide tube layer than in the compact oxide, due to the larger area of the former (see Figure 2.61). On the other hand, the light absorption increases at lower pH's because of a generation of a larger density of localized energy states by proton intercalation. Thus, the protonated oxide tubes act like antennas with an efficient transport of photogenerated minority-charge charges (holes).

The energy band diagram shown in Figure 2.62 resumes the optoelectronic properties of anodically grown high ordered arrays of oxide nanotubes. The electron affinity of TiO_2 was taken from Ref. 173. Hence, it is located 0.43 eV below the reported work function for Ti. This ensures a good ohmic contact metal-oxide. The localized intra-band states arising from proton intercalation are represented by short lines. Here, the supposition of oxide nanotubes with a higher defect concentration than the base compact oxide was adopted. This gives rise to an n-n^+-type junction and hence each change of the electrode potential (by changing the position of the Fermi level) is distributed between the underlying compact oxide (band bending) and the Helmholtz layer at the oxide–electrolyte interface. The electron exchange

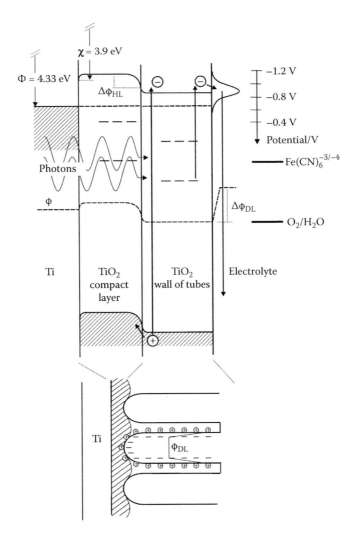

FIGURE 2.62 Energy band diagram of anodic porous titanium oxide at a potential more negative than the flat band potential.

between surface states at the oxide–electrolyte interface and the semiconductor leads to a Fermi level pinning at the negative side of the flat band potential. The effect is enhanced with the surface area of tubes, and a shift of the flat band potential toward more positive potentials is thus observed. The acid–base reactions at the surface, on the other hand modify the Helmholtz-capacity C_H, from which the pH-dependence results. By means of spectroscopy studies, it was demonstrated that surface states arise as a consequence of oxygen vacancies 0.3 eV above the conduction band [174]. The charging of these surface states by photoexcitation of electrons from intraband states leads to an increase of the surface capacity, which appears as an additional peak superposed to the semiconductor response.

The next system to be presented in this chapter is the p-InP–electrolyte interface. This material is the preferred one when designing photoelectrochemical structures for the photogeneration of hydrogen [175,176] due to the position of its energy band edges and the optimal band gap: 1.34 eV. The maximal photovoltage theoretically achievable with p-InP is about 0.9 V (see band diagram in Figure 2.63). The favorable energetic conditions are, however, shadowed by the photoinduced corrosion and the low catalytic activity of an InP-surface for the hydrogen evolution reaction. As already discussed in general terms in Chapter 1, the strategy to be applied consists of the formation of a stabilizing film to isolate the semiconductor from the highly corrosive electrolytes. One alternative is the formation of a conductive oxide by an electrochemical treatment in HCl solution. It has been demonstrated that the potentiodynamic cycling of p-InP in the cathodic region in hydrochloric acid under illumination increases the catalytic activity of the surface for the hydrogen evolution reaction [177]. The enhancement of the catalytic effect in this case has a merely energetic origin related with the formation of a dipole layer at the semiconductor–oxide interface arising after the electrochemical treatment. In contrast to the anodic methods described above, the oxide on p-InP grows by a photoelectrochemical process at potentials where the predominant reaction is the proton reduction. The complex chemical scenario developed at the surface during the electrode cycling leads to a thin oxide with a particular composition and conducting properties as described below. The general surface activation method consists of an initial short chemical etching in a 0.5% bromide methanol solution, which ensures the elimination of a native oxide layer. The second step consists of a cyclic potentiodynamic scanning in HCl solution under illumination. Lewerenz and Schulte [177] have reported the surface activation of p-type InP(111) when cycling between 0.4 V and −0.8 V (SCE). XPS and UPS studies [178] performed after electrochemical conditioning indicate the formation of an oxide layer composed of In_2O_3, InCl, and $In(PO_3)_3$ [179]. The chemical and electronic surface transformation generated by the electrochemical

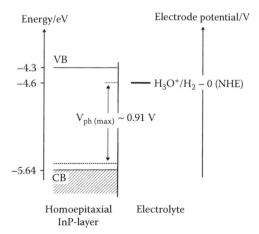

FIGURE 2.63 Idealized energy band diagram of p-InP at the flat band potential in acid electrolytes.

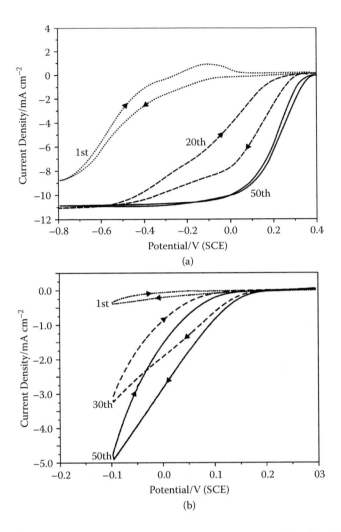

FIGURE 2.64 (a) Cyclic current-voltage curves of p-InP(111) in 0.5 M HCl under illumination. Only curves recorded after several cycles are shown. Scan rate: 20 mV s^{-1} [after Schulte and Lewerenz, 2002] [171]. (b) Cyclic current-voltage curves of an In-rich homoepitaxial p-InP(100) substrate in 0.5 M HCl under illumination. Scan rate: 10 mV s^{-1}

cycling leads to an increasing surface activity. The cathodic photocurrent, which corresponds mainly to the hydrogen evolution, increases with the number of cycles up to a stationary value (see Figure 2.64).

Spectroscopic studies reported in Ref. 178 indicate the increase of the electron affinity of P-type InP(111) by 0.9 eV after the electrochemical conditioning. The energy shift of the semiconductor band edge is to be attributed to the formation of a dipole layer at the p-InP(111)–oxide interface, presumably as a consequence of the formation of In-Cl bonds. This implies the formation of metallic In, which can be formed by the reduction process induced by photogenerated electrons [180–182]:

$$p\text{-InP} + e^-_{CB}(hv) + 3\ H^+ \rightarrow In + PH_{3(g)} \qquad V^0 = -0.6\ V\ (SCE) \qquad (2.89)$$

The formed In particles are oxidized as the potential reaches values anodic from the standard potentials for the following oxidation reactions during the cyclic scans:

$$In + 3\ H_2O \rightarrow In(OH)_3 + 3\ H^+ + 3\ e^- \qquad V^0 = -0.43\ V\ (SCE) \qquad (2.90)$$

$$2\ In + 3\ H_2O \rightarrow In_2O_3 + 6\ H^+ + 6\ e^- \qquad V^0 = -0.412\ V\ (SCE) \qquad (2.91)$$

$$In + Cl^- \rightarrow InCl + e^- \qquad V^0 = -0.38\ V\ (SCE) \qquad (2.92)$$

Furthermore, the substrate oxidation by injection of holes from the valence band occurs during the potential scanning:

$$3\ InP + 15\ H_2O + 24\ h^+ \rightarrow In(PO_3)_3 + 2\ In(OH)_3 + 24\ H^+ \qquad (2.93)$$
$$V^0 = -0.84\ V\ (SCE)$$

$$InP + 4\ H_2O + 8\ h^+ \rightarrow InPO_4 + 8\ H^+ \qquad V^0 = -0.92\ V\ (SCE) \qquad (2.94)$$

It is evident by regarding this complex reaction schema that the spatial distribution of the oxide components in the growing oxide film depends largely on the potential window selected for the potential scan. The atomic structure of the surface also plays a decisive role. For instance, the activation of homoepitaxial layers grown by MOVPE (metallo-organic vapor phase deposition) on p-InP(100), the structure of which is described below, succeeds by selecting a different potential scan window. It should be noted that the current response is also different (see Figure 2.64).

Homoepitaxial films are grown by a controlled injection rate of In- and P-precursors in fixed proportions and a selected temperature program. Using Zn-precursors for semiconductor doping, films up to 3 μm thickness with a donor concentration of 10^{17} cm^{-3} can be grown. The surface structure of In-rich homoepitaxial layers grown by MOVPE on InP(100) substrates shows, in vacuum, a (2 × 4) reconstruction where In is present in a In:P relation of 9:1 (see schematic in Figure 2.65) [183,184]. This is different from the atomic structure of InP(111)-surfaces, which show atomic planes of only In or P atoms depending on the previous treatment. AFM pictures of the new surface show the formation of terraces with 2 nm of height (Figure 2.66). It should be noted, that the lateral growth of terraces is pinned at some points, leading to the formation of pit-like cavities. The origin of these points is unclear but they seem to arise from defects present on the substrate.

Regarding the current-voltage curves obtained during the activation of the In-rich homoepitaxial layer, it can be inferred that the catalytic activity of conditioned (oxidized) In-rich homoepitaxial film is inferior to that shown by the conditioned p-type InP(111) (compare cyclic scans in Figure 2.64). The growth of thin film by MOPVE introduces an interesting concept for the construction of light absorbers with an efficient material usage. The catalytic activity is given by photoelectrodeposition

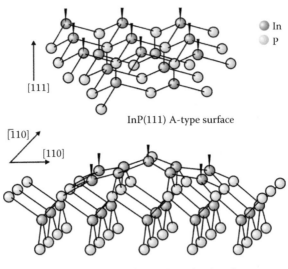

InP(111) A-type surface

(2×4) In-rich InP(001) surface

FIGURE 2.65　Atomic structure of a (111)- and a reconstructed (2×4)-InP surface.

of Rh-nanoparticles at −0.2 V from 5 mM $RhCl_3$ + 0.5 M KCl + 5% isopropanol. Isopropanol is introduced as an additive to improve the flatness of the deposit.

The electrochemical cycling in HCl solution also leaves particles with a lateral size of 40 nm to 70 nm and a height of 10 nm to 20 nm, some of which are located at the step edges. They can a priori be ascribed to In-islands formed by a decomposition reaction (Eq. 2.89).

The surface chemical transformations during the conditioning steps (chemical etching, electrochemical cycling, and Rh-photoelectrodeposition) were analyzed by synchrotron radiation photoelectron spectroscopy (SRPES). The P 2p core level spectra shown in Figure 2.67 can be deconvoluted in three component signals. That at 132.45 eV and 133.20 eV are assigned to $In(PO_3)_3$ and $InPO_4$, respectively. The line at 128.40 eV (InP) arises from the substrate. This indicates that a thin oxide film is already formed after the first chemical etching step. The assignment to indium phosphates is supported by the appearance of lines at 444.75 eV and 445.05 eV in the In 3d core level spectrum shown in Figure 2.68. The chemical pretreatment also leads to the formation of indium oxide, indicated by a line at 444.30 eV. The subsequent photoelectrochemical cycling increases the phosphate/oxide ratio, as indicated by comparing the corresponding signal intensities in the In 3d core level spectrum. The chemical transformation involves the formation of metallic In, as indicated by the appearance of a line at 443.85 eV. The third conditioning step consists of depositing Rh at −0.2 V under illumination. This leads to an increase of the oxidation products $InPO_4$, $In(PO_3)_3$ in relation to In_2O_3. The formation of In_2O_3 implies a local surface alkalization, which arises as a consequence of a strong consumption of protons and is in part due to the blocking effect of hydrogen bubbles attached to the electrode during the photoelectrochemical cycling. The Pourbaix diagram shown in Figure 2.69 shows that metallic In is not stable in the used scan windows and how the speciation

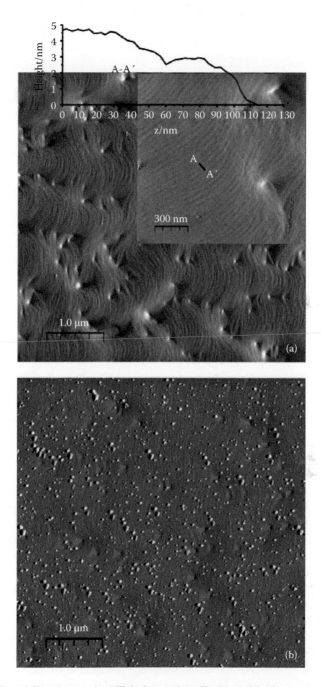

FIGURE 2.66   (a) Tapping-mode AFM-picture (amplitude mode) of a reconstructed (2×4)-InP surface after 30 s etching in 0.5% Br$_2$-methanol solution; (b) the same surface after subsequent electrochemical cycling (electrochemical conditioning).

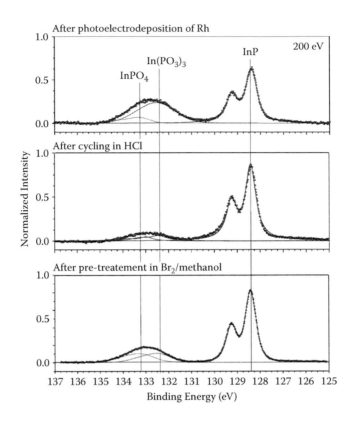

FIGURE 2.67 SRPES P 2p core level signal of a homoepitaxial p-InP(100) layer recorded with an excitation energy of 200 eV after each conditioning step.

of dissolved In(III) changes within the scan windows at the electrode surface. This diagram predicts instability of In_2O_3 in the acid solution. It should be noted that the formed oxide is not a mixing of single phases. The indicated compounds describe solely the local chemical environment of In or P atoms in the oxide. The surface affinity to form indium oxide depends on the concentration of In atoms at the surface. They react with water, thus enabling the introduction of O atoms into the In-P bonds to form the In-O-P bonds [186], which are more thermodynamically stable than In-O-In or P-O-P [187].

It has been demonstrated that the anodization of InP at potentials higher than 5 V leads to the formation of compositionally inhomogeneous films, as for instance on application of a constant current of 15 $\mu A\ cm^{-2}$ to InP(100) in pure water. In this case, an inner layer phosphate layer composed of $In(PO_3)_3$ is formed [188,189]. The outermost part consists of In_2O_3. These findings were ascribed to a preferential migration of In^{3+} from the InP–oxide interface toward the oxide–electrolyte interface upon cation vacancies induced by the electric field. Though the conditions under which these experiments were performed are different from the case discussed here, the comparison may be useful for understanding the transport processes occurring in the photoelectrochemical cycling.

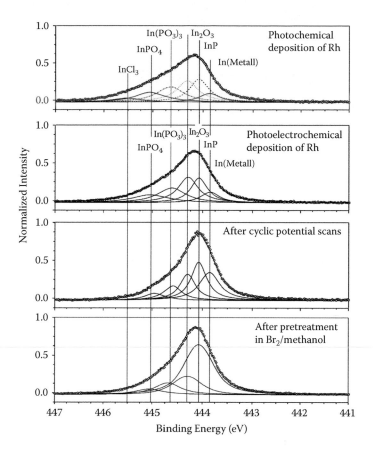

FIGURE 2.68 SRPES In 3d core level signal of a homoepitaxial p-InP(100) layer recorded with an excitation energy of 600 eV after each conditioning step.

SRPES investigations of the oxide film formed by the electrochemical conditioning indicate an increase of the intensity ratio between P 2p lines of $InPO_4$ and $In(PO_3)_3$, that is, $r = (I_{InPO4}/I_{In(PO3)3})$, with the excitation energy: hv = 200 eV, r = 0.62; hv = 335 eV, r = 4.146; hv = 600 eV, r = 6.75. This tendency suggests in principle that the oxide composition can be represented by the sandwich-structure $In(PO_3)_3/InPO_4/$ InP. Such a structure, with an In-rich (P-poor) inner oxide film can be explained in terms of a faster migration of P^{5+} with a ionic radius of 0.35 Å in comparison with In^{3+} with a radius of 0.81 Å [190].

Spectroscopic data (see Figure 2.68) indicates the presence of an In_2O_3-rich phase, the position of which, however, cannot be unambiguously determined. Reported investigations on the thermal oxidation of InP in oxygen atmosphere show the formation of a In_2O_3-rich layer at the outermost part of the oxide film, whereas a preponderant concentration of $InPO_4$ appears at the InP-oxide interface [191–193]. Similar results can be found after the anodization of InP in aqueous electrolytes [194,195].

The migration mechanism of ions by the oxide growth is still a matter of discussion. The transport of In^{3+} occurs by hopping upon oxide defects [196]. The inward

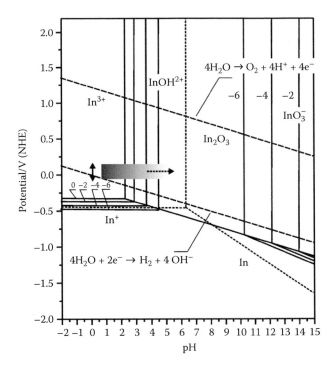

FIGURE 2.69 Stability V-pH diagram for the system Indium-Wasser at 25°C [178].

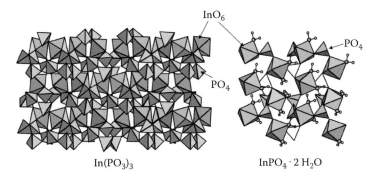

FIGURE 2.70 Atomic structure of $In(PO_3)_3$ and $InPO_4$.

movement of O atoms result from a transport of O^{2-} ions by a translocation of P-O bonds like as it occurs in the silicon oxide growth. Thus, an O-rich phosphate layer $(In(PO_3)_3)$ appears at the outermost part of the growing oxide. The water containing $InPO_4$ is characterized by an open structure where the PO_4 tetrahedrons are connected to InO_6 octahedrons by double coordinated O atoms [197,198] (Figure 2.70). Water is located at the corner of the octahedrons. On the other hand, $In(PO_3)_3$ presents a closed structure, characterized by infinite chains of PO_4 tetrahedrons linked with InO_6 octahedrons. This structure shows a large flexibility due to the wide varying angle of the P-O-P bond. Hence, a small concentration of defects appears in this structure [199,200]. The predominance of $InPO_4$ in the oxide structure of the oxide grown during the photoelectrochemical treatment of the homoepitaxial InP-layer predicts a large concentration of defects and hence, the formation of localized energy levels in the oxide band gap.

The compositional profile introduces a variation in depth of the electronic properties of the film. $In(PO_3)_3$ behaves as an insulator with a band gap of 5.6 eV and is the preferred compound for the fabrication of MIS (metal–insulator–semiconductor) junctions. $InPO_4$ behaves as a high band gap semiconductor with a band gap of 3.8 eV. Similarly, In_2O_3, with a band gap of 3.5 eV, behaves as an n-type semiconductor.

Now, we will analyze the electronic characteristics of the conditioned homoepitaxial p-type InP film in contact with the acid electrolyte. A semiconductor affinity of 4.3 eV was measured by SRPES valence band spectra obtained at low excitation energies as shown below. Thus, the energy band diagram shown in Figure 2.71a was constructed by adopting a flat band potential of 0.515 V (SCE) and the formation of an interfacial dipole of 0.4 eV at the semiconductor-film interface, as determined by UPS measurements (see below). The flat band potential was estimated by comparing

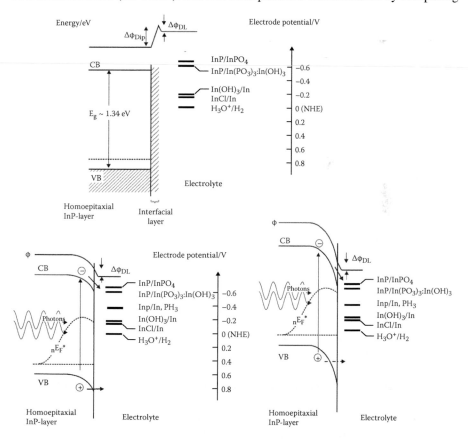

FIGURE 2.71 Energy band diagram of the system p-InP/oxide/electrolyte with the semiconductor (a) at the flat band potential, (b) at 0.3 V, and (c) at −0.1 V.

experimental determined affinity value with that reported by Lewerenz and Schulte [178] for a conditioned p-type InP(111) surface: and a flat band potential of 0.715 V (SCE).

According to this energy diagram, it is to note that the cathodic reactions occurring by injection of photogenerated electrons from the conduction band compete with those taking place by injection of holes in the valence band. The diagrams shown in Figure 2.71b and c represent the energy frame of reactions occurring at each potential limit of the cyclic scan during the photoelectrochemical treatment of p-type homoepitaxial InP film in HCl: −0.3 V and 0.1 V (SCE). Close to the anodic limit (0.1 V), anodic oxidation of InP to $InPO_3$, $In(PO_3)_3$, and $In(OH)_3$ occur preferentially. Close to the cathodic limit, cathodic reduction of InP to In and $PH_3(g)$ and hydrogen evolution occurs, whereas the anodic process is hampered by the formed energy barrier (band bending). Therefore, the relative proportions of $In(OH)_3$, $InPO_4$, and $In(PO_3)_3$ will be determined by the selection of the voltammetric scan limits. This also has consequences for the conductivity of the film. $In(OH)_3$, which stays for hydrated In_2O_3, behaves as a n-type semiconductor, the conduction band of which is close the conduction band of the p-type InP layer. This facilitates the electron transfer from the semiconductor to the electrolyte.

UPS measurements have shown that the electrochemical conditioning step (Figure 2.72) leads to a shift of the secondary electron edge by 0.4 eV. This change is not connected with a shift of the position of the In 4d core level signal between 17 and 18 eV [201], as observed in the valence band spectra for an surface sensitive excitation energy of 150 eV. The spectrum taken after the photoelectrochemical treatment shows also an emission between 3 eV and 4 eV that can be ascribed to the O 2p core level signal of In_2O_3 [202–204]. The broaden emission around 5 eV observed after the photoelectrodeposition of Rh can be ascribed to the structure of the density of states of d-electrons [205–207].

The change of electron affinity by the photoelectrochemical conditioning can be ascribed to the formation of an interfacial dipole, which arises due to the formation of P-O, In-O, and/or In-Cl. The higher electronegativity difference [208] between P and O (1.4) and between In and O (2.0) in comparison with 0.6 between In and P should justify the appearance of a positive partial charge at the semiconductor side of the interface.

The change of the electron affinity by formation of interfacial dipoles can be described in terms of the Helmholtz–Perrin equation [209–211]:

$$\Delta\chi = \frac{4\pi e}{\varepsilon_i \varepsilon_0} N_{dip}\mu(\theta) \qquad (2.95)$$

where N_{dip} is the surface density of dipoles, ε_i is the dielectric constant of the dipole layer, and $\mu(\theta)$ is the individual contribution of dipolar bonds, with a coverage θ. If we try to attribute the formation of dipole layer to the In-O bonds, we should start from a density of In atoms of 4.3×10^{14} cm^{-2} for a (100)-surface. Taking a dipole moment of 1 D (1 D: Debye = 3.34×10^{-28} C cm) for the In-O bond in a In_2O_3 crystal, we calculate a shift of ~21 eV with Eq. (2.95), which appears unrealistic! Now, if we assume that the formation of In-Cl bonds is the cause for the shift of an electron affinity by 0.4

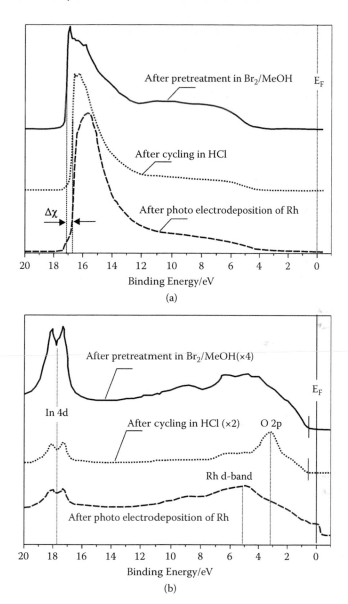

FIGURE 2.72 (a) UPS spectra recorded after each conditioning step of the homoepitaxial p-InP(100): (1) after chemical etching; (2) after photoelectrochemical cycling in HCl, and (3) after photoelectrodeposition of Rh. Excitation: He hv = 21.2 eV; (b) SRPES valence band spectra. Excitation energy: 150 eV.

eV, we can calculate a density of bonds of 2.2×10^{12} cm^{-2}, taking a dipole moment of $\mu = 3.79$ D. This makes 0.5% of the total amount of In atoms at the InP(100) surface. This is a logical reasoning, in view that electron affinity changes just after the treatment in HCl solution and not with the appearance of the first oxide layers. The question is actually whether these bonds are formed on In(III) atoms at the InP-oxide interface or appear as a result of an induced oxidation of clusters of In(0) formed by photoinduced reduction (Eq. 2.89). SRPES spectra show a weak signal at the binding energy region corresponding to the Cl 2p core level signal after the photoelectrochemical treatment in HCl. This signal appears better defined after the photoelectrodeposition of Rh at 198.2 eV, which can be ascribed to InCl [212] (see Figure 2.73). SRPES experiments performed on photoelectrodeposited In in chloride solution show a shift of the secondary electron edge of 0.4 V in comparison with that measured on photoelectrodeposited In from sulfate solutions (see Figure 2.74), thus supporting the idea that the shift of electron affinity arises as a consequence of an electrosorption of Cl⁻ on the formed metallic In at the interface. The electrosorption of In takes place at potentials more positive than the threshold at −1.1 V (SCE) (see Figure 2.75). The standard potential for the formation of $InCl_3$ is situated 0.5 V more positive [213]. The adsorption threshold potential is related with the potential of zero charge, which for In is of −0.89 V, much more negative than that for Pt of −0.26 V, for instance.

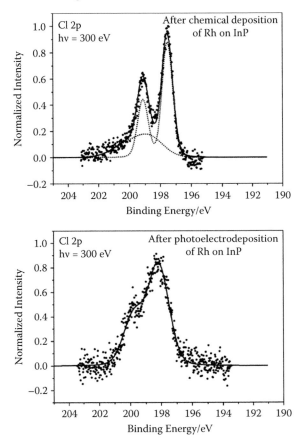

FIGURE 2.73 SRPES Cl 2p core level signal of a homoepitaxial p-InP(100) layer recorded with an excitation energy of 300 eV after the photochemical (open circuit deposition) and photoelectrochemical surface conditioning (photoelectrochemical cycling in HCl and photoelectrodeposition of Rh).

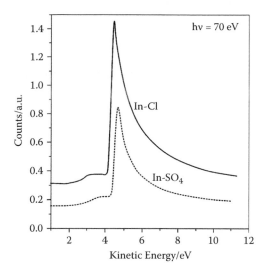

FIGURE 2.74 SRPES secondary electron edge of In particles photoelectrodeposited in chloride and sulfate solutions onto p-Si(111).

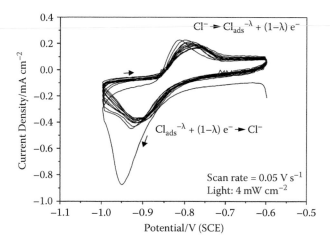

FIGURE 2.75 Cyclic current-voltage curves in 0.5 M KCl of In particles obtained by photoelectrodeposition onto p-Si(111) from a chloride solution.

REFERENCES

1. Khan, S.U., Kainthla R.C. and J.O'M. Bockris. 1987. The redox potential and the Fermi level in solution. *J. Phys. Chem.* 91:5974.
2. Lohman, P. 1967. Fermi-niveau and flachband potential von molekülkristallen aromatischer kohlenwasserstoffe. *Z. Naturforsch.A* 22:843.
3. Trasatti, S. 1977. In Gerischer, H. and C.W. Tobias (Eds.), *Advances in Electrochemistry and Electrochemical Engineering*, Vol. 10, p. 233. New York: Wiley-Interscience.

4. Trasatti, S. 1971. Work function, electronegativity and electrochemical behaviour of metals I. *J. Electroanal. Chem.* 33:351.
5. Trasatti, S. 1972. Work function, electronegativity and electrochemical behaviour of metals II. *J. Electroanal. Chem.* 39:163.
6. Oura, K. Lifshits, V.G., Saranin, A.A., Zotov, A.V., and M. Katayama. 2003. *Surface Science. An Introduction,* Springer, Berlin.
7. Lorenz, W.J. and G. Staikov. 1995. 2D and 3D thin film formation and growth mechanisms in metal electrocrystallization—an atomistic view by in situ STM. *Surf. Sci.* 335:32.
8. Gibbs, J.W., Bumstead H.A., Longley, W.R. and R.G. Van Name. 1928. *The Collected Works of J. Willard Gibbs, Longmanns.* New York: Green and Co.
9. Volmer, M. 1921, *Physik Z.* 22:646.
10. Volmer, M. and A. Weber. 1926. *Z. Phys. Chem., Stoechiom. Verwandtschaftsl.* 119: 277.
11. Stranski, I.N. 1928. *Z. Phys. Chem. Stoechiom. Verwandtschaftsl* 136:297.
12. Stranski, I.N. and R. Kaischev. 1934. *Z. Phys. Chem.* B24:100; Stranski, I.N. and R. Kaischev 1935. Gleichgewichtsform und Wachstumsform der Kristalle. *Ann. Phys.* 415:330.
13. Becker, R. and W. Döring. 1935. Kinetische Behandlung der Keimbildung in übersättigen Dämpfen. *Ann. Phys.* 416:719.
14. Kaischev, R. 1957. Zur kinetischen Ableitung der Kristallkeimbildungsgeschwindigkeit. *Z. Elektrochem.* 61:35.
15. Conway, B.E. and J. O´M. Bockris 1961. On the calculation of potential energy profile diagrams for processes in electrolytic metal deposition. *Electrochim. Acta* 3:340.
16. Gerischer, H. 1960. Electrocrystallization. *Electrochim. Acta* 2:1.
17. Stoyanov, S. 1973. On the atomistic theory of nucleation rate. *Thin Solid Films* 18:91.
18. Milchev, A., Stoyanov, S. and R. Kaischev. 1974. Atomistic theory of electrolytic nucleation: I. *Thin Solid Films* 22:255.
19. Milchev, A., Stoyanov, S. and R. Kaischev. 1974. Atomistic theory of electrolytic nucleation: II. *Thin Solid Films* 22:267.
20. Becker, R. and W. Döring. 1935. Kinetische Behandlung der Keimbildung in übersättigen Dämpfern. *Ann. Phys.* 24:719.
21. Stoyanov S. 1978. Nucleation theory for high and low supersaturations, Current topics in materials science, Vol.3, E. Kaldis (Ed.), North-Holland, Amsterdam.
22. Muñoz, A.G., Salinas, D. R. and J. B. Bessone. 2003. First stages of Ni deposition onto vitreous carbon from sulfate solutions. *Thin Solid Films* 429:119.
23. Fleischmann M. and H.R. Thirsk. 1963. *Advances in Electrochemistry and Electrochemical Engineering,* P. Delahay (Ed.) Vol.3, Wiley, New York.
24. Armstrong R.D. and J.A. Harrison. 1969. Two-dimensional nucleation in electrocrystallisation. *J. Electrochem. Soc.* 116:328.
25. Harrison J.A. and H.R. Thirsk. 1971. *Electroanalytical Chemistry,* A.J. Bard (Ed.), Vol. 5. New York: Marcel Dekker.
26. Saraby-Reintjes A. and M. Fleischmann. 1984. Kinetics of electrodeposition of nickel from watts baths, *Electrochim. Acta* 29:557.
27. Avrami, M. 1939. Kinetics of phase change. I General theory. *J. Chem. Phys.* 7:1103; 1940. Kinetics of phase change. II Transformation–time relations for random distribution of nuclei. *J. Chem. Phys.* 8:212; 1941. Granulation, phase change, and microstructure kinetics of phase change. III. *J. Phys. Chem.* 9:177.
28. Abyaneh, M.Y. 2002. Extracting nucleation rates from current-time transients. Part I: the choice of growth models. *J. Electroanal. Chem.* 530:82.
29. Abyaneh, M.Y and M. Fleischmann. 1981. The electrocrystallization of nickel. Part I. Generalized models of electrocrystallization. *J. Electroanal. Chem.* 119:187.
30. Scharifker, B. and G. Hills. 1981. Electrochemical kinetics at macroscopically small electrodes. *J. Electroanal. Chem.* 130:81.

31. Gunawardena, G., Hills, G., Montenegro, I. and B. Scharifker. 1982. Electrochemical nucleation. Part I. General considerations. *J. Electroanal. Chem.* 138:225.
32. Scharifker, B.R. and J. Mostany. 1984. Three-dimensional nucleation with diffusion controlled growth. Part I. Number density of active sites and nucleation rates per site. *J. Electroanal. Chem.* 177:13.
33. Sluyters-Rehbach, M., Wijenberg, J.H.O.J., Bosco, E. and J.H. Sluyters. 1987. The theory of chronoamperometry for the investigation of electrocrystallization. Mathematical description and analysis in the case of diffusion-controlled growth. *J. Electroanal. Chem.* 236:1.
34. Mirkin, M.V. and A.P. Nilov. 1990. Three-dimensional nucleation and growth under controlled potential. *J. Electroanal. Chem.* 283:35.
35. Heerman, L. and A. Tarallo. 1999. Theory of the chronoamperometric transient for electrochemical nucleation with diffusion-controlled growth. *J. Electroanal. Chem.* 470:70.
36. Scharifker, B. and G. Hills. 1983. Theoretical and experimental studies of multiple nucleation. *Electrochim. Acta* 28:879.
37. Cao, Y., Searson, P.C. and A.C. West. 2001. Direct numerical simulation of nucleation and three-dimensional, diffusion-controlled growth. *J. Electrochem. Soc.* 148:C376.
38. Markov, I., Boynov, A. and S. Toschev. 1973. Screening action and growth kinetics of electrodeposited mercury droplets. *Electrochim. Acta* 18:377.
39. Markov, I. and S. Toschev. 1975. Growth kinetics and screening action of mercury droplets electrodeposited from dilute solutions of electrolyte. *Electrodep. Surf. Treat.* 3:385.
40. Scheludko, A. and M. Todorova. 1952. *Bull. Acad. Sci.Bulg.(Phys.)* 3:61.
41. Milchev, A. and J. Malinowski. 1985. Phase formation, stability and nucleation kinetics of small clusters. *Surf. Sci.* 156:36.
42. Konstantinov, I. and J. Malinowski. 1975. Initial stages of development as nucleation and growth phenomena. *J. Phot. Sci.* 23:145.
43. Hills, G., Pour, A.K. and B. Scharifker. 1983. The formation and properties of single nuclei. *Electrochim. Acta* 28:891.
44. Kinoshita, K. 1998. *Carbon, electrochemical and physicochemical properties.* New York: John Wiley & Sons.
45. Saidman, S.B. and J. B. Bessone. 1997. Activation of aluminium by indium in chloride solutions. *Electrochim. Acta* 42:413.
46. Muñoz, A.G., Saidman, S.B. and J.B. Bessone. 2002. Corrosion of an Al-Zn-In alloy in chloride media. *Corros. Sci.* 44:2171.
47. Piercy, R. and N. A. Hampson. 1974. A study of the differential capacitance of indium in some aqueous solutions. *J. Electroanal. Chem.* 53:271.
48. Muñoz, A.G. and J. B. Bessone. 1998. Effects of different anions on the electrochemical behaviour of In. *Electrochim. Acta* 43:1067.
49. Hautier, G., D'Haen, J., Maex, K. and P.M. Vereecken. 2008. Electrodeposited freestanding single-crystal indium nanowires. *Electrochem. Solid-State Lett.* 11:K47.
50. Milchev, A., Vassileva, E. and V. Kertov. 1980. Electrolytic nucleation of silver on a glassy carbon electrode: Part I. Mechanism of critical nucleus formation. *J. Electroanal. Chem.* 107:323.
51. Nagy, Z., Blaudeau, J.P., Hung, N.C., Curtiss, L.A. and D.J. Zurawski. 1995. Chloride ion catalysis on the copper deposition reaction. *J. Electrochem. Soc.* 142:L87.
52. De Levie, R. 1970. On the electrochemical oscillator. *J. Electroanal. Chem.* 25:257.
53. Oskam, G., van Heerden, D. and P.C. Searson. 1998. Electrochemical fabrication of n-Si/Au Schottky junctions. *Appl. Phys. Lett.* 73:3241.
54. Zambelli, T., Munford, M.L., Pillier, F., Bernard, M.-C. and P. Allongue. 2001. Cu electroplating on H-terminated n-Si(111): properties and structure of n-Si/Cu junctions. *J. Electrochem. Soc.* 148:C614.

55. Tumanski, S 2001. *Thin film magnetoresistive sensors*. Bristol, UK: Institute of Physics Publishing.
56. Mahdi, A.E., Panina, L. and D. Mapps. 2003. Some new horizons in magnetic sensing: high-Tc SQUIDs, GMR and GMI materials. *Sens. Actuators A: Phys.* 105:271.
57. Moina, C.A., de Oliveira-Versic, L. and M. Vazdar. 2004. Magnetic domain states in nanosized Co nuclei electrodeposited onto monocrystalline silicon. *Mater. Lett.* 58:3518.
58. Moina, C.A and M. Vazdar. 2001. Electrodeposition of nanosized nuclei of magnetic Co-Ni alloys onto n-Si(100). *Electrochem. Commun.* 3:159.
59. Lopez-Díaz, L., Klaui, M., Rothmann, J. and J.A.C. Bland. 2002. Fast and controllable switching in narrow ring nanomagnets. *J. Magn. Magn. Mater.* 242–245:553.
60. Muñoz, A.G. 2006. Photoelectrodeposition of thin Co films on p-Si from sulfate solutions. *Surf. Coat. Tech.* 201:3030.
61. Lohrengel, M.M. 1993. Thin anodic oxide layers on aluminium and other valve metals: high field regime. *Mater. Sci. Eng. R* 11:243.
62. Karwas, C. and T. Hepel. 1989. Morphology and composition of electrodeposited cobalt-zinc alloys and the influence of boric acid. *J. Electrochem. Soc.* 136:1672.
63. Radisic, A., Oskam, G. and P.C. Searson. 2004. Influence of oxide thickness on nucleation and growth of copper on tantalum. *J. Electrochem. Soc.* 151:C369.
64. Muñoz, A.G. and G. Staikov. 2006. Electrodeposition of Co on oxide modified p-Si surfaces. *Electrochim. Acta* 51:2836.
65. Bard, A.J. and L.R. Faulkner. 2001. *Electrochemical Methods: Fundamentals and Applications*, 2nd ed. New York: Wiley.
66. Jungblut, H. and H.J. Lewerenz. 2000. Photo-induced ultrathin electropolishing layers on silicon: formation, composition and structural properties. *Appl. Surf. Sci.* 168:194.
67. Llopis, J.F. and F. Colom. 1976. Platinum, in *Encyclopedia of Electrochemistry of the Elements*, Vol. 6, Ch. VI-4, A.J. Bard (Ed.). New York: Marcel Dekker.
68. Rao, C.R.K. and D.C. Trivedi. 2005. Chemical and electrochemical depositions of platinum group metals and their applications. *Coord. Chem. Rev.* 249:613.
69. A.G. Muñoz and H.J. Lewerenz. 2010. Model experiments on electrochemical formation of nanodimensioned noble metal–oxide–semiconductor junctions at Si(111) surfaces. *Electrochim. Acta* 55:7772.
70. Lau, A.L.Y. and A.T. Hubbard. 1970. Study of the kinetics of electrochemical reactions by thin-layer voltammetry: III. Electroreduction of the chloride complexes of platinum(II) and (IV). *J. Electroanal. Chem. Interfacial Electrochem.* 24:237.
71. Gregory, A.J., Levason, W., Noftle, R.E., Le Penven, R. and D. Pletcher. Studies of platinum electroplating baths Part III. The electrochemistry of $Pt(NH_3)_4^{-x(H2O)2+2}$ and $PtCl_4^{-x(H2O)(2-x)-x}$. 1995. *J. Electroanal. Chem.* 399:105.
72. Gröning, Ö. and L.I. Elding. 1989. Water exchange of trans-dichlorodiaquaplatinum(II) and tetraaquaplatinum(II) studied by an oxidative-addition quenching technique. Isotopic shifts and platinum-195 NMR chemical shifts for mixed chloro-aqua complexes of platinum(II) and platinum(IV). *Inorg. Chem.* 28:3366.
73. Pletcher, D. and R.I. Urbana. 1997. Electrodeposition of rhodium. Part 1. Chloride solutions. *J. Electroanal. Chem.* 421:137.
74. Keller, F., Hunter, M.S. and D.L. Robinson. 1953. Structural features of oxide coatings on aluminum. *J. Electrochem. Soc.* 100:411.
75. O'Sullivan, J. P. and G.C. Wood. 1970. Morphology and mechanism of formation of porous anodic films on aluminium. *Proc. R. Soc. Lond. A* 317:511.
76. Lee, W., Schwirn, K., Steinhart, M., Pippel, E., Scholz, R. and U. Gösele. 2008. Structural engineering of nanoporous anodic aluminium oxide by pulse anodization of aluminium. *Nature: Nanotechnology* 3:234.
77. Yin, A.J., Li, J., Jian, W., Bennett, A. J. and J.M. Xu. 2001. Fabrication of highly ordered metallic nanowire arrays by electrodeposition. *Appl. Phys. Lett.* 79:1039.

78. Masuda, H., Yotsuya, M. and M. Ishida. 1998. Spatially selective metal deposition into a hole-array structure of anodic porous alumina using a microelectrode. *Jpn. J. Appl. Phys.* 37:L1090.
79. Saito, M., Shibasaki, M., Nakamura, S. and M Miyagi. 1994. Optical waveguides fabricated in anodic alumina films. *Opt. Lett.* 19:710.
80. Lazzara, T.D., Aaron Lau, K.H. and W. Knoll. 2010. Mounted nanoporous anodic alumina thin films as planar optical waveguides. *J. Nanosci. Nanotech.* 10:4293.
81. Sieber, I., Hildebrand, H., Friedrich, A. and P. Schmuki. 2005. Formation of self-organized niobium porous oxide on niobium. *Electrochem. Commun.* 7:97.
82. Muñoz, A.G. and G. Staikov. 2006. Electrodeposition of metals on anodized thin Nb films. *J. Solid State Electrochem.* 10:329.
83. Gomes, M.A.B. and L.O. de Souza Bulhões. 1990. Diffusion coefficient of H^+ at Nb_2O_5 layers prepared by thermal oxidation of niobium. *Electrochim. Acta* 35:765.
84. Cabanel, R., Chaussy, J., Mazuer, J., Delabouglise, G., Joubert, J.C., Barral, G. and C. Montella. 1990. Electrochromism of Nb_2O_5 thin films obtained by oxidation of magnetron-sputtered NbNx. *J. Electrochem. Soc.* 137:1444.
85. Schwitzgebel, G. and T. Unruh. 1995. Electrochemical investigations of polycrystalline $HxNb_2O_5$. *J. Solid State Chem.* 115:260.
86. Hayashi, Y., Miyakoshi, T. and M. Masuda. 1991. Colouration of oxidized niobium surfaces by electrochemical hydrogen charging. *J. Less Common Met.* 172:851.
87. Hayashi, Y., Arita, M., Koga, K. and M. Masuda. 1995. Photoelectrochemical properties of hydrogen in anodically oxidized niobium. *J. Alloys Compd.* 231:702.
88. Komura, T., Nakanori, T. and K. Takahashi. 1993. Electrochemical insertion of hydrogen into diniobium pentaoxide. *Bull. Chem. Soc. Jpn* 10:2858.
89. Schultze, J.W. and M.M. Lohrengel. 2000. Stability, reactivity and breakdown of passive films. Problems of recent and future research. *Electrochim. Acta* 45:2499.
90. Torresi, R.M. and F.C. Nart. 1988. Growth of anodic niobium oxide films. *Electrochim. Acta* 33:1015.
91. Heusler, K.E. and M. Schulze. 1975. Electron-transfer reactions at semiconducting anodic niobium oxide films. *Electrochim. Acta* 20:237.
92. Macagno, V. and J.W. Schultze. 1984. The growth and properties of thin oxide layers on tantalum electrode. *J. Electroanal. Chem.* 180:157.
93. Schultze, J.W. and C. Bartels. 1983. Electron-transfer reactions on metal-coated passive titanium. *J. Electroanal. Chem.* 150:583.
94. de Sá, A.I., Rangel, C.M., Skeldon, P. and G.E. Thompson. 2002. Growth of anodic oxides on sputtered Al-Nb alloys. *Key Engineering Materials* 230:44.
95. Magnussen, N., Quinones, L., Dufner, D.C., Cocke, D.L. and E.A. Schweikert. 1989. Analysis of anodic oxide films on niobium. *Chem. Mater.* 1:220.
96. Schwartz, R.J., Chiou, Y.L. and H.W. Thompson Jr. 1970. Electrical conduction in niobium-niobium oxide-gold diodes. *Thin Solid Films* 6:81.
97. D'Alkaine, C.V., de Souza, L.M.M. and F.C. Nart. 1993. The anodic behaviour of niobium-I. The state of the art. *Corros. Sci.* 34:109.
98. Schultze, J.W. and L. Elfenthal. 1986. Electron-transfer reactions on pure and modified oxide films. *J. Electroanal. Chem.* 204:153.
99. Schultze, J.W. and V.A. Macagno. 1986. Electron-transfer reaction on passive tantalum electrodes. *Electrochim. Acta* 31:355.
100. Dignam, M.J. 1981. *The Kinetics of the Growth of Oxides: Comprehensive Treatise of Electrochemistry*, J.O'M. Bockris (Ed.), Vol. 4. New York: Plenum Press.
101. Cabrera, N. and N.F. Mott. 1949. Theory of the oxidation of metals. *Rep. Prog. Phys.* 12:163.

102. In the case of metals, the potential drop in the oxide is considered as the difference between the applied potential and the standard potential for oxide formation, that is, $(V - V_{ox}^0)$. In the case of semiconductors, the same applies only if $C_{ox} \ll C_{sc}$.

103. Burstein G.T. and A.J. Davenport. 1989. The current-time relationship during anodic oxide film growth under high electric field. *J. Electrochem. Soc.* 136:936.

104. Young, L. and F.G. Zobel. 1966. An ellipsometric study of steady-state high field ionic conduction in anodic oxide films on tantalum, niobium and silicon. *J. Electrochem. Soc.* 113:277.

105. Lohrengel, M.M. 1993. Thin anodic oxide layers on aluminium and other valve metals: high field regime. *Mat. Sci. Eng.* R11:243.

106. Lohrengel, M.M., Moehring, A. and M. Pilaski. 2000. Electrochemical surface analysis with the scanning droplet cell. *Fresenius J. Anal. Chem.* 367:334.

107. Lohrengel, M.M. 1991. Principles and applications of pulse techniques to corrosion studies, in M.G.S. Ferreira, C.A. Melendres (Eds.), *Electrochemical and optical techniques for the study and monitoring of metallic corrosion*, pp. 69–121. Dordrecht: Kluwer.

108. Muñoz, A.G., Moehring, A. and M.M. Lohrengel. 2002. Anodic oxidation of chemically hydrogenated Si(100). *Electrochim. Acta* 47:2751; Muñoz, A.G., and M.M. Lohrengel. 2002. *J. Solid State Electrochem.* 6:513.

109. de Wit, H.J., Wijenberg, C. and C. Crevecoeur. 1979. Impedance measurements during anodization of aluminium. *J. Electrochem. Soc*, 126:779.

110. Piazza, S., Sunseri, C. and F. Di Quarto. 1990. A photocurrent spectroscopy study of the initial stages of anodic oxide film formation on niobium. *J. Electroanal. Chem.* 293:69.

111. Rüße, S., Lohrengel, M.M. and J.W. Schultze. 1994. Ion migration and dielectric effects in aluminium oxide films. *Solid State Ionics* 72:29.

112. Deal, B.E. 1974. The current understanding of charges in the thermally oxidized silicon structure. *J. Electrochem. Soc.* 121:198C.

113. Chazalviel, J.-N. 1992. Ionic processes through the interfacial oxide in the anodic dissolution of silicon. *Electrochim. Acta* 37:865.

114. Schmuki, P., Böhni, H. and J.A. Bardwell. 1995. In situ characterization of anodic silicon oxide films by AC impedance measurements. *J. Electrochem. Soc.* 142:1705.

115. Clark, K.B., Bardwell, J.A. and J.M. Baribeau. 1994. Physical characterization of ultrathin anodic silicon oxide films. *J. Appl. Phys.* 76:3114.

116. O'Reilly, E.P. and J. Robertson. 1983. Theory of defects in vitreous silicon oxide. *Phys. Rev. B* 27:3780.

117. Jonscher, A.K. 1981. A new understanding of the dielectric relaxation of solids. *J. Mat. Sci.* 16:2037.

118. Jonscher, A.K. 1999. Dielectric relaxation in solids. *J. Phys. D: Appl. Phys.* 32:R57.

119. Helms, C.R. and E.H. Poindexter. 1994. The silicon-silicon-dioxide system: its microstructure and imperfections. *Rep. Prog. Phys.* 57:191.

120. D'Alkaine, C.V., de Souza, L.M.M. and F.C. Nart. 1993. The anodic behaviour of niobium-III. Kinetics of anodic film growth by potentiodynamic and galvanostatic techniques-General models, equations and their applications. *Corros. Sci.* 34:129.

121. Cohen, M.H., Fritzsche, H. and, S.R. Ovishinsky. 1969. Simple band model for amorphous semiconducting alloys. *Phys. Rev. Lett.* 22:1065.

122. Mott, N.F. and E.A. Davis. 1979. *Electronic process in non-crystalline materials*. Oxford: Clarendon Press.

123. Fritzsche, H. 1972. Electronic phenomena in amorphous semiconductors. *Ann. Rev. Mater. Sci.* 2:697.

124. Quarto, F. Di, La Mantia, F. and M. Santamaria. 2005. Physicochemical characterization of passive films on niobium by admittance and electrochemical impedance spectroscopy studies. *Electrochim. Acta* 50:5090.

125. Cohen, J.D. and D.V. Lang. 1982. Calculation of the dynamic response of Schottky barriers with a continuous distribution of gap states. *Phys. Rev. B* 25:5321.
126. Archibald, W. and R.A. Abram. 1983. A theory of the admittance of an amorphous silicon Schottky barrier. *Phil. Mag. B* 48:111.
127. Archibald, W. and R.A. Abram. 1986. More theory of the admittance of an amorphous silicon Schottky barrier. *Phil. Mag. B* 54:421.
128. Da Fonseca, C., Ferrera, M.G. and M. Da Cunha Belo. 1994. Modeling of the impedance behaviour of an amorphous semiconductor Schottky barrier in high depletion conditions. Applications to the study of the titanium anodic oxide/electrolyte junction. *Electrochim. Acta* 39:2197.
129. Heidelberg, A., Rozenkranz, C., Schultze, J.W., Schäpers, Th. and G. Staikov. 2005. Localized electrochemical oxidation of thin Nb films in microscopic and nanoscopic dimensions. *Surf. Sci.* 597:173.
130. Li, Y.-M. and L. Young. 2000. Niobium anodic oxide films: effect of incorporated electrolyte species on DC and AC ionic current. *J. Electrochem. Soc.* 147:1344.
131. Muñoz, A.G. and G. Staikov. 2006. Electrodeposition of metals on anodized thin Nb films. *J. Solid State Electrochem.* 10:329.
132. Wold, A. 1993. Photocatalytic properties of TiO_2. *Chem. Mater.* 5:280.
133. Hashimoto, K., Irie, H. and A. Fujishima. 2005. TiO_2 photocatalysis: A historical overview and future prospects. *Jpn. J. Appl. Phys.* 44:8269.
134. Mathieu, J.B., Mathieu, H.J. and D. Landolt. 1978. Electropolishing of titanium in perchloric acid-acetic acid solutions. *J. Electrochem. Soc.* 125:1039.
135. Schultze, J.W., Lohrengel, M.M. and D. Ross. 1983. Nucleation and growth of anodic oxide films. *Electrochim. Acta* 28:973.
136. Antonelli, D.M. and J.Y. Ying. 1995. Synthese von hexagonal gepacktem, mesoporösem TiO_2 mit einer modifizierten Sol-Gel-Methode. *Angew. Chem.* 107:2202.
137. Pecchi, G., Reyes, P., Sanhuesa, P. and J. Villaseñor. 2001. Photocatalytic degradation of pentachlorophenol on TiO_2 sol-gel catalysts. *Chemosphere* 43:141.
138. Miao, Z., Xu, D., Ouyang, J., Guo, G., Zhao, X. and Y. Tang. 2002. Electrochemically induced sol-gel preparation of single-crystalline TiO_2 nanowires. *Nano Lett.* 2:717.
139. Akbar, S.A. and L.B. Younkman. 1997. Sensing mechanism of a carbon monoxide sensor based on anatase titania. *J. Electrochem. Soc.* 144:1750.
140. Savage, N.O., Akbar, S.A. and P.K. Dutta. 2001. Titanium dioxide based high temperature carbon monoxide selective sensor. *Sens. Actuators B* 72:239.
141. Sakthivel, S. and H. Kisch. 2003. Photocatalytic and photoelectrochemical properties of nitrogen-doped titanium dioxide. *ChemPhysChem* 4:487.
142. Asahi, A., Morikawa, T., Ohwaki, T., Aoki, K. and Y. Taga. 2001. Visible-light photocatalysis in nitrogen-doped titanium dioxide. *Science* 293:269.
143. Choi, Y., Umebayashi, T. and M. Yoshikawa. 2004. Fabrication and characterization of C-doped anatase TiO2 photocatalysts. *J. Mater. Sci.* 39:1837.
144. Gong, D., Grimes, C.A., Varghese, O.K., Hu, W., Singh, R.S., Chen, Z. and E.C. Dickey. 2001. Titanium oxide nanotube arrays prepared by anodic oxidation. *J. Mater. Res.* 16:3331.
145. Mor, G.K., Varghese, O.K., Paulose, M., Mukherjee, N. and C.A. Grimes. 2003. Fabrication of tapered, conical-shaped titania nanotubes. *J. Mater. Res.* 18:2588.
146. Macak, J.M., Tsuchiya, H. and P. Schmuki. 2005. High aspect ratio TiO2 nanotubes by anodization of titanium. *Angew. Chem. Int. Ed.* 44:2100.
147. Macak, J.M., Tsuchiya, H., Taveira, L., Aldabergerova, S. and P. Schmuki. 2005. Smooth anodic TiO2 nanotubes. *Angew. Chem. Int. Ed.* 44:7463.
148. Mor, G.K., Varghese, O.K., Paulose, M., Shankar, K. and C.A. Grimes. 2006. A review on highly ordered, vertically oriented TiO2 nanotube arrays: fabrication, material properties, and solar energy applications. *Solar Energy Mater. Solar Cells* 90:2011.

149. Li, A. P., Müller, F., Birner, A., Nielsch, K. and U. Gösele. 1998. Hexagonal pore arrays with a 50–420 nm interpore distance formed by self-organization in anodic alumina. *J. Appl. Phys.* 84:6023.
150. Muñoz, A.G. 2007. Semiconducting properties of self-organized TiO2 nanotubes. *Electrochim. Acta* 52:4167.
151. Muñoz, A.G., Chen, Q. and P. Schmuki. 2007. Interfacial properties of self-organized TiO_2 nanotubes studied by impedance spectroscopy. *J. Solid State Electrochem.* 11:1077.
152. Retter, U. and H. Lohse. 2002. Electrochemical impedance spectroscopy, in *Electroanalytical Methods. Guide to Experiments and Applications*, F. Scholz (Ed.). Berlin-Heidelberg: Springer Verlag.
153. van de Krol, R., Groossens, A. and J. Schoonman. 1997. Mott–Schottky analysis of nanometer-scale thin film anatase TiO2. *J. Electrochem. Soc.* 144:1723.
154. Lee, E.-J. and S.-I. Pyun. 1992. Analysis of nonlinear Mott–Schottky plots obtained from anodically passivating amorphous and polycrystalline TiO2 films. *J. Appl. Electrochem.* 22:156.
155. Simons, W., Hubin, A. and J. Vereecken. 1999. The role of electrochemical impedance spectroscopy (EIS) in the global characterisation of the reduction kinetics of hexacyanoferrate on anodised titanium. *Electrochim. Acta* 44:4373.
156. Dean, M.H. and U. Stimming. 1989. The electronic properties of disordered passive films. *Corros. Sci.* 29:199.
157. Lausmaa, J., Kasemo, B., Mattson, H. and H. Odelius. 1990. Multi-technique surface characterization of oxide films on electropolished and anodically oxidized titanium. *Appl. Surf. Sci.* 45:189.
158. Marino, C.E.B., Nascente, P.A.P., Baggio, S.R., Rocha-Fihlo, R.C. and N. Bocchi. 2004. XPS characterization of anodic titanium oxide films grown in phosphate buffer solutions. *Thin Solid Films* 468:109.
159. Ghicov, A., Tsuchiya, H., Macak, J.M. and P. Schmuki. 2005. Titanium oxide nanotubes prepared in phosphate electrolytes. *Electrochem. Commun.* 7:505.
160. Boschloo, G.K., Goossens, A. and J. Schoonman. 1997. Photoelectrochemical study of thin anatase TiO2 films prepared by metallorganic chemical vapour deposition. *J. Electrochem. Soc.* 144:1311.
161. Boschloo, G.K. and A. Goossens. 1996. Electron trapping in porphyrin-sensitized porous nanocrystalline TiO2 electrodes. *J. Phys. Chem.* 100:19489.
162. Schindler, P.W. and H. Gamsjäger. 1971. H.P. Boehm, Acidic and basic properties of hydroxylated metal oxide surfaces. *Discuss. Faraday Soc.* 52:286.
163. Lyon, L.A. and J.T. Hupp. 1999. Energetics of the nanocrystalline titanium dioxide/ aqueous solution interface: approximate conduction band edge variations between $H_0 = -10$ and $H_- = +26$. *J. Phys. Chem. B* 103:4623.
164. Kavan, L., Kratochvilová, K. and M. Grätzel. 1995. Study of nanocrystalline TiO_2 (anatase) electrode in the accumulation regime. *J. Electroanal. Chem.* 394:93.
165. Ghicov, A., Tsuchiya, H., Hahn, R., Macak, J.M., Muñoz, A.G. and P. Schmuki. 2006. TiO2 nanotubes: H+ insertion and strong electrochromic effects. *Electrochem. Commun.* 8:528.
166. Sakai, N., Fujishima, A., Watanabe, T. and K. Hashimoto. 2001. Highly hydrophilic surfaces of cathodically polarized amorphous TiO2 electrodes. *J. Electrochem. Soc.* 148: E395.
167. Wang, H., He, J., Boschloo, G., Lindström, H., Hagfeldt, A. and S.-E. Lindquist. 2001. Electrochemical investigations of traps in a nanostructured TiO2 film. *J. Phys. Chem. B* 105:2529.
168. Hahn, R., Ghicov, A., Tsuchiya, H., Macak, J.M., Muñoz, A.G. and P. Schmuki. 2007. Lithium-ion insertion in anodic TiO2 nanotubes resulting in high electrochromic contrast. *Phys. Status Solidi (a)* 204:1281.

169. Cronemeyer, D.C. 1959. Infrared absorption of reduced rutile TiO2 single crystals. *Phys. Rev.* 113:1222.
170. Sakata, K. 1983. Study of electronic structure of reduced TiO2 and VxTi1-xO2 crystals by ESCA and optical absorption. *Phys. Status Solidi (b)* 116:145.
171. Redmond, G., Fitzmaurice, D. and M. Grätzel. 1993. Effect of surface chelation on the energy of an intraband surface state of a nanocrystalline TiO2 film. *J. Phys. Chem.* 97:6951.
172. Howe, R.F. and M. Grätzel. 1985. EPR observation of trapped electrons in colloidal titanium dioxide. *J. Phys. Chem.* 89:4495.
173. Rothenberger, G., Fitzmaurice, D. and M. Graetzel. 1992. Spectroscopy of conduction band electrons in transparent metal oxide semiconductor films: optical determination of the flatband potential of colloidal titanium dioxide films. *J. Phys. Chem.* 96:5983.
174. Ramamoorthy, M., King-Smith, R.D. and D. Vanderbilt. 1994. Defects on TiO2 (110) surfaces. *Phys. Rev. B* 49:7709.
175. Kobayashi, H., Mizuno, F., Nakato, Y. and H. Tsubomura. 1991. Hydrogen evolution at a platinum-modified indium phosphide photoelectrode: improvement of current-voltage characteristics by hydrogen chloride etching. *J. Phys. Chem.* 95:819.
176. Kobayashi, H., Mizuno, F. and Y. Nakato. 1994. Surface states in the band-gap for Pt-deposited p-InP photoelectrochemical cells. *Appl. Surf. Sci.* 81:399.
177. Schulte, K.H. and H.J. Lewerenz. 2002. Combined photoelectrochemical conditioning and photoelectron spectroscopy analysis of InP photocathodes. I. The modification procedure. *Electrochim. Acta* 47:2633.
178. Lewerenz, H.J. and K.H. Schulte. 2002. Combined photoelectrochemical conditioning and surface analysis of InP photocathodes II. Photoelectron spectroscopy. *Electrochim. Acta* 47:2639.
179. Segar, P.R., Koval, C.A., Koel, B.E. and S.C. Gebhard. 1990. A comprehensive investigation of HCl- and Br2/NH3(aq)-etched p-InP interfaces. *J. Electrochem. Soc.* 137:544.
180. Etcheberry, A., Gonçalves, A.-M., Mathieu, C. and M. Herlem. 1997. Cathodic decomposition of n-InP during hydrogen evolution in liquid ammonia. *J. Electrochem. Soc.* 144:928.
181. Bogdanoff, P., Friebe, P. and N. Alonso-Vante. 1998. A new inlet system for differential electrochemical mass spectroscopy applied to the photocorrosion of p-InP(111) single crystals. *J. Electrochem. Soc.* 145:576.
182. Quinlan, K.P., Yip, P.W., Rai, A.K. and Th.N. Wittberg. 1996. Oxidation of n-InP and indium in the negative potential region at pH 5. *J. Electrochem. Soc.* 143:524.
183. Schmidt, W.G., Briggs, E.L., Bernholc, J. and F. Bechstedt. 1999. Structural fingerprints in the reflectance anisotropy spectra of InP(001)-2×4 surfaces. *Phys. Rev. B* 59:2234.
184. Irene, E. A. 1995. A comparison of the oxidation and passivation of Si, Ge and InP. *Mater. Sci. Forum* 185–188:37.
185. Losev, V.V. and A.I. Modolov. 1976. *Encyclopedia of Electrochemistry of the Elements*, Vol. VI, A.J. Bard, H. Lund (Eds.). New York: Marcel Dekker.
186. Chen, G., Visbeck, S.B., Law, D.C. and R.F. Hicksa. 2002. Structure-sensitive oxidation of the indium phosphide (001) surface. *J. Appl. Phys.* 91:9362.
187. Lide, D.R. (Ed.). 2000. *CRC Handbook of Chemistry and Physics 1999–2000: A Ready-Reference Book of Chemical and Physical Data*, 81st ed. Boca Raton, FL: CRC Press.
188. Robach, Y., Joseph, J., Bergignat, E. and G. Hollinger. 1989. Anodic Oxidation of InP in Pure Water. *J. Electrochem. Soc.* 136:2957.
189. Hollinger, G., Joseph, J., Robach, Y., Bergignat, E., Commere, B., Viktorovitch, P. and M. Froment. 1987. On the chemistry of passivated oxide-lnP interfaces. *J. Vac. Sci. Technol. B* 5:1108.
190. Weast, R.C. (Ed.). 1978. *CRC Handbook of Chemistry and Physics,* 59th ed. Cleveland, OH: Chemical Rubber Co.

191. Wager, J.F. and C.W. Wilmsen. 1982. Plasma-enhanced chemical vapor deposited SiO2/ lnP interface. *J. Appl. Phys.* 53:5789.

192. Wager, J.F., Makky, W.H., Wilmsen, C.W. and L.G. Meiners. 1982. Oxidation of InP in a plasma-enhanced chemical vapour deposition reactor. *Thin Solid Films* 95:343.

193. Wager, J.F., Geib, K.M., Wilmsen, C.W. and L.L Kazmerski. 1983. Native oxide formation and electrical instabilities at the insulator/lnP interface. *J. Vac, Sci. Technol. B* 1:778.

194. Pakes, A., Skeldon, P., Thompson, G.E., Hussey, R.J., Moisa, S., Sproule, G.I., Landheer, D. and M.J. Graham. 2002. Composition and growth of anodic and thermal oxides on InP and GaAs. *Surf. Interface Anal.* 34:481.

195. Echeverria, F., Skeldon, P., Thompson, G.E., Habazaki, H. and K. Shimizu. 2001. Morphology and composition of layered anodic films on InP. *J. Mater. Sci.* 36:1253.

196. Faur, M., Faur, M., Jayne, D.T., Goradia, M. and C. Goradia. 1990. XPS Investigation of Anodic Oxides Grown on p-Type InP. *Surf. Interface Anal.* 15:641450.

197. Haushalter, R. C. and L. A. Mundi. 1992. Reduced molybdenum phosphates: octahedral-tetrahedral framework solids with tunnels, cages, and micropores. *Chem. Mater.* 4:31.

198. Tang, X. and A. Lachgar. 1998. The missing link: synthesis, crystal structure, and thermogravimetric studies of InPO4.H2O. *Inorg. Chem.* 37:6181.

199. Robach, Y., Besland, M.P., Joseph, J., Hollinger, G., Viktorovitch, P., Ferret, P., Pitaval, M., Falcou, A. and G. Post. 1992. Passivation of InP using In(PO)3-condensed phosphates: from oxide growth properties to metal-insulator-semiconductor field-effect-transistor devices. *J. Appl. Phys.* 71:2981.

200. Benyama, J., Durand, J. and L. Cot. 1988. Structure cristalline du trimetaphosphate d'lndium(I I I). *Z. Anorg. Allg. Chem.* 556:227.

201. Sun, Y., Liu, Z., Machuca, F., Pianetta, P. and W.E. Spicer. 2003. Preparation of clean InP (100) surfaces studied by synchrotron radiation photoemission. *J. Vac. Sci. Technol. A* 21:219.

202. Klein, A. 2000. Electronic properties of In2O3 surfaces. *Appl. Phys. Lett.* 77:2009.

203. Klein, A. 2001. Role of surfaces and interfaces for the electronic properties of conducting oxides. *Mat. Res. Soc. Symp. Proc.* 666: F1.10.1.

204. Chye, P.E., Su, C.Y., Lindau, I., Graner, C.M., Pianetta, P. and W.E. Spicer. 1979. Photoemission studies of intial stages of oxidation of GaSb and InP(001). *Surf. Sci.* 88:439.

205. Smith, N.V., Wertheim, K.G., Hüfner, S. and M.M. Traum. 1974. Photoemission spectra and band structures of d-band metals. IV. X-ray photoemission spectra and densities of states in Rh, Pd, Ag, Ir, Pt, and Au. *Phys. Rev. B* 10:3197.

206. Bringer, J.S. 1976. Applying electron spectroscopy for chemical analysis to industrial catalysis. *Acc. Chem. Res.* 9:86.

207. Gagarin, G.S. 1979. Valence-band electronic structure of deposited metals. *Theoretical and Experimental Chemistry* 15:543.

208. Allred, A.L. and E.G. Rochow. 1958. A scale of electronegativity based on electrostatic force. *J. Inorg. Nucl. Chem.*5:264.

209. Helmholtz, H. 1879. Studien über elektrische Grenzschichten. *Ann. Phys.* 7:337.

210. Perrin, J. 1904. Mecanisme de l′electrisation de contact et solutions. *J. Chim. Phys* 2:601.

211. Campbell, C.T. 1990. *Annu. Rev. Phys. Chem.* 41:775.

212. Freeland, B.H., Habeeb, J.J. and D.G. Tuck. 1977. Coordination compounds of indium. Part XXXIII. X-Ray photoelectron spectroscopy of neutral and anionic indium halide species. *Can. J. Chem.* 55:1527.

213. Muñoz, A.G. and J. B. Bessone. 1998. Effects of different anions on the electrochemical behaviour of In. *Electrochim. Acta* 43:1067.

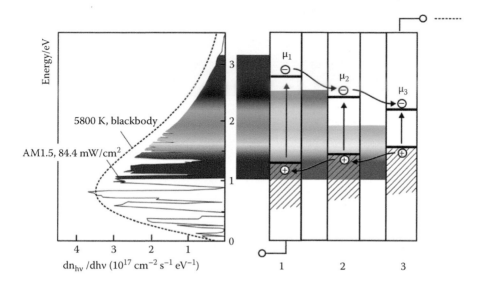

FIGURE 1.12 Schematic showing the working principle of a tandem-cell. The solar spectrum is indicated as the energy density of photons as a function of photon energy.

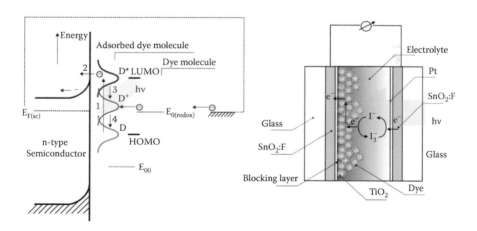

FIGURE 1.18 Basic electronic processes taking place at a semiconductor-electrolyte interface in a dye-sensitized photoelectrochemical cell. The energy band diagram (*left*) represents a cell in the short circuit case. The photon absorption brings the dye-molecule (D) to its excited state (D*). The electron transfer to the semiconductor conduction band (2) leaves an oxidized dye-molecule (D⁺) (3), which is returned to its original state by reaction with the redox-electrolyte (I_3^-/I^-) (4). *Right:* construction details of a dye-sensitized cell.

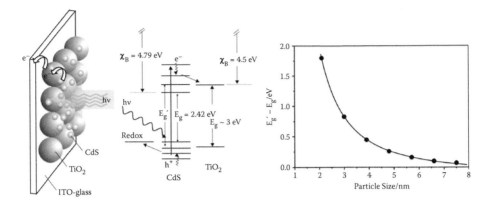

FIGURE 1.19 *Left:* Working principles of a photoelectrochemical cell based on wide band gap semiconductor sensitized with nano-sized semiconducting crystals. *Right:* experimentally measured band gap as a function of the size of CdS particles (adapted from ref. 97).

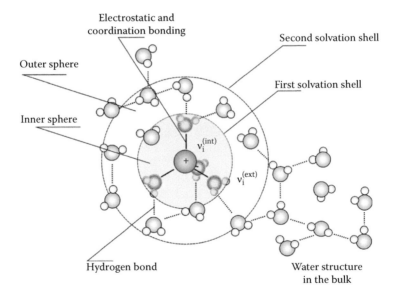

FIGURE 4.12 Schematic depicting the inner and outer solvation shells of a cation. The internal frequency modes are represented by v_i. The extent of coupling of ion fluctuations with the surrounding water structure is given by the interaction constant g_i.

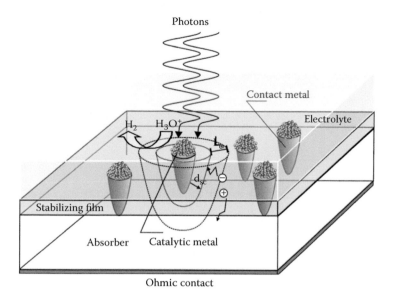

FIGURE 4.34 Schematic of a nanoemitter photoelectrode based on noncatalytic metals covered by a thin film of a metal of the platinum group.

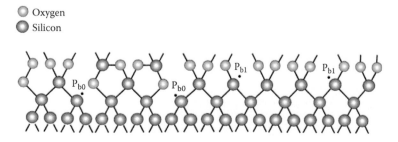

FIGURE 5.13 Schematic depicting the Si-SiO$_2$ atomic structure and the associated electronic defects P$_{b0}$ and P$_{b1}$.

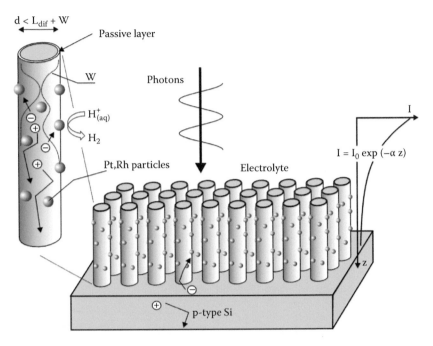

FIGURE 6.4 Schematic of a rodlike photoelectrocatalytic structure where photon absorption and excess carrier separation occur in an orthogonalized configuration; the catalytically active nanoparticles are supposed to form rectifying junctions with the p-type semiconductor upon contact with the respective redox electrolyte with a semispherical space charge region modulated by the curvature of the rod circumference; $I(z)$: absorption profile, α: absorption coefficient, d: diameter of the rods, W: length of space charge layer.

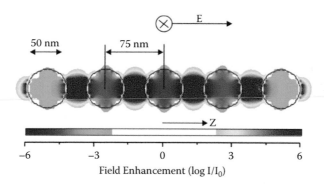

FIGURE 6.7 Distribution of the component of the electric field parallel to the z-direction for an array of 5 Au nanoparticles in air excited at the collective resonance frequency by a propagating plane wave with the electric field polarized in the z-direction (adapted from Maier et al. (8)).

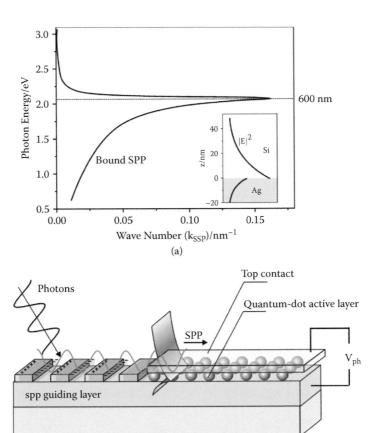

FIGURE 6.11 (a) Dispersion diagram showing the relationship between frequency of incident light and wave vector $(2\pi/\lambda)$ of surface-plasmon polaritons on a Ag/Si interface. The bound SPP mode appears at energies below the surface plasmon resonance energy of 2.07 V for Ag. The inset schematizes the profile of the SPP mode along the Si/Ag interface; (b) plasmonic quantum-dot solar cells designed for enhanced light absorption in ultra-thin quantum-dot layers by coupling to SPP modes propagating in the plane of the interface Ag/Si.

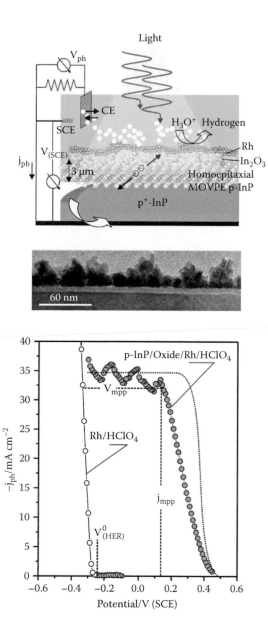

FIGURE 6.13 Schematic of a light conversion EMOS junction in 1 M HClO4 and their output power characteristic formed electrochemically on thin homoepitaxial films of p-InP. Light intensity: 100 mWcm^{-2} (W-I lamp); open circles: Rh I-V curve; V_{mpp}, j_{mpp} denote voltage and current at the maximum power point. (.....) expected I-V characteristic for improved charge transfer.

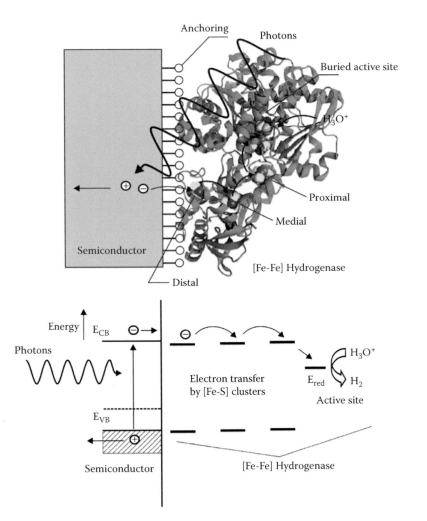

FIGURE 6.17 Electron transfer at a photocathode constituted by a semiconducting absorber covered by electrocatalytic hydrogenase attached to the surface by an organic anchor.

3 Physics of Electrolyte–Metal–Oxide–Semiconductor (EMOS) Contacts in Nanodimensions

The understanding of the electronic process taking place in the energy conversion devices based on nanodimensioned electrolyte–metal–oxide–semiconductor contacts requires a familiarization with basic aspects of the thermodynamics and electronic of the formed contact. The first is necessary to determine the energetic frame established upon electronic equilibrium among the different phases constituting the contact. The conversion process involves essentially the trapping of minority charge carriers generated upon illumination of the semiconductor. These reach charge carriers by diffusion, the limits of the electric field created at the semiconductor–oxide–metal contact and are further transported to the metal–electrolyte interface through the stabilizing film (oxide). Maximal conversion efficiencies require the selection of materials and processes that achieve an optimal semiconductor band bending (electric field) and an efficient electron transfer at the different interfaces and through the oxide. The final step, the electron transfer at the metal–electrolyte interface, depends on the electrocatalytic properties of the metal islands and will be treated in Chapter 4.

3.1 THE ELECTRON TRANSFER IN NANODIMENSIONED SCHOTTKY CONTACTS

The equality of the electrochemical potentials is the *condition sine qua non* to describe the electronic equilibrium among various phases in contact. Each isolated phase is characterized by an electrochemical potential that is determined by inherent properties (Figure 3.1) as defined below. Here, surface states are considered as a hypothetical phase for the sake of simplicity in the analysis. After contacting two or more phases, there is an electron exchange until the electrochemical potential of electrons reaches the same value in the different phases. The charging of surface states leads to potential changes at the interface between two phases in contact. Surface states are characterized by an electroneutrality level, E^0_{ss}. The charge of surface states is thus negative (positive) if the equilibrium position of the Fermi level is above (below) E^0_{ss}.

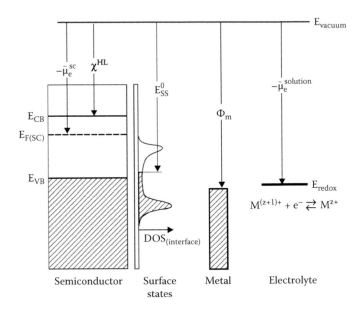

FIGURE 3.1 Electronic properties determining the electrochemical potential in different phases.

Under electronic equilibrium, we have

$$\tilde{\mu}_e^{sc} = \tilde{\mu}_e^{m} = \tilde{\mu}^{s} = \tilde{\mu}_e^{ss} \tag{3.1}$$

where $\tilde{\mu}_e^i$ is the electrochemical potential of electrons in the phase i (sc: semiconductor, m: metal, s: solution, ss: surface states). The electrochemical potential is given by $\tilde{\mu}_e^i = \mu_e^i - e\varphi$, where μ_e^i is the chemical potential and φ is the Volta potential. This latter is given by the sum of mere electrostatic, ϕ, or Galvani potential and a dipole contribution, χ: $\varphi = \phi + \chi$. The dipole contribution arises as a result of a spatial charge separation occurring at each contact interface. In metals, dipoles appear due to structural factors such as atomic corrugation and preferential orientation of solvent molecules. Upon contact with other phase, structural and chemical changes arise at the metal surface that leads to change of the surface dipole:

$$\tilde{\mu}_e^M = \mu_e^M - e\chi^M - e\delta\chi^M - e\varphi = -\Phi^M - e\varphi \tag{3.2}$$

where Φ^M is the work function of the metal, which includes the dipole effects. Moreover, the electrochemical potential of a semiconductor is identified as the Fermi level. This equivalence is not trivial and is derived after describing the electron distribution by the Maxwell-Boltzmann statistic and the supposition of a noninteracting gas of electrons. Therefore, this equality should be taken just as a good approximation [1]. Thus, we can write

$$\tilde{\mu}_e^{sc} = E_F \tag{3.3}$$

The position of the Fermi level is determined in the vacuum energy scale by the electron affinity of the semiconductor, an experimental measurable quantity: χ^{sc}. Thus, $E_F = \chi^{sc} + E_g/2 - \Delta E_F$, where E_g is the semiconductor band gap and ΔE_F represents the shift of the Fermi level due to the semiconductor doping.

The concept of Fermi level in solutions containing a redox pair was introduced for the first time by Gerischer in 1960 (see Ref. 147 in Chapter 1). In principle, this concept is artificial and used just as a tool to describe the electronic equilibrium of metal and semiconductors in contact with a solution. To understand it, let us consider the following general redox reaction:

$$M^{(z+1)+}_{(aq)} + e^-_{(vacuum)} \rightarrow M^{z+}_{(aq)} \tag{3.4}$$

The redox potential for this reaction is given by

$$e\varphi_{redox(vacuum)} = -(\mu_{M^{z+}} - \mu_{M^{(z+1)+}}) \tag{3.5}$$

Now, consider the electrochemical reaction:

$$M^{(z+1)+}_{(aq)} + e^-_{(m)} \rightarrow M^{z+}_{(aq)} \tag{3.6}$$

for which one has that in equilibrium

$$\tilde{\mu}_{M^{(z+1)+}} + \tilde{\mu}_e^m = \tilde{\mu}_{M^{z+}} \tag{3.7}$$

Thus, from relation 3.5 and 3.7 we have

$$\mu_{M^{(z+1)+}} - \mu_{M^{z+}} = e\varphi_{redox(vacuum)} = e(\varphi_m - \varphi_s) - \mu_e^m \tag{3.8}$$

where μ_e^m is the chemical potential of electrons in the metal. Upon electronic equilibrium between metal and solution, we can write

$$\tilde{\mu}_e^m = \tilde{\mu}_e^s = \mu_e^s - e\varphi^s \tag{3.9}$$

and

$$e\varphi^m - e\varphi^s = \mu_e^m - \mu_e^s \tag{3.10}$$

where $\tilde{\mu}_e^s$ represents the electrochemical potential of electrons in the solution, which by definition we set equal to the hypothetical Fermi level of the solution. From Eq. (3.7) to (3.10) we have

$$E_F^s = \tilde{\mu}_e^s = \mu_e^s - e\varphi^s = e\varphi_{redox(vacuum)} - e\varphi^s \tag{3.11}$$

It should be noted that the usual assignment of the Fermi level of the solution, E_F^s, to the redox potential in the vacuum scale $\varphi_{redox(vacuum)}$ is just an approximation.

Moreover, it must be made clear that the term *Fermi level*, which describes a Fermi–Dirac statistics, is not totally adequate for the description of bounded electrons in a redox system.

The Fermi level of redox systems is not an experimental measurable quantity. They are positioned in the vacuum scale by taking the standard hydrogen electrode as a reference for which a value was determined by different experimental and theoretical methods [2–5] to be −4.6 ± 0.1 eV.

Back to our initial discussion about the electronic equilibrium of contacting phases, we have that it is achieved by electron exchange, that is, variation of the electron density and consequent change of potential up to reach the equality of phase Fermi levels. Figure 3.2 shows the band diagrams exemplifying different cases.

The first case represents an ideal n-type semiconductor-metal contact, also known as Schottky-contact, where the $\Phi_m > E_F^{sc}$. Because of much larger density of states for free electrons in metals (in the order of 10^{22}–10^{23} cm^{-3}) than that of the semiconductors at the conduction or valence band (~10^{19} cm^{-3}), the electron exchange does not imply large changes of potential in the metal. The alignment of the Fermi level in the semiconductor is achieved by the spatial distributed charge from the surface up to a given depth to compensate the exchanged charge. The length of this area is known a space charge layer. The potential drop in this region leads to a band bending, which is given by

$$e\Delta\varphi_{bb} = \Phi^M - \chi^{sc} - (E_F - E_{CB}^0) \tag{3.12}$$

E_{CB}^0 is the energy of the conduction band edge in the bulk.

The electronic equilibrium between a semiconductor and a redox system in the electrolyte requires equalization of the oxidation and reduction processes, that is, zero charge accumulation under an atomic dynamical equilibrium. The charge separation at the semiconductor–electrolyte interface brings about additional electrostatic and

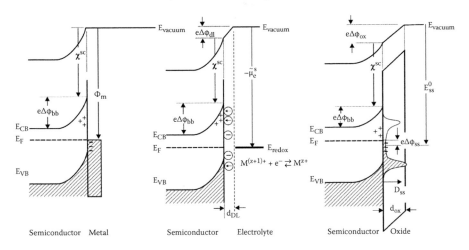

FIGURE 3.2 Energy band diagrams describing the equilibrium of semiconductor junctions with different phases.

chemical transformations. If we disregard for now each chemical transformation arising from the reactivity of the semiconductor surface with the electrolyte species, we
have in principle a change of the potential at the double layer. This results merely
from electrostatic interactions, which involves modification of the molecular solution
structure respective the bulk close to the contact surface. The potential drop is given by

$$\Delta\varphi^{dl} = \frac{q_{sc}}{C_{dl}} = \frac{q_{dl}d_{dl}}{\varepsilon_0\varepsilon_{dl}} \tag{3.13}$$

where q_{dl} refers to the charge built at both sides of the interface, d_{dl} is the thickness of
the double layer, where structural solution changes take place, and C_{dl} is the double
layer capacity. The length of the double layer is typically about ~ 0.3 nm. After
equalizing the electrochemical potentials of semiconductor and electrolyte, the band
bending of the semiconductor can be calculated as

$$e\Delta\varphi_{bb} = -\tilde{\mu}_e - \left[\chi^{sc} + (E_{CB}^0 - E_F)\right] - e\Delta\varphi_{dl} \tag{3.14}$$

Possible chemical reactions occurring at the semiconductor–electrolyte interface leads
for instance to the formation of surface dipole layers or oxide films, as already discussed in Chapter 2 for the electrochemical conditioning of p-type InP photoelectrodes.

As a third case, we regard the semiconductor–oxide contact. For the sake of simplicity we consider an electronic insulating oxide, as for instance SiO_2 or Al_2O_3.
The semiconductor–oxide interface is inherently connected with the formation of
interfacial surface states. With this denomination, we refer to localized energy levels
which arise as a result of an interruption of the lattice periodicity of the crystalline semiconductor, that is, uncompleted coordinated atoms or the formation of new
chemical bonds as well as stressed lattice bonds. These energy states are connected
with particular spatial structural features of the surface and act as electron traps.
They are characterized by a distinct energy distribution of their surface density and
a so-called electroneutrality level. This means that a negative or positive charge will
be built up depending on whether the position of the Fermi level of the semiconductor is shifted above or below this level. Depending on the density of surface states,
the electron exchange between semiconductor and surface states leads to a *Fermi
level pinning*. Thus, each shift of the position of the Fermi level results in partial or
no variation of the semiconductor band bending. This is compensated by a potential
drop in the adjacent phase, the double layer, or, as in the now analyzed case, the
oxide film. The charging of surface states polarizes the dielectric oxide, so that an
electric field arises. The charge at the surface states can be calculated by

$$q_{ss} = e^2 \int_{E_{ss}^0}^{E_F} D_{ss}d\varphi \tag{3.15}$$

where D_{ss} [eV^{-1} cm^{-2}] is the density of states (DOS). Accordingly, the electric field in
the dielectric film is given by

$$\Delta\varphi_{ox} = \frac{d_{ox}}{\varepsilon_0\varepsilon_{ox}}e^2\int_{E_{ss}^0}^{E_F}D_{ss}d\varphi \tag{3.16}$$

The band bending of the semiconductor can be calculated as

$$\Delta\phi_{bb} = (E_{ss}^0 - e\phi_{ss}) - e\Delta\phi_{ox} - \left[\chi^{sc} + (E_{CB}^0 - E_F)\right] \tag{3.17}$$

where $\Delta\varphi_{ss}$ is the difference between the Fermi and electroneutrality level.

The behavior of semiconductor-oxide-metal contacts is a combination of the aforementioned first and third cases. Unless the transfer process is controlled by tunneling through the oxide, the current-voltage characteristic of semiconductor-oxide-metal contacts presents a Schottky-type form:

$$j = A_{eff}T^2\exp\left(-\frac{\Phi_B}{kT}\right)\left[\exp\left(\frac{e}{nkT}V\right) - 1\right] \tag{3.18}$$

where Φ_B [eV] is the Schottky barrier height and n is the ideality factor. V is the difference between the Fermi level at metal and semiconductor; in equilibrium $V = 0$. In this case, the barrier height is not only determined by the energy distance between the electron affinity of the semiconductor and the work function of the metal, but also by the density of surface state at the oxide-semiconductor interface. Figure 3.3 shows a detailed energy band diagram for an s-o-m contact. In the hypothetical case of a constant density of surface state, D_{ss}, the charging of surface states is given by

$$q_{ss} = eD_{ss}\left[\Phi_B - (E_{ss}^0 - \chi^{sc})\right] \tag{3.19}$$

Moreover, the space charge that forms in the depletion layer of the semiconductor is given by

$$q_{sc} = \left(2\varepsilon_{sc}N_D\left[\Phi_B - \left(E_{CB}^0 - E_F\right) - kT\right]\right)^{1/2} \tag{3.20}$$

with N_D as the doping concentration. According to Gauss's law and under the condition that an exactly equal and opposite charge q_m appears at the metal surface, that is, $q_m = -(q_{sc} + q_{ss})$, the potential drop in the oxide, $\Delta\varphi_{ox}$, can be calculated as

$$\Delta\varphi_{ox} = -\frac{q_m}{\varepsilon_{ox}\varepsilon_0}d_{ox} = \frac{(q_{ss} + q_{sc})}{\varepsilon_{ox}\varepsilon_0}d_{ox} \tag{3.21}$$

After introduction of Eq. (3.19) and (3.20) in (3.21), we have

$$\Phi_B = (\Phi_m - \chi^{sc}) + \frac{e^2D_{ss}}{\varepsilon_{ox}\varepsilon_0}d_{ox}\left[\Phi_B - (E_{ss}^0 - \chi^{sc})\right]$$

$$+ \left[\frac{2e^3\varepsilon_{sc}}{\varepsilon_{ox}\varepsilon_0}N_Dd_{ox}^2\left(\frac{\Phi_B}{e} - \frac{(E_{CB}^0 - E_F)}{e} - \frac{kT}{e}\right)\right]^{1/2} \tag{3.22}$$

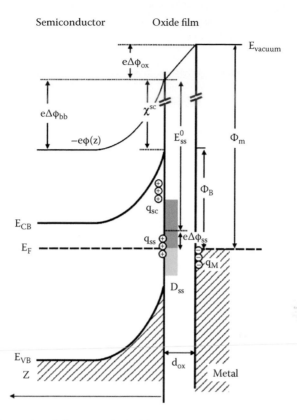

FIGURE 3.3 Energy band diagram of a MOS contact in electronic equilibrium.

The term between square brackets is on the order of some ten of mV for $N_D < 10^{18}$ cm^{-3}, so that it can be neglected. After some rearrangement, Eq. (3.22) reduces to the following linear relationship of the barrier height with the work function of metal:

$$\Phi_B \approx s\left(\Phi_m - \chi^{sc}\right) + (1-s)\left(E_{ss}^0 - \chi^{sc}\right) = s\Phi_m + C \tag{3.23}$$

The proportionality constant s is called the interfacial coefficient and C is a constant including interfacial properties. The electroneutrality level, E_{ss}^0, and the mean density of states, D_{ss}, can be determined experimentally after having measured the height barriers for a given semiconductor with different metals as follows:

$$E_{ss}^0 = \chi^{sc} - \frac{C + s\chi^{sc}}{1-s} \tag{3.24}$$

$$D_{ss} = \frac{(1-s)}{s} \frac{\varepsilon_{ox}\varepsilon_0}{e^2 d_{ox}} \tag{3.25}$$

This procedure is usually applied for contacts formed by physical methods (sputtering, vapor deposition), where the surface properties of the semiconductor are not largely modified. The method is not suitable for electrochemically achieved contacts due to the different interfacial reaction products arising from the electrochemical deposition process.

Two limiting cases can be recognized in Schottky-type contacts. When $D_{ss} \to \infty$, $s \to 0$. Hence, the Fermi level is pinned at the interface states at $e\Delta\varphi_{ss}$ below the electroneutrality level or the particular case of an n-type semiconductor and the barrier height becomes independent of the metal work function. The other limit behavior appears in the absence of surface states, that is, $D_{ss} \to 0$. The contact shows an ideal Schottky behavior with $s = 1$, usually known as the Barden limit.

The electron affinity of some semiconductors can be tailored by controlled modification of the chemical surface composition. The functionalization of the silicon surface with organic groups, such as Si–CH$_3$ [6], Si-phenyl [7,8] and other alkyl terminations [9], for instance, is being currently investigated by many researchers in order to modify the stability and surface electronic properties with potential applications in the construction of chemical and biochemical sensors [9,10], organic transistors [11] and electrochemical solar cells [12,13]. The covalent bond linkage with organic adsorbates can also mediate the connection of diverse biochemical polymers such as DNA and proteins [14,15]. In spite of the very low recombination velocity of well hydrogen-terminated Si surfaces, they oxidize in water containing milieus within first 30 min. UV photoelectron spectroscopy measurements performed by Hunger et al. [16] under ultra-high vacuum conditions indicate that the electron affinity for Si(111)-CH$_3$ surfaces is 0.5 eV less than that for H:Si(111) surfaces. The shift of energy bands toward the energy of vacuum implies an increase of the barrier height of the Schottky-type contact, and hence of the photovoltage when these modified surfaces are used as photoanodes in photoelectrochemical conversion devices. Maldonado and Lewis [6] reported the electrochemical preparation of Schottky-type contacts of methylated n-type Si(111) with Cd and Pb. They observed an increase of the height barrier of about 0.41 eV and 0.23 eV, respectively, in comparison with the similar contacts with H:Si(111) surfaces. But one of the most interesting features of methylated Si surfaces is the low recombination velocity, comparable to that of hydrogen-terminated surfaces [17]. The shift of the electron affinity is well described by the Helmholtz equation:

$$\Delta\chi = \frac{4\pi}{\varepsilon_i \varepsilon_0} N_{ad}\mu(\theta) \tag{3.26}$$

where N_{ad} is the surface concentration of adsorbed species, ε_i is the dielectric permittivity of the dipole layer, and $\mu(\theta)$ is the dipole moment contribution normal to the surface, which is a function of the surface coverage θ [18].

The Fermi level pinning is caused by the formation of surface or interface states, the origin of which is discussed in several formulated theories as discussed below. The extent of Fermi level pinning is quantified in terms of the deviation of the ideal Schottky behavior by the interfacial coefficient s [19]. Experimental values of the barrier height are usually presented as a function of the metal electronegativity rather than their work function. The electronegativity is an atomic property, whereas the work function depends on the chemistry and surface structure of metal. Thus, the former is preferred to characterize the behavior of semiconductor–metal contacts. The work function of polycrystalline metals correlates linearly with the corresponding electronegativities as [20,21]

$$\Phi_m = 2.27 X_m + 0.34 \tag{3.27}$$

where X_m is the Mulliken electronegativity [22].

The first theories to explain the Fermi level pinning phenomenon were formulated by Heine [23] and Tersoff [24]. They postulated the formation of gap states induced by the contacting metal. Thus, this theory is known as metal-induced surface states or MIGS. Interface states arise as a result of the penetration of the electron wave function of the metal about some monolayers into the semiconductor. They interact with the virtual states in the semiconductor band gap deriving from the complex band structure of the semiconductor. The solution of the Schrödinger equation predicts the formation of states in the band gap with a complex wave vector $\bar{k} = \mathbf{k}_r + i\mathbf{k}_i$. Hence, the wave function $\psi(r)$ will be damped by the imaginary part of the wave vector: $\exp(-\mathbf{k} \cdot \mathbf{r})$. Thus, the coupling of the protruding wave functions of the metal with the damped wave function in the band gap of the semiconductor leads to the formation of new wave functions (i.e., states) that can take up charges. Figure 3.4 shows the course of charge density with the distance to the metal–semiconductor interface by charged surface states as calculated by Louie, Chelikowsky, and Cohen [25,26] for several semiconductors. The Jellium model for the metal and a pseudo-potential formalism were applied for the calculations.

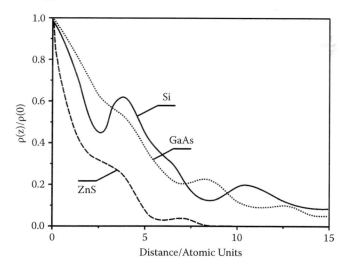

FIGURE 3.4 Normalized density of interface states in the semiconductor band gap calculated as a function of the distance to the interface (after Louie, Chelikowsky, and Cohen [25]). Length unity is given in Bohr radius (0.053 nm).

Without specifying the origin of surface states the following expression for the s-coefficient can be derived by considering only electrostatic arguments [19]:

$$s = \left(1 + \frac{e^2}{\varepsilon_i \varepsilon_0} D_{is} \delta_i\right)^{-1} \tag{3.28}$$

where ε_i and δ_i are the dielectric constant and the thickness of the interface, respectively. According to the calculations reported by Louie et al. [25], a density of surface states $D_{ss} = 4.5 \times 10^{14}$ eV^{-1}cm^{-2} for Si, 5×10^{14} eV^{-1}cm^{-2} for GaAs and 1.4×10^{14} eV^{-1}cm^{-2} for ZnS are to be taken. The thickness of the interface δ_{ss} is obtained from the distance at which $\rho(z)/\rho(0) = 0.27$ in Figure 3.4. Thus, values of 0.3 nm for Si, 0.28 nm for GaAs and 0.09 nm for ZnS result. Putting these values into Eq. (3.28), we obtain values of $s = 0.075$ for Si, 0.073 for GaAs and 0.55 for ZnS assuming a dielectric constant $\varepsilon_i = 2$ [27].

On the other hand, the MIGS model offers a correlation between the s-coefficient and the static dielectric constant, derived from the relation between this latter with the semiconductor band gap as follows:

$$\varepsilon_\infty = 1 + \left(\frac{\hbar\omega_p}{\langle E_g \rangle}\right)^2 \tag{3.29}$$

where $\hbar\omega_p$ represents the plasmonic energy of valence band electrons. Thus, we have [28,29]

$$s = \frac{1}{1 + 0.29 \dfrac{(\varepsilon_\infty - 1)^2}{\varepsilon_i}} \tag{3.30}$$

Other theories were formulated to explain the origin of surface states such as the dangling-bond theory [30], the unified defect model from Spicer [31], the surface disorder model (DIGS) from Hasegawa and Sawada [32] and the effective work function model (EWF) postulated by Woodall and Freeouf [33,34].

The DIGS model, in particular, postulates the formation of surface states as a result of a perturbation of the surface structure after oxide growth or metal deposition. According to this theory, the fluctuating bond lengths at the surface lead to the formation of bonding and anti-bonding states, whose energy distributions are characterized by an electroneutrality level. From these considerations, the following expression for the s-coefficient is derived:

$$s = \text{sech}\left[\delta_{dis}\left(\frac{e^2 N_{DIGS}(E_0)}{\varepsilon_0 \varepsilon_\infty}\right)^{1/2}\right] \tag{3.31}$$

where δ_{dis} is the thickness of the disordered layer and $N_{DIGS}(E_0)$ is the density of defects per volume and energy units at the electroneutrality level E_0. The validation of this theory by experimental evidence is, however, very difficult, since the changes of density of defects with the deposited metal cannot be explained in a convincing way.

3.2 ANALYSIS OF COMPLEX SYSTEMS: THE EMOS INTERFACE

The interpretation of the electronic behavior of the electrolyte–metal–oxide–semiconductor (EMOS) interface is sometimes very complex due to the presence of various interfaces through which the photogenerated electrons have to be transported before the electrochemical reaction can occur. We have analyzed the behavior of semiconductor–oxide–metal interfaces, where the current voltage behavior can be studied by simply applying an external potential difference which moves the Fermi levels (or electrochemical potentials) of the semiconductor and metal away from each other. One of the first grades of difficulty appearing after the introduction of the electrolyte as an additional phase is how the Fermi level in the solution is defined, and the experimental access to Fermi level changes of the electrode. The control of the Fermi level in the photoelectrode requires the introduction of an additional electrolyte interface. From a thermodynamic analysis of an electrochemical cell formed by two inert electrodes in a solution containing a redox pair, Kahn et al. [36] came to the conclusion that the absolute potential of each electrode can be defined as

$$V_{abs} = (\varphi^m - \varphi^s) - \frac{\mu_e^m}{ne} \tag{3.32}$$

where μ_e^m is the chemical potential of electrons in the metal, φ^s and φ^m are the Volta potential in the solution and in the metal, respectively, and n is the number of electrons exchanged in the redox reaction. According to this definition, we have

$$\tilde{\mu}_e^m = \mu_e^m - e\varphi^m = -e(\varphi^m - \varphi^s) + \frac{\mu_e^m}{n} - e\varphi^s = -V_{abs} - e\varphi^s \tag{3.33}$$

Provided that there is an electronic equilibrium at the metal–electrolyte interface, we have

$$\tilde{\mu}_e^m = \tilde{\mu}_e^s \equiv E_F^s = -eV_{abs} - e\varphi^s \tag{3.34}$$

For a discharged interface, the Galvani potential $\phi = 0$, $\varphi_s = \chi_s$, and Eq. (3.34) reduces to

$$\tilde{\mu}_e^s = -eV_{abs} - e\chi^s \tag{3.35}$$

This is the definition of the Fermi level of electrons in the solution formulated by Gerischer and Eckardt [37]. As already mentioned, its quantity offers a definition consistent with the work function of metals in spite of the controversial use of the name "Fermi level" for electrons governed by Boltzmann statistics. Thus, we arrive at an expression very similar to Eq. (3.11). Comparing both equations, we have that the absolute potential $V_{abs} = \varphi_{redox\ (vacuum)}$, a nontrivial conclusion. Thus, the Fermi level for an electrode constituted by a semiconductor–oxide–metal interface can be determined with respect to that of the solution. Note that, different from Schottky-type contacts, the electron transfer at EMOS-contacts is driven by the different electrochemical potential between MOS and electrolyte, provided there is electronic equilibrium at the MOS contact.

If we immerse a metal in an electrolyte solution containing a redox pair, the different electrochemical potentials of electrons in both phases leads to an electron exchange and hence to an interfacial charging until equilibrium is achieved: that is, $\tilde{\mu}_e^s = \tilde{\mu}_e^m$. The process is schematized in Figure 3.5 for the particular case when $\Phi^m < -\tilde{\mu}_e^s$. The equilibration process implies a restructuring of the double layer and a related change of the potential at the double layer.

The electron transfer during the charging process can be analyzed by the semi-classical treatment following the concepts of Marcus and Gerischer [38], according to which the discharge current density is written as

$$j = -ev^- c_{ox} \int_{-\infty}^{\infty} \rho(E) f(E) W_{ox}(E) dE + ev^+ c_{red} \int_{-\infty}^{\infty} \rho(E)[1 - f(E)] W_{red}(E) dE \qquad (3.36)$$

where $\rho(E_F)$ is the electron density at the metal surface at the Fermi level, $f(E)$ is the Fermi–Dirac distribution function, λ is the reorganization energy of the redox species, and c_{ox} is the concentration of the oxidized species (electron acceptor). v^- and v^+ are frequency factors accounting for the probability of electron transfer, which includes the transmission coefficient. W_{ox} and W_{red} refer to the probability of finding an empty or filled redox state in the solution, which are described by a Gaussian distribution function with a maximum at E_{ox} and E_{red}.

$$W_{ox} = \frac{1}{\sqrt{4\lambda_{ox}kT}} \exp\left[-\frac{(E - E_{ox})^2}{4\lambda kT} \right] \qquad (3.37a)$$

$$W_{red} = \frac{1}{\sqrt{4\lambda_{red}kT}} \exp\left[-\frac{(E - E_{red})^2}{4\lambda kT} \right] \qquad (3.37b)$$

The redox level of the electrons can be defined thermodynamically regarding the reaction $Ox_{(aq)} + e_{redox} = Red_{(aq)}$ for which we can write

$$\tilde{\mu}_{e(redox)} = \tilde{\mu}_{e(redox)}^0 + kT \ln\left(\frac{c_{red}}{c_{ox}} \right) \qquad (3.38)$$

Per definition, we have that the electrochemical potential of electrons is equal to the Fermi level of redox electrons, that is, $\tilde{\mu}_{e(redox)} = E_{F(redox)}$. According to Gerischer's interpretation, we find $E_{F(redox)}$ from the equality: $c_{ox} W_{ox}(E_{F(redox)}) = c_{red} W_{red}(E_{F(redox)})$. Introducing expressions (3.37a,b) into the last equality and after a bit mathematics, we obtain

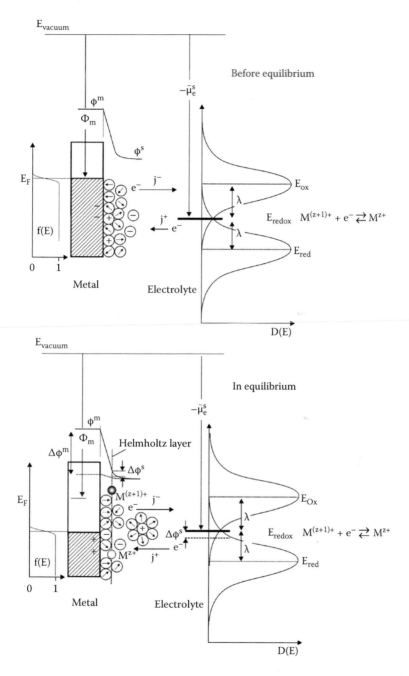

FIGURE 3.5 Energy band diagram representing a metal-electrolyte junction in the presence of a redox couple in the solution before and after setting on electronic equilibrium.

$$W_{ox} = \frac{1}{\sqrt{4\lambda_{ox}kT}}\left(\frac{c_{red}}{c_{ox}}\right)^{1/2}\exp\left[-\frac{\lambda_{ox}}{4kT}\right]\exp\left[\frac{\alpha(E-E_{F(redox)})^2}{kT}\right] \quad (3.39a)$$

$$W_{red} = \frac{1}{\sqrt{4\lambda_{red}kT}}\left(\frac{c_{ox}}{c_{red}}\right)^{1/2}\exp\left[-\frac{\lambda_{red}}{4kT}\right]\exp\left[-\frac{(1-\alpha)(E-E_{F(redox)})^2}{kT}\right] \quad (3.39b)$$

provided λ_{ox}, $\lambda_{red} \gg E^0_{F(redox)}$. The coefficient α is the equivalent of the charge transfer coefficient in the Butler–Volmer expression (see below) and represents, according to Gerischer's interpretation, a symmetry factor associated with λ_{ox} and λ_{red}. The density of occupied states of the redox system can be expressed as

$$D_{ox}(E) = \left[c_{ox}W_{ox}(E)+c_{red}W_{red}(E)\right][1-f(E,E_{F(redox)})] = \quad (3.40a)$$

$$= c_{ox}W_{ox}(E_{F(redox)})\exp\left[\frac{\alpha(E-E_{F(redox)})}{kT}\right]$$

$$D_{red}(E) = \left[c_{ox}W_{ox}(E)+c_{red}W_{red}(E)\right]f(E,E_{F(redox)}) = \quad (3.40b)$$

$$= c_{red}W_{red}(E_{F(redox)})\exp\left[-\frac{(1-\alpha)(E-E_{F(redox)})}{kT}\right]$$

Replacing Eqs. (3.39) and (3.40) in (3.36), the following current-potential relation can be obtained after some rearranging:

$$j = j_0\left\{\exp\left[\frac{(1-\alpha)e\eta}{kT}\right]-\exp\left[-\frac{\alpha e\eta}{kT}\right]\right\} \quad (3.41)$$

where the exchange current density is given by

$$j_0 = \frac{e\pi kT}{2\sin[(1-\alpha)\pi]}\nu(E_{F(m)})\rho_m(E_{F(m)})\left[c_{ox}W_{ox}(E_{F(redox)})+c_{red}W_{red}(E_{F(redox)})\right] \quad (3.42)$$

and $e\eta = E_{F(redox)} - E_{F(m)}$ with $E_{F(m)}$ as the Fermi level at the metal. Here, it was additionally supposed that $\nu^+ = \nu^- = \nu(E_{F(m)}) = \nu(E_{F(redox)})$.

In the absence of a redox pair in the solution, there are no electron exchanges between metal and ion in solution, except for partial electron transfer with possible adsorbing species. In this case, electrostatic interactions dominate the interface charging. In the case of a preferential orientation of water molecules with the oxygen toward the metal surface, the metal charges slightly positive and solvated cations accumulate at the outer Helmholtz plane, the closest approximation distance of solvated ions. As a consequence, an electric field of 10^6–10^8 V cm^{-1} appears. After the introduction of a redox pair in the solution, an electron exchange occurs between metal and redox species until $j = 0$. The restructuring of the metal–electrolyte interface leads also to changes of the solution potential, $\Delta\varphi^s$, and so of the electrochemical potential of electrons in the solution: $\Delta\tilde{\mu}_e^s$. The potential in the solution drops on distance from the electrode given by:

$$d = \left(\frac{8\pi e^2 N_A}{\varepsilon_0 \varepsilon_s kT} I_\pm\right)^{-1/2} \sim 0.0855 \times I_\pm^{-1/2} \; [\text{nm}] \qquad (3.43)$$

where N_A is Avogadro's number, ε_s is the dielectric constant of the solvent, and $I_\pm$ is the ionic strength, given by

$$\frac{1}{2}\sum_j z_j^2 \, c_j.$$

In concentrated solutions (i.e., for $c > 0.1$ M), the charge at the metal surface is completely compensated in the Helmholtz layer, that is, within 0.3 nm. Let us consider the following example. We take a platinum electrode of large surface area, $A > 1$ cm^2, and introduce it into a solution containing 0.1 M $FeCl_2$ and 0.1 M $FeCl_3$. For this solution, we have $I_\pm = 0.9$ and hence, the potential drop φ^m-φ^s should be placed within a distance of 0.3 nm from the metal surface; i.e., the Helmholtz layer. The Fermi level of Pt is situated at -5.65 eV $\pm \chi_s$. The redox level for the redox pair Fe^{+2}/Fe^{+3} is found at -5.27 eV. This makes an energy difference of 0.34 eV, which is compensated by a potential drop at the Helmholtz layer after the system has achieved equilibrium. This requires the charging of the metal surface by electron exchange and electrostatic interactions given by $\Delta q^m = \Delta\varphi^m \times C_H$ (C_H: Helmholtz layer capacity ~ 5–20 μF cm^{-2}). Thus, we have $\Delta q^m = 1.7 - 6.8$ μC cm^{-2}. The charging of the metal surface brings about a reorientation of surface dipoles, most water molecules. The charge induced by surface dipoles can be calculated by

$$\Delta q_{dip} = n_{dip}\frac{\mu_{dip}^2}{3kT}\Delta|\vec{E}| \qquad (3.44)$$

where n_{dip} [cm^{-3}] is the dipole concentration, μ_{dip} [D] is the corresponding dipole moment, and $|\vec{E}|$ [V cm^{-1}] is the interfacial field strength. According with our given example the change of field strength is about 3.4×10^7 V cm^{-1} (taking a distance of 0.1 nm). Taking a dipole moment 1.85 D for the water molecule, a concentration of oriented dipoles of 1.6×10^{21} cm^{-3} can be estimated by Eq. (3.44) assuming a surface charge of 1.7 μC cm^{-2}. This corresponds to a density of reoriented dipoles of ca. 1.5×10^{13} cm^{-2}, that is, a fraction of 0.04 of the total density of water molecules inside the Helmholtz layer: $n_w = 55.5 \times N_A \times 10^{-3}$ [cm^{-3}] $\times d_H$.

After having briefly analyzed the properties of metal-electrolyte interfaces, we want to consider the more complex EMOS interface. Figure 3.6 represents the process of linking a MOS with a metal-electrolyte interface, both in electronic equilibrium. In the first case, the difference between the semiconductor electron affinity and the work function is absorbed by the semiconductor band bending and a dielectric potential drop at the oxide film.

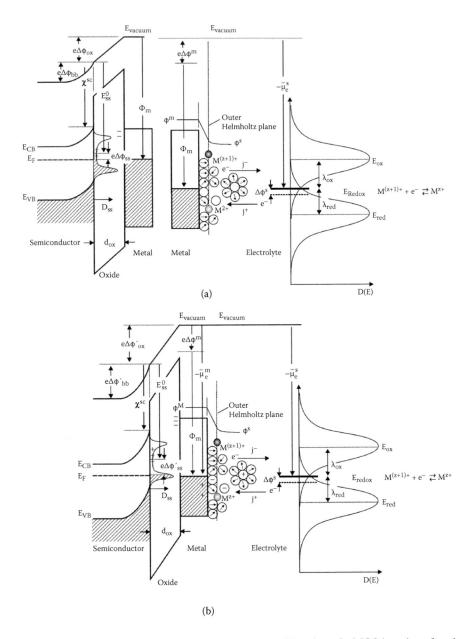

FIGURE 3.6 Energy band diagrams depicting the equilibration of a MOS junction after the contact with the metal–electrolyte junction.

This contrasts with the metal–electrolyte interface, where the energy difference is absorbed by the metal by surface charging and a change of potential drop at the Helmholtz layer. If we contact now both interfaces, a new charge exchange among the different phases takes place until all phases reach the same electrochemical potential. This brings about mainly changes of the Volta potential in the semiconductor and in the oxide film. Practically no changes should be observed at the metal–electrolyte interface. The position of the Fermi level is governed by the redox electrolyte. This can be seen regarding the characteristic capacitance of each interface. Considering the exchanged charge after contacting the MOS with the ME interface, we have

$$\Delta q = (C_{sc} + C_{ss})\Delta\varphi_{bb} = C_{ox}\Delta\varphi_{ox} = C_H\Delta\varphi_H \tag{3.45}$$

where C_{ss} refers to the capacity of surface states, which are seen as connected in parallel with that of the space charge layer. The capacities of the semiconductor and Helmholtz layer are functions of the applied potential and the chemistry of the interface. Thus, it is more convenient to speak of differential capacity: $C_i = dq_i/d\varphi$. The Helmholtz capacity typically shows values of 10–50 μF cm^{-2}. This value is 10-fold larger than that typical for oxides, < 2 μF cm^{-2}, or those of semiconductors in depletion: $C_{HL} + C_{ss} < 5$ μF cm^{-2}. Thus, we have $(\Delta\varphi_{ox} + \Delta\varphi_{bb}) > \Delta\varphi_H$. Should the semiconductor potential be driven in the accumulation region, its capacity increases exponentially with potential, and the potential changes are practically absorbed by the oxide film. The relation $(\Delta\varphi_{bb}/\Delta\varphi_{ox})$ is of fundamental importance in the case of photoelectrodes based on EMOS contacts, since the photoinduced retraction of the band bending defines the maximal achievable photopotential (see band diagram in Figure 3.7). In ideal conditions, the photopotential changes to the same extent as the potential of the redox couple, that is, $|d\Delta\varphi_{ph}/d\Delta\varphi_{redox}| = 1$. A slope <1 is an indication of Fermi level pinning.

It should be noted that $\Delta\varphi_{ox}$ is mainly induced by charging of the states at the semiconductor–oxide interface. Let us consider the case of a Si-SiO$_2$ interface with an oxide thickness of 1 nm. The potential drop induced by charging of interface states in the oxide is given by

$$\Delta\varphi_{ox} = \frac{en_{ss}d_{ox}}{\varepsilon_0\varepsilon_{ox}} \tag{3.46}$$

where n_{ss} is the surface density of rechargeable interface states. The maximal achievable photopotential for an n-type Si with a doping concentration of $\sim 10^{15}$ cm^{-3} is about 0.85 V. Hence, according to Eq. (3.46) and adopting $\varepsilon_{SiO2} = 3.9$, a concentration of interface states of 1.83×10^{13} cm^{-2} should lead to a complete loss of the photoeffect. Such a concentration is close to that measured in very defective silicon dioxide films as discussed in details in Chapter 5.

In the ideal case of a complete absence of interface states, the photopotential is given by

$$eV_{ph}^m = e\Delta\varphi_{bb} = \tilde{\mu}_e^s - \chi^{sc} - (E_{CB}^0 - E_F) = -e\varphi_{redox}^{vac} - \chi^s - \chi^{sc} - \Delta E_N \tag{3.47}$$

It should be noted, that the maximal theoretical achievable photopotential is directly proportional to the standard potential of the redox pair. The presence of surface states brings about a partial or total Fermi level pinning so that $(dV_{ph}/d\varphi_{Redox}) < 1$ or $(dV_{ph}/d\varphi_{Redox}) = 0$, respectively.

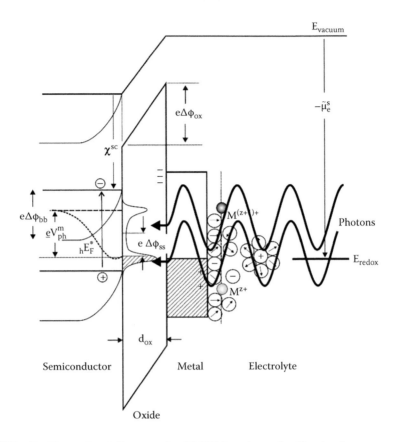

FIGURE 3.7 Energy band diagram of an EMOS junction under illumination.

The electron transfer in MOS contacts will be discussed regarding the band diagram in Figure 3.8. In this case, we deal with a two-electrode system where the circulating current is controlled by applying externally a tension ΔV between the semiconductor and the metal: a solid-state Schottky-type contact. This requires an ohmic contact at the rear side of the semiconductor, usually achieved by depositing a metal with a work function close to the Fermi level of the semiconductor. An ohmic contact is made by smearing eutectic GaIn or evaporating Al onto n-type Si or by evaporating Au onto p-type Si. In the case of p-type InP, the formation of an ohmic contact requires a bit more work. A usual procedure consists of the formation of a very highly doped ($> 10^{18}$ cm^{-3}) layer of semiconductor adjacent to the contact metal [39]. The depletion region is thus so thin that field emission or tunneling of charge carrier can take place. The formation of a highly doped surface can be achieved via two methods. One of them consists of growing a highly doped epitaxial layer by, for example, chemical vapor deposition on the semiconductor prior to metal deposition. The epitaxial layer is usually a material with a smaller band gap such as In$_{0.53}$Ga$_{0.47}$As (E_g: 0.75 eV) lattice matched to InP or a graded In$_x$Ga$_{1-x}$As layer to InAs at the surface [40]. The other method uses an external dopant—for instance, Zn (p-type) or Ge (n-type). The dopant metal is deposited

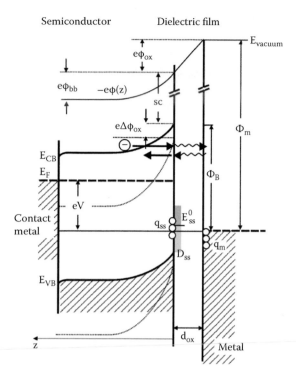

FIGURE 3.8 Energy band diagram of a MIS junction after applying a forward bias V. The band bending for the junction in equilibrium (V = 0) is indicated as reference.

as a thin layer of a few to tens of nanometers by beam evaporation and driven into the semiconductor by diffusion during a heating program. The dopant is part of a multilayer metallization scheme including metallic adhesion layers, diffusion barriers, and capping layers to prevent oxidation during annealing. Common ohmic contacts are listed in Table 3.1.

The oxide film in our analyzed MOS contact behaves as a dielectric and its thickness is narrow enough so that electrons may tunnel. In equilibrium, E_F^{sc} equals Φ_m. This demands a semiconductor band bending, $e\varphi_{bb}$, and an electric field in the oxide film: $\bar{E} = e\varphi_{ox} / d_{ox}$. Thus $\chi^{sc} + (E^s_{CB} - E_F) + e\,\varphi_{ox} = \Phi_m$. If we apply tension between both phases $\Delta V \neq 0$, it will be absorbed by semiconductor and oxide film: $\Delta V = e\,\Delta\varphi_{bb} + e\,\Delta\varphi_{ox}$.

Under these conditions, the electron transfer through the SO interface occurs by electron emission. Transferred electrons reach the MO interface by tunneling. Providing the tunneling rate transfer does not limit the electron transfer, the current-voltage behavior can be described by the thermionic emission theory, provided the following conditions are met: (i) $\Phi_B \gg kT$, (ii) there is thermic equilibrium at the emission level, and (iii) the equilibrium is not disturbed by the current flow. According to this theory, the current flow is given by integration of the product of the electron concentration with energy larger than the potential barrier by their velocity in the transport direction:

TABLE 3.1

Selected Ohmic Contacts to III-V Semiconductors

Contact	Doping level [cm^{-3}]	Contact resistance [Ω cm^2]	Annealing temp. [°C]	Ref.
	n-type			
Pd-InP	1×10^{18}	7×10^{-5}	300–350	41
Pd/Ge/Au-InP	1×10^{17}	2.5×10^{-6}	300–375	42
Pd/Ge-InP	1×10^{17}	6×10^{-6}	400–450	42
Pd/Ge-GaAs	5×10^{18}	4×10^{-7}	400–415	43
	p-type			
Pd/Pt/Au/Pd-InGaP/GaAs	3×10^{19}	$< 10^{-6}$	415–440	44
Zn/Pd/Pt/Au-AlGaAs/GaAs	5×10^{18}	8.5×10^{-7}	440	45
Ti/Pt/Au-InGaAs	5×10^{18}	6×10^{-7}	450	46
Ti/Pt-InGaAs	1.5×10^{19}	4×10^{-6}	450	47
Ti/Pt-InGaAs	2×10^{18}	3.4×10^{-8}	450	48
Pd/Zn/Pd/Au/-InP		7×10^{-5}	420–425	49

$$j_{thermionic} = \int_{E_F + \Phi_B}^{\infty} ev_z \, dn = \underbrace{\left(\frac{4 \, em_e^* k^2}{h^3} \right)}_{A^*} T^2 \exp\left[-\frac{\Phi_B}{kT} \right] \exp\left[\frac{e(V - \varphi_{ox})}{kT} \right] \tag{3.48}$$

where $\Phi_B = (E_{CB}^0 - \Phi_M)$, $V = E_F - \Phi_m$, and A^* is the effective Richardson constant [50]. This has a value of 120 A cm^{-2} K^{-2} for free electrons. In the case of semiconductors, the effective electron mass in k-space must be considered for the calculation. The effective mass is a tensor of the form

$$\left(\frac{1}{\mathbf{m}^*} \right) = \hbar^2 \nabla_k (\nabla_k E) \tag{3.49}$$

where E is the band energy in the k-space. In the case of Si, the constant energy surfaces in the k-space for the bottom of the conduction band show six ellipsoids along the $\langle 100 \rangle$ axes perpendicular to the (100) square surfaces of the Brillouin zone boundary. Thus, effective masses for electrons perpendicular and parallel to the axis (100), $m_{e\perp}^*$ and $m_{e\parallel}^*$ respectively, can be defined. Because of the practically spherical form of the valence band, however, the directional dependence of the effective mass disappears. The upper two bands of Si can be approximately fitted by two parabolic bands with different curvatures. Hence, heavy ($m_{h,h}^*$) and light holes ($m_{l,h}^*$) appears corresponding to the wider and narrower bands, respectively. Therefore, different effective Richardson constants are to be considered for n-type Si for the emission in $\langle 111 \rangle$ and $\langle 100 \rangle$ surface directions:

$$\frac{A^{*}_{n-Si\langle 111\rangle}}{A} = \frac{6}{m_e}\left[\frac{m^{2}_{e,\perp} + 2m^{*}_{e,//}m^{*}_{e,\perp}}{3}\right] = 2.2 \qquad (3.50a)$$

$$\frac{A^{*}_{n-Si\langle 100\rangle}}{A} = 2\frac{m^{*}_{e,\perp}}{m_e} + 4\frac{\sqrt{m^{*}_{e,//}m^{*}_{e,\perp}}}{m_e} = 2.1 \qquad (3.50b)$$

For p-type Si, we have

$$\frac{A^{*}_{p-Si}}{A} = \frac{m^{*}_{l,h} + m^{*}_{h,h}}{m_e} = 0.66 \qquad (3.51)$$

At $V = 0$, the electron flux S $\rightarrow$ M equals that of M $\rightarrow$ S. This latter does not depend on applied potential and can be obtained from expression (3.48) by making $V = 0$. Hence, the total current is found from the difference of the countercurrents:

$$j_{thermionic} = A^{*}T^{2}\exp\left[-\frac{\Phi_B}{kT}\right]\cdot\left[\exp\left[\frac{e(V-\varphi_{ox})}{kT}\right]-1\right] \qquad (3.52)$$

$$= j_S\left[\exp\left[\frac{e(V-\varphi_{ox})}{kT}\right]-1\right]$$

If the length of the space charge layer is thin enough, electrons may tunnel through the potential barrier of the contact as occurs for high-doped semiconductors, for instance, $N_D > 10^{17}$ cm^{-3} in Si, or as the semiconductor is driven into deep depletion. The current through the potential barrier in semiconductors in depletion can be calculated by regarding the quantum transmission coefficient, $T(E)$, multiplied by the probability of occupied states in the semiconductor and the probability of finding unoccupied states in the metal. Hence, we have

$$j_{T(sc\rightarrow m)} = \frac{A^{*}T}{k}\int_{0}^{\infty}T(\varepsilon)\exp\left[-\frac{e\varphi_{bb}+(E_{CB}-E_F)+\varepsilon}{kT}\right]d\varepsilon + \qquad (3.53)$$

$$+\frac{A^{*}T}{k}\int_{0}^{e\varphi_{bb}}f_{sc}(\eta)T(\eta)[1-f_m(\eta)]d\eta$$

where $f_{sc,m}$ represents the Fermi–Dirac distribution function. ε and η are the energy coordinates upward and downward from the potential maximum of the energy barrier. The first term in this expression corresponds to the flux of electrons injected by a thermionic mechanism and will reduce to Eq. (3.48) by making $T(\varepsilon) = 1$. The second integral represents the flux of tunneled electrons. A similar expression can be obtained for the current flowing from the metal to the semiconductor:

$$j_{T(m \to sc)} = -\frac{A^*T}{k} \exp\left[-\frac{\Phi_B}{kT}\right] \int_0^\infty T(\varepsilon) \exp\left(-\frac{\varepsilon}{kT}\right) d\varepsilon - \qquad (3.54)$$

$$-\frac{A^*T}{k} \int_0^{e\varphi_{bb}} f_m(\eta) T_t(\eta)[1 - f_{sc}(\eta)] d\eta$$

The quantum-mechanical transmission coefficient for energies $E < E_{CB}{}^s$ can be calculated by applying the Wentzel–Kramers–Brillouin approximation [51]:

$$T(E) = \exp\left(-\frac{\varepsilon\varepsilon_0 \overline{E}_m^2}{2N_D E_{00}}\left[\sqrt{c} - (1-c)\tanh^{-1}\sqrt{c}\right]\right) \qquad (3.55)$$

with

$$E_{00} = \frac{e\hbar}{2}\sqrt{\frac{N_D}{m_e^* \varepsilon\varepsilon_0}} \quad \text{and} \quad c = \frac{2N_D}{\varepsilon\varepsilon_0 \overline{E}_m^2}(E_{CB}^s - E)$$

Equation (3.55) can be reasonably well approximated by

$$T(E) \approx \exp\left[\frac{\left(E_{CB}^s - E\right)^2}{E_{00}e\varphi_{bb}}\right] \qquad (3.56)$$

Thus, after the introduction of this expression in the second integral of Eq. (3.54), one obtains

$$j_{tunnel} = \frac{A^*T}{k}\frac{1}{2}\sqrt{\pi E_{00}e\varphi_{bb}} \exp\left(-\frac{\Phi_B - eV}{kT} + \frac{E_{00}e\varphi_{bb}}{4(kT)^2}\right) \qquad (3.57a)$$

$$\times\left[erf\left(\frac{\sqrt{E_{00}e\varphi_{bb}}}{2kT}\right) - erf\left(\sqrt{\frac{e\varphi_{bb}}{E_{00}}}\right)\right]$$

for $\Phi_B - eV > e\varphi_{bb}$ and

$$j_{tunnel} = \frac{A*T}{k}\frac{1}{2}\sqrt{\pi E_{00}e\varphi_{bb}}\,\exp\left(-\frac{\Phi_B - eV}{kT} + \frac{E_{00}e\varphi_{bb}}{4(kT)^2}\right)\times \qquad (3.57b)$$

$$\times\left[erf\left(\frac{\sqrt{E_{00}e\varphi_{bb}}}{2kT}\right) - erf\left(\frac{\sqrt{E_{00}e\varphi_{bb}}}{2kT} - \frac{\Phi_B - eV}{\sqrt{E_{00}e\varphi_{bb}}}\right)\right] +$$

$$+\left(\frac{\Phi_B - eV}{kT} - 1\right)\left[erf\left(\frac{\Phi_B - eV}{\sqrt{E_{00}e\varphi_{bb}}}\right) - erf\left(\sqrt{\frac{e\varphi_{bb}}{E_{00}}}\right)\right] +$$

$$+\frac{\sqrt{E_{00}e\varphi_{bb}}}{\sqrt{\pi}kT}\left[\exp\left(\frac{(\Phi_B - eV)^2}{E_{00}e\varphi_{bb}}\right) - \exp\left(-\frac{e\varphi_{bb}}{E_{00}}\right)\right]$$

for $\Phi_B - eV > e\varphi_{bb}$. The deviation from the Schottky behavior (thermionic emission) introduced by electron tunneling is generally accounted for by the introduction of an ideality factor n. Hence, we have

$$j_T = A*T^2\exp\left[-\frac{\Phi_B}{kT}\right]\cdot\exp\left(-\frac{e(V - \varphi_{ox})}{nkT}\right)\cdot\left[\exp\left(\frac{e(V - \varphi_{ox})}{kT}\right) - 1\right] \qquad (3.58)$$

The fraction of current by tunneling increases at low temperatures and with doping concentration. This has consequences in the case of nanodimensioned contacts. The solution of the Poisson equation for metal-semiconductor contacts formed by nanodimensioned metal hemispheres onto semiconductors predicts a reduction of the space charge layer to dimensions close to the radius of the particle [52–54]. Equation (3.59) gives the Poisson equation expressed in spheroidal coordinates (see Figure 3.9).

$$\frac{1}{r^2(\xi^2 + \eta^2)}\left\{\frac{\partial}{\partial\xi}\left[(1 + \xi^2)\frac{\partial\varphi}{\partial\xi}\right] + \frac{\partial}{\partial\eta}\left[(1 - \eta^2)\frac{\partial\varphi}{\partial\eta}\right]\right\} = -\frac{eN_D}{\varepsilon_{sc}\varepsilon_0} \qquad (3.59)$$

The spheroidal coordinates are related to the Cartesian coordinates by $z = r_0\sqrt{(\xi^2 - 1)(1 - \eta^2)}\sin\varphi$, $x = r_0\sqrt{(\xi^2 - 1)(1 - \eta^2)}\cos\varphi$, $y = r_0\eta\xi$.

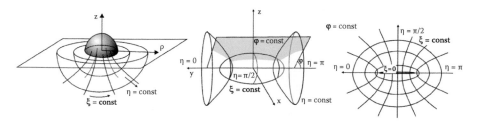

FIGURE 3.9   Schematic showing the spheroidal coordinate system.

In the particular case of the integration of Eq. (3.59) for a metal sphere imbedded in a n-type semiconductor, with the boundary condition that the charge on the sphere cancels the total charge in the space charge region, the following radial potential function results:

$$\varphi(r) = \frac{kT}{e} \frac{1}{2L_D^2} \left[(r_o + W)^2 - \frac{2(r_o + W)^3}{3(r_o + r)} - \frac{(r_o + r)^2}{3} \right] \qquad (3.60)$$

where W is the length of the space charge layer, given by $L_D = \sqrt{(\varepsilon_{sc}\varepsilon_0 kT)/e^2 N_D}$. The zero-point of the potential is chosen in the semiconductor bulk. The value of W is fixed by the second boundary condition $\varphi(r_o) = \varphi_s$, where φ_s is the total potential drop over the space charge region. Equation (3.60) offers a simple useful relation to estimate the threshold beyond which size influences the Schottky behavior. Figure 3.10 shows a graphical representation of Eq. (3.60) in units of the characteristic length $l_c = L_D\sqrt{2e\varphi_{sc}/kT} = \sqrt{2\varepsilon_{sc}\varepsilon_0\varphi_{sc}/eN_D}$, which allows deciding if a given contact should be considered as small or large.

The solution of the Poisson equation for a contact formed by a metal disk on a semiconductor requires the application of numerical methods. Figure 3.11 shows the results [54] for different diode sizes for a semiconductor with a doping concentration of 10^{15} cm^{-3}, a band bending $\varphi_{sc} = 0.4$ V, and a dielectric constant $\varepsilon_{sc} = 11.7$, where the following relations of spherical to cylindrical coordinates were used:

$$\rho = r\sqrt{(1+\xi^2)(1-\eta^2)};$$

$$z = r\xi\eta; \quad 0 \le \xi \le \infty; \quad -1 \le \eta \le 1$$

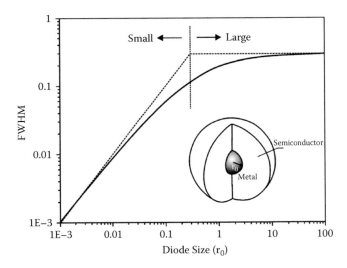

FIGURE 3.10 Plot of the calculated barrier full width at half maximum (FWHM) as a function of diode size (both in units of *lc*). The dashed lines represent the asymptotic values for $r_0 \gg l_c$ (semi-infinite diode). Inset: schematic showing the model system: a metallic sphere embedded in a semiconductor (after Smit et al. [52]).

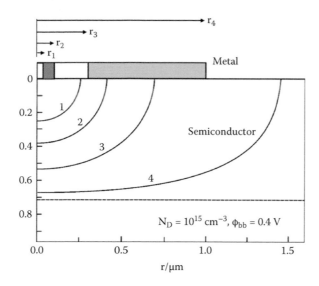

FIGURE 3.11 Contours of the barrier FWHM generated at the contact of disk shaped metal islands with a semiconductor surface as calculated by solving the Poisson equation (3.59).

It can be seen that the length of the space charge layer reduces to the dimensions of the contact for disks smaller than 1µm. This is a common case for photoelectrons formed by electrochemical deposition of metal particles, the size of which oscillates between some tens of nm to 500 nm. Contacts larger than 1.5 µm can be treated as infinite, for which the length of the space charge layer, that is, the electric field, is given by $W = \sqrt{(2\varepsilon_{sc}\varepsilon_0\varphi_{bb})/eN_D}$.

The shrinkage of the potential barrier at nanodimensioned metal-semiconductor contacts leads to an increase of the tunneling contribution to the flowing current. The contribution of tunnel conductance in small contacts can be estimated by applying Eq. (3.57a), where the size effect is introduced as an apparent modification of the doping concentration on reduction of the space charge layer, W. Thus, taking $N_D = (2\varepsilon_{sc}\varepsilon_0\varphi_{bb})/(eW^2)$, we have $E_{00} = (e\hbar/W)\sqrt{2\varphi_{bb}/(m_e^*e)}$. Introducing this term into expression (3.57a) for the calculation of the quantum transmission coefficient, the relation between tunneling and thermionic current can be calculated as a function of the length of the space charge layer as shown in Figure 3.12 for different values of the semiconductor band bending. It should be noted, that a considerable increase of the tunnel contribution is possible for space charge layers narrower than 200 nm. If we look at Figure 3.11, it is evident that these conditions are reached for contacts in the range of 10 to 50 nm.

Theoretical predictions of a reduction of the space charge layer with contact size were corroborated experimentally by Smit et al. [55]. They performed measurements of the current-voltage characteristic by contacting with a STM tip Co_2Si islands formed by annealing of Co layers deposited onto reconstructed 7 × 7 reconstructed Si(111) surfaces. Figure 3.13 shows their results on a silicon sample with a resistivity of 10 Ωcm. According to the doping concentration, the threshold size to consider size effects in this case is about 1 µm. Thus, the islands with a size of about 30 nm

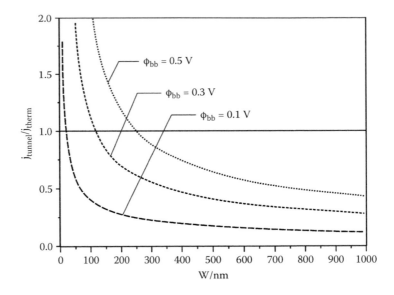

FIGURE 3.12   Relationship between tunnel and thermionic current as a function of the space charge layer length of a Schottky contact calculated for different values of band bending.

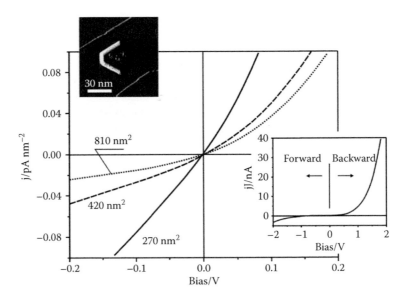

FIGURE 3.13   Current-bias curves performed on Co_2Si nanodimensioned contacts of different sites deposited on silicon of a resistance of 10 Ω cm. Inset at bottom right-side corner: full current-bias curve for a 810 nm^2 island; upper left-side corner: STM image of a typical island. Partially reprinted with permission from Smit et al. *Appl. Phys. Lett.* 80 (2002) 2568. Copyright 2002, American Institute of Physics.

are expected to show clear size effects. It was shown that the differential resistance at zero bias is 10^4 times lower than that expected for a planar Schottky contact. Considering the fact that $j_{thermionic}/j_{tunnel} \sim 10^{10}$ for a planar contact, an increase of the tunnel contribution by a factor of $\sim 10^{14}$ was inferred. Although the tendency is in agreement with the general theory, the extreme increase of the tunnel current infers the presence of edge effects, not addressed in their theory. The fact that a large bias, the reverse current shows a more accentuated current increase than at the forward bias, implies that small diodes act as ballistic points where charges are emitted by a Fowler–Nordheim tunneling mechanism. The application of the Poisson equation for solving the electronic structure of contact of such dimensions is however questionable, if one regards the distance among doping elements, estimated in about 100 nm for a doping concentration of 10^{15} cm^{-3}. This question was explored by Smit et al. [56]. They found that the scatter of conductance measurements at zero bias increases toward smaller contact sizes. The spread was modeled by Poisson statistics given by $sd(N)/\langle N \rangle = 1/\sqrt{\langle N \rangle}$. Regarding $\langle N \rangle$ as proportional to the contact area, a relation of the standard deviation of the conductance with area of the type $sd(\sigma) = C/\sqrt{A}$ is predicted, it being in agreement with the experimental results. The investigations reported by Smit et al. [56] confirm the reduction of the space charge layer toward smaller contacts for undoped silicon (without dispersion). The Coulomb wall of randomly distributed dopant atoms disturbs the conduction band profile and leads to a lowering of the local barrier; electrons cross the space charge layer by resonance tunneling.

The decoration of n-type silicon with nanodimensioned Pt particles is a prototype photoelectrode attempting to exploit the advanced electronic characterization of this material. The first successful photoelectrochemical experiments carried out by Nakato and coworkers [57–59] were explained in terms of a spatial modification of the electron affinity of silicon under the deposited metal islands. Thereby, it was assumed that the photoelectrochemical behavior was governed by the formation of local Schottky contacts. The possibility of Fermi level pinning, however, was not taken into account. Thus, according to the presented theory, electrostatic interactions at the semiconductor–metal interface were to be considered the origin of shifted band edges. It was found that electrodes formed by Pt islands in an oxidized surface matrix generate higher photovoltages than a surface covered by a continuous thin metal layer. Figure 3.14 compares the photopotentials obtained at the open circuit with different redox couples for the Pt-covered n-type Si surface and after etching it in an alkaline solution. The Fermi level pinning can be clearly observed in the latter case, where the photopotential practically does not change with the potential of the redox couple. This is probably due to the generation of MIGS, since these contacts were fabricated by electron-beam evaporation onto silicon surfaces etched with 10% HF prior to deposition. The oxide surface, the structure of which is supposed to be that schematized in Figure 3.14 b, shows a linear relation between the photopotential and the difference between the flat band and redox potential. At $V_{ph} = 0$, $|V_{fb}-V_{redox}| \sim$ 0.3 V, which can be ascribed to the potential drop in the surface oxide (see energy band diagram in Figure 3.14c). At $|V_{fb}-V_{redox}| > 0.9$ V, the Fermi level for minority charge carriers, $_nE^*_F$, is pinned at the VB edge and photopotential does not change.

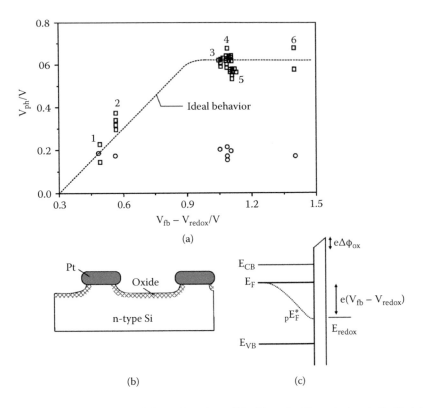

FIGURE 3.14 (a) Photopotential as a function of V_{fb}-V_{redox} for Pt-coated and alkali-etched n-type silicon (squares) and Pt-coated n-type silicon (circles). Redox couples: 1: $Fe(C_2O_4)^{4-}$/ $Fe(C_2O_4)^{3-}$, pH 6.2; 2: I^-/I_3^- (7 M H^+); 3: Fe^{2+}/Fe^{3+}, pH 0.5; 4: Br^-/Br_2 (8.6 M H^+), 5: $Fe(CN)_6^{4-}$/ $Fe(CN)_6^{3-}$, pH 7.6; 6: Br^-/Br_2, pH 2.9; (b) Schematic of the surface structure after Pt coating a and alkaline etching; (c) Energy band diagram showing the retraction of the band bending upon illumination (after Nakato et al. [57]).

The barrier height of Si-metal contacts obtained by physical methods was found to follow [50]: $\Phi_{B(Si)} = 0.27\ \Phi_m\ -0.55$. Hence, a density of interface states $D_{ss} = 2.7\pm0.7 \times 10^{13}$ eV^{-1} cm^{-2} can be calculated. This value is two orders of magnitude higher than that found in completely hydrogenated surfaces (~10^{11} eV^{-1} cm^{-2}) or after thermal oxidation [60]. A large density of surface states in the range of 10^{13} eV^{-1} cm^{-2} is also generated at silicon-oxide interfaces as a thin oxide film is formed by wet chemical or electrochemical treatments. The more effective collection of photogenerated charges in metal contacts in an oxide covered surface matrix can be explained in terms of a modulation of the space charge layer. The space charge layer created around the localized metal contacts is thought to introduce radial electric fields, which act as collectors of photogenerated charge carriers in the semiconductor bulk. The nanoemitter solar cell concept developed by Lewerenz and coworkers [61,62] and was realized by depositing catalytic metals into fingers formed by localized photo-etching of silicon surfaces passivated by porous anodic oxides. Partly self-organized pores are formed by anodization of H-terminated Si in fluoride containing

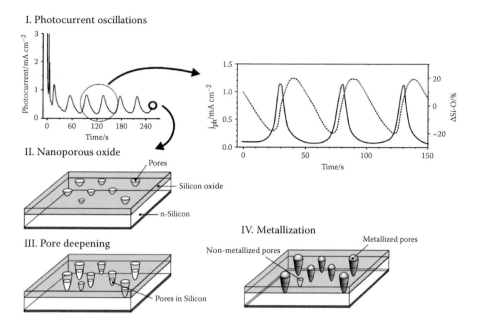

FIGURE 3.15   Steps for in situ preparation of a nanoemitter solar cell. After electrodeposition, the metallic nanoemitters protrude somewhat out of the oxide surface indicated in step IV. Inset: phase shift between current oscillations and mean oxide thickness measured by changes of the Si-O infrared adsorption.

solutions at 4–6 V. In this potential region, a dynamic process consisting in the formation of pores by etching in the oxide and their subsequent repassivation arises. This leads to current oscillations synchronized with the etching and repassivation of the oxide. Figure 3.15 shows a typical photocurrent oscillation of n-type Si by applying a potential of 6 V (vs. SCE) in NH_4F 0.1 M (pH 4) under illumination (light power: 100 mW cm^{-2}). The frequency of the oscillations depends on the solution pH (2 to 4) and applied potential (4 V to 7 V), so that one can tailor the density of pores.

The oxidation occurs by injection of holes from the valence band, photo-generated in n-type or induced by applying potential in a p-type Si:

$$Si + 2 H_2O + 4 h^+_{VB} (V,h) \rightarrow SiO_2 + 4 H^+ \tag{3.61}$$

Pores are formed by etching via HF or the bridge compound HF_2^- at pH between 2 and 4:

$$SiO_2 + 6 HF \rightarrow SiF_6^{2-} + 2 H^+ + 2 H_2O \tag{3.62}$$

$$SiO_2 + 2 HF_2^- \rightarrow SiF_6^{2-} + H_2O + OH^- \tag{3.63}$$

Regarding the stoichiometry of both reactions, it can be inferred that the etching rate by Eq. (3.63) is three times larger than that by Eq. (3.62). The former leads to a local alkalization that limits further reaction, as expressed by the kinetic expression for the dissolution of SiO_2 proposed by Knotter [63]:

$$\frac{d(SiO_2)}{dt} = k_{exp} \frac{K[H^+][HF_2^-]}{1+K[H^+]} \tag{3.64}$$

where k_{exp} is the experimental rate constant and K is an equilibrium constant involving the formation of active sites.

The inset in Figure 3.15 shows the time dependency of the change of the Si-O vibrational absorption between $980\ cm^{-1}$ and $1260\ cm^{-1}$ during photocurrent oscillations obtained with a specially designed cell with an adapted ATR-MIR configuration [64]. From this result, the oscillation of the integral thickness between 8 and 12 nm was calculated [65]. It was observed that the largest porosity of the oxide film is achieved at the point where the integral oxide thickness reaches its minimum at 1/5 of the maximum of the oscillating current. Thereafter, pores are deepened by a polarization of the electrode at −0.85 V (SCE) in the dark for 20 min. Thus, pores of up to 20 nm of depth and 70 nm of width can be etched. The presence of oxide on the walls of pores is removed by a short dipping in 50% HF prior to electrochemical metal deposition. Deposition is performed by polarizing the surface at −0.75 V for 30 s in a solution containing 1 mM H_2PtCl_6 and 0.1 M K_2SO_4. Figure 3.16a shows a secondary electron mode SEM picture of pores after the first stages of metal filling, which accounts for the selectivity of deposition inside the pores. A transmission microscopy picture of a cross-sectional film of this structure shows that the particles are in direct contact with the silicon substrate. Such a structure has shown a conversion efficiency of 11.2% using I_3^-/I^- as a redox pair and a W-I lamp [62].

This type of structure can also be applied for the construction of photocathodes for light-induced hydrogen evolution. As shown in Figure 3.16b, the etching of oxide pores leaves square-shaped pores, in the wall of which preferential electrochemical metal deposition occurs. Metal deposition was performed from a chloride complex of noble metals $PtCl_6^{2-}$, $IrCl_6^{3-}$, and $RhCl_6^{3-}$ as will be discussed in Chapter 5.

The insertion of metallic collectors of photogenerated minority carriers in a passivated substrate improves the photoresponse in comparison with the performance of simple metal–semiconductor contacts obtained by electrodeposition. Figure 3.17a compares the photopotentials obtained in an n-type Si surface covered with Rh deposits with those obtained from a nanoemitter cell using Pt as metal for collectors. It can be seen that the photopotential for the nanoemitter device increases according to the relation $\Delta V_{ph} = 0.46\ V_{redox} + 0.25$ V, whereas that for the Rh-Si contact does not change in a significant way. In the dark, however, the open circuit potential of the Rh-Si contact changes linearly, with a slope close to 1, with the redox potential. This fact indicates that the change of potential under illumination does not contribute to a semiconductor band bending. For the nanoemitter device, the flat band potential varies according to $V_{fb} = 0.83\ V_{redox} - 0.50$ V, which accounts for a strong Fermi level pinning. The band banding of the electrode is thus given by $V_{fb} - V_{redox} = 0.17$ $V_{redox} + 0.50$ V. By analogy with the Schottky contact, the density of interface states can be calculated according to Eqs. (3.23) and (3.25) by substituting Φ_m by V_{redox}.

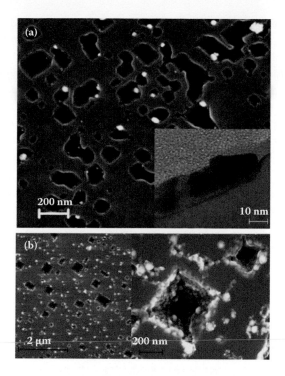

FIGURE 3.16　(a) Secondary electron mode SEM picture of a nanoemitter electrode obtained by polarization of n-type Si(100) at +6 V in 0.1 M NH_4F solution of pH 4.0 upon illumination at 20 mW cm^{-2}, alkaline etching in 4 M NaOH at −0.85 V in the dark and 5 s dipping in HF 50% and finally Pt deposition at −0.8 V from 1 mM H_2PtCl_6 + 0.1 M K_2SO_4; Inset: high-resolution transmission microscopy picture of a cross-sectional cut over a pore; (b) SEM picture of nano-emitter type cell with incomplete filled pores: p-type semiconductor: metal: Rh.

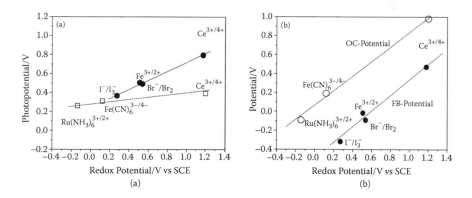

FIGURE 3.17　(a) Photopotential measured in different redox solution of a nanoemitter cell constructed with n-type Si(100) and Pt deposits (black circles) and of a n-type Si(111) covered with an electrodeposited Rh film (squares); (b) open circuit potential of an n-type Si(111) covered with Rh and flat band potential of the nanoemitter cell in different redox solutions in the dark.

Assuming a value of the interfacial oxide thickness of ~ 1 nm and a dielectric constant of 6, a density of surface states of 1.62×10^{14} cm^{-2} eV^{-1} can be calculated. It should be noted that the high density of interface states is brought about by the bad electronic quality of the metal–silicon contact of nanoemitters.

The electrochemical deposition of noble metals is accompanied by the formation of an interfacial amorphous film, which can be ascribed to the formation of an oxide film (see Chapter 5) and amorphous Si caused by intercalation of hydrogen. The formation of a uniform film can be well identified in high-resolution transmission microscopy (HR-TEM) pictures of cross-sectional sheets cut after deposition of Pt, Rh, and Ir (Figure 3.18).

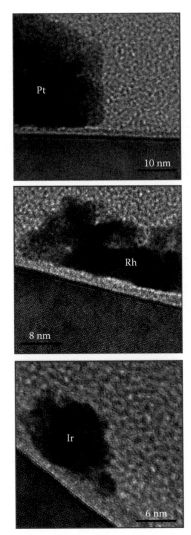

FIGURE 3.18   High-resolution TEM pictures of cross-sectional cuts performed on n-type Si after electrodeposition of different noble metals: Pt at –0.35 V, Ir at –0.9 V, and Rh at –0.4 V.

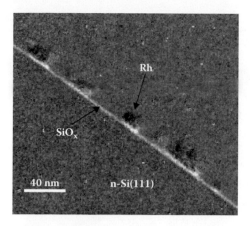

FIGURE 3.19 Three-window mode energy-filtered TEM image (EFTEM) at the O K-edge of cross-sectional cut of n-type Si(111) after electrodeposition of Rh.

The width of the film ranges about 3 nm, this being larger than that predicted for SiO_2 regarding spectroscopic measurements (see Chapter 5). Part of the amorphous film is constituted by amorphous silicon generated by hydrogen intercalation during the deposition. The presence of an oxide film is pointed out on an energy-filtered TEM picture (EFTEM) [66] corresponding to the O K-edge of the electron energy loss spectrum (EELS) (see Figure 3.19). The bright line beneath the Rh particles indicates indubitably the formation of oxide. The brighter appearance under the particles indicates a higher O density, that is, a more compact oxide film.

Evidence for hydrogen intercalation can be well noticed after prolonged deposition times. This is a natural consequence of the catalytic activity of the formed metal islands. Figure 3.20 shows HRTEM images of cross-sectional sheets of n-Si(111):H after 20 s of Rh deposition from 5 mM $RhCl_3$ + 0.5 M NaCl at −0.6 V (SCE). It can be seen that the amorphous interfacial film increased up to 50 nm, which can be undoubtedly ascribed to amorphization of the underlying substrate by intercalation in depth of atomic hydrogen generated at the deposited metal particles. The inward diffusion of H was studied by Allongue et al. [67] under cathodic and anodic bias in fluoride solutions by applying nuclear reaction analysis (NRA) and in situ capacitance methods. They observed that the incorporation of H takes places by volume diffusion, for which a diffusion coefficient $D_H \sim 10^{-13}$ cm^{2}s in n-type Si was estimated. Regarding the temporal evolution of the flat band potential, they also inferred the formation of a p$^+$-doped layer. Incorporated H atoms in p-type Si compensate the acceptors, so that the apparent doping concentration of the contaminated layer reduces practically to zero. In the present case, the TEM analysis indicates a large extent of amorphization. According to this experimental evidence, one could infer that the inward diffusion of H leads to the formation of a transition region from p-doped toward a highly disordered Si beneath the metal island.

If we consider that the contaminated layer increases according to $\delta_H = (D_H \, t)^{1/2}$, a thickness of 3.1 nm and 14.1 nm for 1 s and 20 s of diffusion, respectively, can be calculated. According to the results published by Allongue [67], the incorporation of

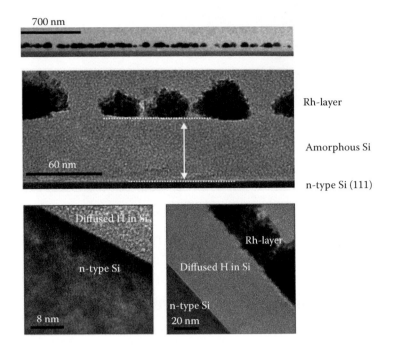

FIGURE 3.20   High-resolution TEM pictures of cross-sectional cuts of n-type Si (111) after longtime deposition (>20 s) at −0.6 V from 5 mM RhCl$_3$ + 0.5 M NaCl.

H gives rise to a concentration of defects of about 10^{20} cm^{-3}, so that defect concentrations in the range of 10^{14} cm^{-2} can be estimated for a hydrogenated layer of 10 nm.

The intercalation of H in the interstitial site of the silicon structure during inward diffusion leads to the formation of a deep acceptor level E_a at 0.06 eV from the conduction band in n-type Si and a deep donor level E_d at 0.52 eV over the valence band in p-type Si [68,69]. Donor levels near the midgap react with the doping levels to form H$^+$. This causes a compensation of majority charge carriers and the shift of the Fermi level toward the midgap. In the case of n-type Si, the formation of a deep donor level close to the conduction band was attributed to the placement of H at an antibonding site of a Si nearest neighbor to a substitutional dopant atom: H-Si···P [70,71].

The inward diffusion of H and the occupancy of interstitial places lead to the breakdown of Si-Si bonds. Amorphous silicon is characterized by a U-like continuous energy distribution of the density of states. The lack of periodicity in the Si lattice brings about an undefined splitting of the energy states of electrons and a large amount of states appears inside the forbidden zone defined for crystalline Si. Instead of band edges, conduction and valence band edges are defined, which instead set an energy threshold between delocalized and localized charge carriers. Hence, electrons with $E > E_{CB}$ and holes with $E < E_{VB}$ are mobiles. Here, it is usual to name the difference E_{CB}-E_{VB} as *mobility band*. This difference is in a-Si:H about 1.7 eV (depending on the H concentration). It was reported elsewhere that the Fermi level of hydrogenated silicon is at 4.3 eV [67]. Let us now analyze the contact of n-type Si

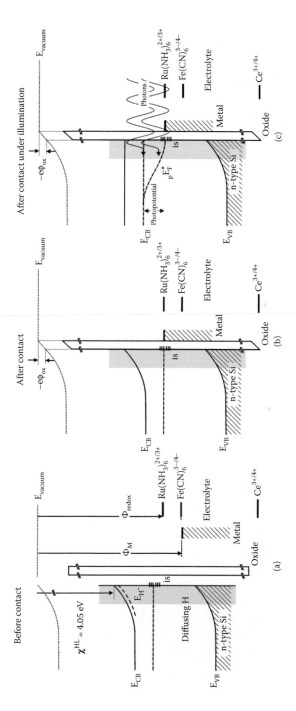

FIGURE 3.21 (a) and (b) Energy band diagrams representing a gedankenexperiment where the n-type Si with a H-contaminated surface is put into contact with metal and electrolyte; (c) the behavior of the interface upon illumination.

with the hydrogenated layer, the metal, and the redox electrolyte. Figure 3.21a and b shows the energy band diagrams of the contact. Here, the assumptions of a transition region and the formation of a thin oxide layer beneath the metal island are made. A *np*-junction is formed at the transition from n-type Si to the H containing zone. Because of the high doping concentration of the hydrogenated layer (10^{18}–10^{20} cm^{-3}), practically no potential changes are expected in the amorphous layer.

The representation of the energy band diagram for the hydrogenated layer follows the Fritzsche–Ovishinsky model (Ref. 121 in Chapter 2), according to which the middle of the band gap indicates the energy limit of occupied states. The Fermi level of the hydrogenated layer remains pinned at the interface states at the contact with the oxide layer. Hence, each modification of the position of the Fermi level is absorbed by a potential change in the thin oxide layer, $\Delta\varphi_{ox}$, and the band bending of the semiconductor remains constant. As a consequence, the photopotential does not change with the redox potential.

The stabilizing oxide formed by electrochemical treatment at a homoepitaxial film of p-type InP on an InP(100) substrate is constituted by a mixture of In_2O_3, $InPO_4$ and $In(PO_3)_3$ with a predominance of the first compound as already discussed in Chapter 2. Figure 3.22 shows a HR-TEM picture of a cross section of such an interface after having deposited Rh by electrochemical reduction of its chloride complex. It can be noted that the amorphous structure of the film contrasts with the crystalline nature of the substrate. It should be noted that the 10 nm thick oxide layer does not hinder the charge transport as inferred from the high performance of this type of photoelectrodes for the light-induced hydrogen evolution. An affinity of 4.5 eV was reported for In_2O_3. Thus, the conduction band of the oxide is located 0.1 eV below the conduction band edge of the substrate. The presence of In_2O_3 in the interfacial film was indicated by surface photoelectron emission spectroscopic analysis, from which also $InPO_4$ and $In(PO_3)_3$ can be identified. The spectroscopic signals indicate the formation of chemical environments as found in corresponding

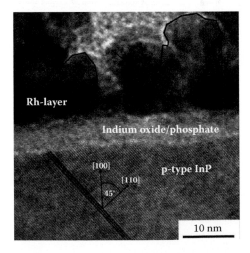

FIGURE 3.22 HR-TEM picture of a cross-sectional cut performed on homoepitaxial (MOVPE) p-type InP film after electrochemical cycling in HCl and Rh deposition.

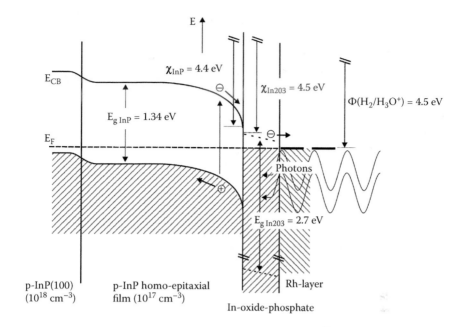

FIGURE 3.23 Energy band diagram showing the electron transfer mechanism on illuminated p-type InP/oxide/Rh junction in acid electrolytes.

pure compounds. Hence, it is valid to affirm that localized states at energies close to the conduction and valence band of indium oxide are present in the otherwise dielectric film. Figure 3.23 shows an energy band diagram for the EMOS interface in equilibrium with the hydrogen evolution reaction. This diagram indicates a maximal achievable photopotential of 0.65 V respective of the hydrogen evolution reaction, as found experimentally.

REFERENCES

1. Harvey, W.W. 1962. The relation between the chemical potential of electrons and energy parameters of the band theory as applied to semiconductors. *J. Phys. Chem. Solids* 23:1545.
2. Reiss, H. and A. Heller. 1985. The absolute potential of the standard hydrogen electrode: A new estimate. *J. Phys. Chem.* 89:4207.
3. Gomer, R. and G. Tryson. 1977. An experimental determination of absolute half-cell EMF's and single ion free energies of salvation. *J. Chem. Phys.* 66:4413.
4. Gurevich, Y.Y. and Y.V. Pleskov. 1982. *Sov. Electrochem.* 18:1315.
5. Trasatti, S. 1986. The absolute electrode potential: An explanatory note. *Pure & Appl. Chem.* 58:955.
6. Maldonado, S. and N.S. Lewis. 2009. Behavior of electrodeposited Cd and Pb Schottky junctions on CH₃-terminated n-Si(111) surfaces. *J. Electrochem. Soc.* 156:H123.
7. Henry de Villeneuve, C., Pinson, J., Bernard, M.C. and P. Allongue. 1997. Electrochemical formation of close-packed phenyl layers on Si(111). *J. Phys. Chem. B* 101:2415.

8. Faucheux, A., Chantal Gouget-Laemmel, A., Henry de Villeneuve, C., Boukherroub, R., Ozanam, F., Allongue, P. and J.-N. Chazalviel. 2006. Well-defined carboxyl-terminated alkyl monolayers grafted onto H-Si(111): Packing density from a combined AFM and quantitative IR study. *Langmuir* 22:153.

9. Buriak, J. M. 2002. Organometallic chemistry on silicon and germanium surfaces. *Chem. Rev.* 102:1271.

10. Schöning, M. J. and H. Lüth. 2001. Novel concepts for silicon-based biosensors. *Phys. Status Solidi A* 185:65.

11. Gartsman, K., Cahen, D., Kadyshevitch, A., Libman, J., Moav, T., Naaman, R., Shanzer, A., Umansky, V. and A. Vilan. 1998. Molecular control of a GaAs transistor. *Chem. Phys. Lett.* 283:301.

12. Lewis, N.S. 1991. An analysis of charge transfer rate constants for semiconductor/liquid interfaces. *Annu. Rev. Phys. Chem.* 42:543.

13. Tan, M.X., Laibinis, P.E., Nguyen, S.T., Kesselman, J.M., Stanton, C.E. and N.S. Lewis. 1994. Principles and applications of semiconductor photoelectrochemistry. *Prog. Inorg. Chem.* 41:21.

14. Harnett, C. K., Satyalakshmi, K. M. and H.G. Craighead. 2000. Low-energy electron-beam patterning of amine-functionalized self-assembled monolayers. *Appl. Phys. Lett.* 76:2466.

15. Livache, T., Bazin, H. and G. Mathis. 1998. Conducting polymers on microelectronic devices as tools for biological analyses. *Clin. Chim. Acta* 278:171.

16. Hunger, R., Fritsche, R., Jaeckel, B., Jaegermann, W., Webb, L.J. and N.S. Lewis. 2005. Chemical and electronic characterization of methyl-terminated Si(111) surfaces by high-resolution synchrotron photoelectron spectroscopy. *Phys. Rev. B* 72:45317.

17. Royea, W.J., Juang, A. and N.S. Lewis. 2000. Preparation of air-stable, low recombination velocity Si(111) surfaces through alkyl termination. *Appl. Phys. Lett.* 77:1988.

18. Weber, R., Winter, B., Hertel, I.V., Stiller, B., Schrader, S., Brehmer, L. and N. Koch. 2003. Photoemission from azobenzene alkanethiol self-assembled monolayers. *J. Phys. Chem. B* 107:7768.

19. Mönch, W. 1990. On the physics of metal-semiconductor interfaces. *Rep. Prog. Phys.* 53:221.

20. Michaelson, H.B. 1978. Relation between an atomic electronegativity scale and the work function. *IBM J. Res. Develop.* 22:72.

21. Gordy, W. and W.J.O. Thomas. 1956. Electronegativities of the elements. *J. Chem. Phys.* 24:439.

22. Mulliken, R.S. 1935. Electronic structures of molecules XI. Electroaffinity, molecular orbitals and dipole moments. *J. Chem. Phys.* 3:573.

23. Heine, V. 1965. Theory of surface states. *Phys. Rev.* A138:1689.

24. Tersoff, J. 1984. Schottky barrier heights and the continuum of gap states. *Phys. Rev. Lett.* 52:465.

25. Louie, S.G., Chelikowsky, J.R. and M.L. Cohen. 1977. Ionicity and the theory of Schottky barriers. *Phys. Rev. B* 15:2154.

26. Louie, S.G. and M.L. Cohen. 1976. Electronic structure of a metal-semiconductor interface. *Phys. Rev. B* 13:2461.

27. Walter, J.P. and M.L. Cohen. 1970. Wave-vector-dependent dielectric function for Si, Ge, GaAs, and ZnSe. *Phys. Rev. B* 2:1821.

28. Mönch, W. 1986. *Festkörperprobleme: Advances in solid state physics,* Vol. 26, P. Grosse (Ed.). Braunschweig: Vieweg.

29. Mönch, W. 1996. Chemical trends of barrier heights in metal-semiconductor contacts: on the theory of the slope parameter. *Appl. Surf. Sci.* 92:367.

30. Sakurai, T. and T. Sugano. 1981. Theory of continuously distributed trap states at Si-SiO$_2$ interfaces. *J. Appl. Phys.* 52:2889.

31. Spicer, W.E., Chye, P.W., Skeath, P.R., Su, C.Y. and I. Lindau. 1979. New and unified model for Schottky barrier and III-V insulator interface states formation. *J. Vac. Sci. Technol.* 16:1422.
32. Hasegawa, H. and H. Ohno. 1986. Unified disorder induced gap state model for insulator-semiconductor and metal-semiconductor interfaces. *J. Vac. Sci. Technol. B* 4:1130.
33. Woodall, J.M. and J.L. Freeouf. 1981. GaAs metallization: some problems and trends. *J. Vac. Sci. Technol.* 19:794.
34. Freeouf, J.L. and J.M. Woodall. 1981. Schottky barriers: an effective work function model. *Appl. Phys. Lett.* 39:727.
35. Hasegawa, H., Sato, T. and T. Hashizume. 1997. Evolution mechanism of nearly pinning-free platinum/n-type indium phosphide interface with a high Schottky barrier height by in situ electrochemical process. *J. Vac. Sci. Technol. B* 15:1227.
36. Khan, S.U.M., Kainthla, R.C. and J.O'M. Bockris. 1987. The redox potential and the Fermi level in solution. *J. Phys. Chem.* 91:5974.
37. Gerischer, H. and W. Eckardt. 1983. Fermi levels in electrolytes and the absolute scale of redox potentials. *Appl. Phys. Lett.* 43:393.
38. Gerischer, H. 1961. *Advances in Electrochemistry and Electrochemical Engineering,* Vol. 1, P. Delahay, Ch. W. Tobias (Ed.). New York: Interscience.
39. Ivey, D.G. 1999. Platinum metals in ohmic contacts to III-V semiconductors. *Platinum Metals Rev.* 43:2.
40. Katz, A. 1992. *Ohmic Contacts to InP and Related Materials in Indium Phosphide and Related Materials: Processing, Technology and Devices.* A. Katz (Ed.). Boston: Artech House.
41. Stremsdoerfer, G., Calais, C., Martin, J.R., Clechet, P. and D. Nguyen. 1990. Low resistance ohmic contacts onto n-InP by palladium electroless bath deposition. *J. Electrochem. Soc.* 137:835.
42. Jian, P., Ivey, D.G., Bruce, R. and G. Knight. 1994. Ohmic contact formation in palladium-based metallizations to n-Type InP. *J. Electron. Mater.* 23:953.
43. Ivey, D.G., Eicher, S., Wingar, S. and T.P. Lester. 1997. Performance of Pd–Ge based ohmic contacts to n-type GaAs. *J. Mater. Sci.: Mater. Electron.* 8:63.
44. Ivey, D.G., Zhang, R., Abid, Z., Eicher, S. and T.P. Lester. 1997. Microstructural analysis of Pd/Pt/Au/Pd ohmic contacts to InGaP/GaAs. *J. Mater. Sci.: Mater. Electron.* 8:281.
45. Ivey, D.G., Jian, P., Eicher, S. and T.P. Lester. 1996. Microstructural analysis of a Au/Pt/Pd/Zn ohmic contact to an AlGaAs/GaAs heterojunction bipolar transistor. *J. Electron. Mater.* 25:1478.
46. Katz, A., Weir, B.E. and W.C. Dautremont-Smith. 1990. Au/Pt/Ti contacts to p-$In_{0.53}Ga_{0.47}As$ and n-InP layers formed by a single metallization common step and rapid thermal processing. *J. Appl. Phys.* 68:1123.
47. Katz, A., Chu, S.N.G., Weir, B.E., Abernathy, C.R., Hobson, W.S., Pearton, S.J. and W. Savin. 1992. Rapid isothermal processing of Pt/Ti contacts to p-type III-V binary and related ternary materials. *IEEE Trans. Electron. Devices* 39:184.
48. Katz, A., Dautremont-Smith, W.C., Chu, S.N.G., Thomas, P.M., Koszi, L.A., Lee, J.W., Riggs, V.G., Brown, R.L., Napholtz, S.G. and J.L. Zilko. 1989. Pt/Ti/p-$In_{0.53}Ga_{0.47}As$ low-resistance nonalloyed ohmic contact formed by rapid thermal processing. *Appl. Phys. Lett.* 54:2306.
49. Ivey, D.G., Jian, P., Wan, L., Bruce, R., Eicher, S. and C. Blaauw. 1991. Pd/Zn/Pd/Au ohmic contacts to p-Type InP. *J. Electron. Mater.* 20:237.
50. Sze, S.M. 1981. *Physics of Semiconductor Devices.* New York: John Wiley & Sons.
51. Schroeder, D. 1991. *An Analytical Model of Non-Ideal Ohmic and Schottky Contacts for Device Simulation. Simulation of Semiconductor Devices and Processes,* Vol. 4, pp. 313. W. Fichtner and D. Aemtner (Eds.). Zurich: Hartung-Gorre Verlag.

52. Smit, G.D., Rogge, S. and T.M. Klapwijk. 2002. Scaling of nano-Schottky-diodes. *Appl. Phys. Lett.* 81:3852.
53. Giannazzo, F., Roccaforte, F., Raineri, V. and S. F. Liotta. 2006. Transport localization in heterogeneous Schottky barriers of quantum-defined metal films. *Europhys. Lett.* 74:686.
54. Donolato, D. 2004. Aproximate analytical solution to the space charge problem in nano-sized Schottky diodes. *J. Appl. Phys.* 95:2184.
55. Smit, G.D.J., Rogge, S. and T.M. Klapwijk. 2002. Enhanced tunneling across nanometer-scale metal-semiconductor interfaces. *Appl. Phys. Lett.* 80:2568.
56. G.D.J. Smit, S. Rogge, J. Caro and T.M. Klapwijk. 2004. Conductance distribution in nanometer-sized semiconductor devices due to dopant statistics. *Phys. Rev. B* 69:035338.
57. Nakato, Y. and H. Tsubomura. 1992. Silicon photoelectrodes modified with ultrafine metal islands. *Electrochim. Acta* 37:897.
58. Yae, S., Kitagaki, M., Hagihara, T., Miyoshi, Y., Matsuda, H., Parkinson, B.A. and Y. Nakato. 2001. Electrochemical deposition of fine Pt particles on n-Si electrodes for efficient photoelectrochemical solar cells. *Electrochim. Acta* 47:345.
59. Nakato, Y., Ueda, K., Yano, H. and H. Tsubomura. 1988. Effect of microscopic discontinuity of metal overlayers on the photovoltages in metal-coated semiconductor-liquid junction photoelectrochemical cells for efficient solar energy conversion. *J. Phys. Chem.* 92:2316.
60. Angermann, H. 2002. Characterization of wet-chemically treated silicon interfaces by surface photovoltage measurements. *Anal. Bioanal. Chem.* 374:676.
61. Aggour, M., Skorupska, K., Stempel Pereira, T., Jungblut, H., Grzanna, J. and H.J. Lewerenz. 2007. Photoactive silicon-based nanostructure by self-organized electrochemical processing. *J. Electrochem. Soc.* 154:H794.
62. Stempel, T., Aggour, M., Munoz, A., Skorupska, K. and H.J. Lewerenz. 2008. *Electrochem. Comm.* 10:1184.
63. Knotter, D.M. 2000. Etching mechanism of vitreous silicon dioxide in HF-based solutions. *J. Am. Chem. Soc.* 122: 4345.
64. Lewerenz, H.J. 1997. Surface scientific aspects in semiconductor electrochemistry. *Chemical Society Reviews* 26:239
65. Rappich, J. and H. J. Lewerenz. 1996. *Thin Solid Films* 276:25.
66. Fultz, B. and J.M. Howe. 2001. *Transmission Electron Microscopy and Diffractometry of Materials.* Berlin: Springer-Verlag.
67. Allongue, P., Henry de Villeneuve, C., Pinsard, L. and M.C. Bernard. 1995. Evidence for hydrogen incorporation during porous silicon formation. *Appl. Phys. Lett.* 67:941.
68. Rizk, R., de Mierry, P., Ballutaud, D., Aucouturier, M. and D. Mathiot. 1991. Hydrogen diffusion and passivation processes in p- and n-type crystalline silicon. *Phys. Rev. B* 44:6141.
69. Pantelides, S.T. 1987. Effect of hydrogen on shallow dopants in crystalline silicon. *Appl. Phys. Lett.* 50:995.
70. Tavendale, A.J., Pearton, S.J. and A.A. Williams. 1990. Evidence for the existence of a negatively charged hydrogen species in plasma treated n-type Si. *Appl. Phys. Lett.* 56:949.
71. Zhu, J., Johnson, N.M. and C. Herring. 1990. Negative-charge state of hydrogen in silicon. *Phys. Rev. B* 41:12354.

4 Electrocatalysis on Nanodimensioned EMOS Contacts

The electron transfer step of photoelectrochemical reactions in light conversion devices designed on the basis of metal–oxide–semiconductor systems depends to a large extent on the complex molecular and electronic structure of the metal–electrolyte interface. As already mentioned, this step is the last of a series of consecutive transfer processes from the semiconductor to the electrolytes upon different interfaces. A comprehensive understanding of transfer processes at metal–electrolyte interfaces requires a description of the metal–electrolyte interface, beginning with the classical concepts, to arrive at a brief revision of the more complex quantum mechanical picture. Reactions of special technological interest, such as the hydrogen evolution and the reduction of CO_2 in carbon fuels, will be treated.

4.1 FUNDAMENTAL ASPECTS OF THE METAL–ELECTROLYTE INTERFACE

A classical representation of the metal–electrolyte interface in aqueous electrolytes is depicted in Figure 4.1. The electrostatic interactions of water molecules with the metal and ions leads to a rupture of the water structure because of a reorientation of the water dipoles. As a consequence, an interfacial gradient of the water electrical permeability appears. The water dipoles in direct contact with ions are completely radial oriented and form the so-called inner solvation shell. There is a transition region between the completely oriented dipoles and the bulk water structure, the extension of which is defined by the Debye-length, and which is referred as the outer solvation shell. The inner and outer solvation shells are characterized by different vibration modes that contribute to the total vibration energy of the charged particle, which determines the energy barrier height in electron transfer processes as explained later. The inner solvation shell limits the approaching distance of the ion to the metal surface to a minimum of 0.2 to 0.5 nm. Thus, completely solvated ions are positioned at the so-called outer Helmholtz plane. In some cases, the solvation energy is overcome by attractive forces of an electrostatic nature, and hence ions get into direct contact with the metal surface. The ionic radius defines the so-called inner Helmholtz plane. This type of ion approach is known as specific adsorption, to distinguish it from approaches involving partial charge transfer or formation of chemical bonding.

The electrostatic interaction of the metal with solvated ions and the water dipoles leads to a charge separation at the interface of a distance of some angstroms and thus

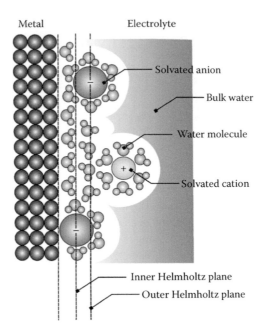

Metal   Electrolyte

— Solvated anion

— Bulk water

— Water molecule

— Solvated cation

— Inner Helmholtz plane
— Outer Helmholtz plane

FIGURE 4.1 Classical representation of the metal–electrolyte interface.

to the appearance of electric fields of up to 10^8 V cm^{-1}. The interface adopts a determined structure of the water and ion distribution to compensate for the charge separation: the double layer. The change of chemical energy derived from the structural modifications is compensated by the electrical work, as required from an invariability of the electrochemical potential in an equilibrated system.

One of the first descriptions of the metal–electrolyte interface was made by Graham in 1947 [1]. According to this model, the interfacial excess of charge at the metal surface is compensated by decaying excess concentration of ions toward solution bulk: the diffuse layer. The region between the metal surface and the closest ion plane (outer Helmholtz plane) is called the compact layer [2]. Therefore, the double layer capacity can be considered as a series connection of two planar capacitors, which represents the electrical behavior of the compact and the diffuse layer, respectively.

A relation between potential and ion concentration in the diffuse layer is given by the Poisson–Boltzmann equation:

$$\nabla^2 \varphi = -4\pi \frac{\rho(\vec{r})}{\varepsilon \varepsilon_0} \tag{4.1}$$

where $\rho(\vec{r})$ is the charge density as a function of the space vector $\vec{r}$, which can be found by applying a Boltzmann-type distribution:

$$\rho(\vec{r}) = \sum_i e C_i(\vec{r}) z_i = \sum_i e z_i C_i^0 e^{-\frac{e z_i}{kT} \varphi(\vec{r})} \tag{4.2}$$

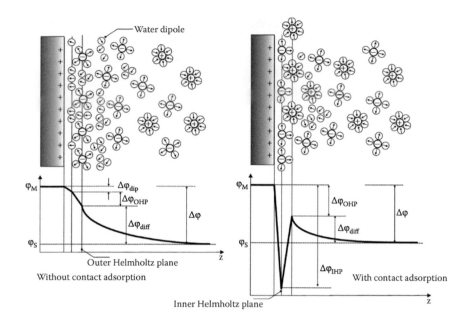

FIGURE 4.2   Potential distribution of a charged metal–electrolyte interface without and with specific adsorption of ions.

where $C_i(\vec{r})$ is the local concentration of the i-ion of charge z_i. The differential equation (4.1) can be solved for the one-dimensional case by using the following identity:

$$\frac{d^2\phi}{dz^2} = \frac{d}{d\phi}\left(\frac{d\phi}{dz}\right)^2 = -\frac{8\pi}{\varepsilon\varepsilon_0}\sum_i e z_i C_i^0 \, e^{-\frac{e z_i}{kT}\phi(z)} \tag{4.3}$$

With the boundary conditions

$$\phi(z \to \infty) = 0; \quad \left(\frac{d\phi}{dz}\right)_{z\to\infty} = 0$$

we have

$$\left(\frac{d\phi}{dz}\right)^2 = -\frac{8\pi kT}{\varepsilon\varepsilon_0}C_i^0\left[e^{\frac{ze}{2kT}\phi(z)} - e^{-\frac{ze}{2kT}\phi(z)}\right]^2 \tag{4.4}$$

$$\left(\frac{d\phi}{dz}\right) = -\sqrt{\frac{32\pi kT C_i^0}{\varepsilon\varepsilon_0}}\,senh\left(\frac{ze}{2kT}\phi(z)\right)$$

which after integration gives

$$\phi(z) = \phi(z=0)e^{-\kappa z}; \quad \kappa = \sqrt{\frac{8\pi C^0(ze)^2}{\varepsilon\varepsilon_0 kT}} \tag{4.5}$$

The term $(1/\kappa)$ can be interpreted as the separation of plates of an equivalent planar capacitor, provided the potential is small enough so that $\sinh(x) \approx x$. The total excess charge in the solution is calculated by applying Gauss's law:

$$q_d = -q_m = \frac{\varepsilon\varepsilon_0}{4\pi}\left(\frac{d\phi}{dz}\right) = -2\sqrt{\frac{\varepsilon\varepsilon_0 kTC_i^0}{2\pi}}\sinh\left(\frac{ze}{2kT}\phi(z)\right) \tag{4.6}$$

where q_d and q_m are the excess charges at the solution and the metal side of the interface, respectively. Thus, we arrive at the following expression for the interfacial capacity:

$$C_d = \frac{\partial q_d}{\partial\phi} = \sqrt{\frac{\varepsilon\varepsilon_0 e^2 C^0}{2\pi kT}}\cosh\left(\frac{ze}{kT}\phi(z)\right) \tag{4.7}$$

Graham's treatment takes in its formulation a combination of the Helmholtz and Gouy–Chapman models. The former describes the interface as a compact double layer. The latter considers a charge compensation by a diffuse layer [3–5]. The representation of the interfacial capacity as a series connection of the capacity of the compact Helmholtz layer and that of the diffuse layer is known as the Stern model [6]. Thus, the total capacity is given by

$$\frac{1}{C} = \frac{1}{C_H} + \frac{1}{C_d} \tag{4.8}$$

For concentrations higher than 0.01 M, a condition met in most practical cases, $C_d \gg C_H$ and the measured capacity corresponds to that of the compact of Helmholtz layer: $C \approx C_H$. The relevance of Eq. (4.7) is found in the prediction of minimum at the so-called potential of zero charge (pzc), at which $q_m = 0$. Today, the interfacial capacitance is measured by means of electrochemical impedance at a high enough frequency so that the response is dominated by the capacitive component of the double layer. This technique replaced the old Lippmann method [7] based on the change of surface tension with the electrode potential and electrolyte concentration, which was also limited to mercury surfaces. As an example, the interfacial capacitance curves reported by Grahame [1], shown in Figure 4.3, were obtained on mercury by using the electrocapillary method in dilute solutions. The curves show a parabolic potential dependence of the capacitance with a minimum at the potential of zero charge. In principle, this constitutes the success of the theory of a diffuse layer, since it allows the determination of the pzc. Hereby, two other aspects should be noted. The first one consists of the appearance of a practically constant capacity at enough negative potential with respect to the pzc. In this potential region, we have $C_d \gg C_H$, and hence $C \approx C_H$. The second aspect is the appearance of a shoulder at potentials more positive than the pzc. This behavior reflects the contact adsorption of anions. Note that the effect appears preponderantly in chloride solutions, a fact that is expected considering the weak solvation of chloride in comparison with the strong solvated F^-. The contact adsorption implies electronic interaction between the anion orbitals and

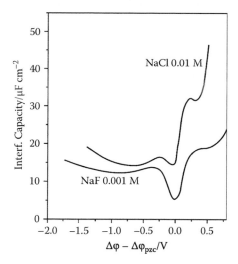

FIGURE 4.3 Capacity-voltage curves of Hg–electrolyte interface in diluted NaCl and NaF solutions [after Grahame (1)].

the metal surface, which introduces important distortions of the potential course at the metal–electrolyte interface, and thus to an increase of the capacity.

The adsorption process also involves a partial elimination of the solvation shell and the expulsion of water molecules adsorbed at the metal surface. This implies that ions with a high hydration number, as for instance F^-, need to overcome a large hydration free energy ($\Delta G_{hydr} = 484$ kJ mol^{-1}) and hence, they are impeded to adsorb. This contrasts with larger ions such as Cl^- ($\Delta G_{hydr} = 341$ kJ mol^{-1}) and I^- ($\Delta G_{hydr} = 273$ kJ mol^{-1}) which easily adsorb.

The specific adsorption occurs with a partial charge transfer and the formation of an interfacial dipole. As already discussed in Chapter 2 for the particular case of Cl^- adsorption on In, the transfer process can be studied by potentiodynamic techniques. Thermodynamically, the adsorption is described by the relation

$$-d\gamma = q_m d\Delta\phi + \Gamma_i d\mu_i = q_m d\Delta\phi + \Gamma_i kTd\ln c_i \qquad (4.9)$$

where γ represents the surface tension and Γ_i is the surface excess concentration of ions, i, at the interface. The change of surface tension with ion adsorption is known as surface pressure π, which for a constant ion concentration can be calculated as

$$\pi = \gamma_{\theta=0} - \gamma_\theta = \int_{\Delta\phi 1}^{\Delta\phi 2} \left[q_m(\theta) - q_m(\theta = 0) \right] d\Delta\phi \qquad (4.10)$$

where $q_m(\theta)$ and $q_m(\theta = 0)$ are the charges transferred to the metal during adsorption and without adsorbing ions, respectively. Thus, the surface ion excess can be calculated from the expression

$$\Gamma_i = \frac{1}{kT}\left[\frac{\partial \pi}{\partial \ln c_i}\right]_{\Delta\phi,\mu_{j\neq i}} \tag{4.11}$$

The charge $q_m(\theta)$ can be experimentally determined by integration of the potentio-dynamic current-voltage curve:

$$q_m = \frac{1}{v_{scan}}\int_{\Delta\phi1}^{\Delta\phi2} i(\Delta\phi)d\Delta\phi \tag{4.12}$$

where v_{scan} is the potential scan rate [8]. The partial charge transfer can be represented by the reaction [9]

$$A^- \rightarrow A^{-(1-qa)} + q_a\,e^- \tag{4.13}$$

The charge q_a is known as the electrosorption number and it can be calculated by

$$q_a = \frac{1}{F}\left(\frac{\partial q_m}{\partial \Gamma_{A^-}}\right)_{\Delta\phi} \tag{4.14}$$

As an example, let us consider the adsorption of Cl$^-$ on Pt(111) surfaces. Figure 4.4 shows cyclovoltammograms performed in potassium perchlorate solution with and without chloride. Perchlorate anions are known not to adsorb specifically, and hence this solution is normally used as a base electrolyte for adsorption studies. The corresponding calculated surface excess and the electrosorption number are shown below as a function of potential. It can be seen that the increase of surface excess in the potential region $0\ V < V < 0.35\ V$ is characterized with $q_a \sim -1$. This indicates a complete transfer of the ion charge during adsorption. At potentials more positive than 0.35, the surface excess decreases together with a stepped change of the electrosorption number from -1 to $+1$. This result reflects the expulsion of adsorbed ions from the surface by a competing formation of an oxide layer according to

$$Pt_s + OH^- \rightarrow Pt\text{-}OH + e^- \tag{4.15}$$

This reaction can be recognized by a wide anodic peak at $V > 0.1\ V$ without chloride. The retarded oxide formation due to the presence of adsorbed chloride leads to a sharp anodic peak at 0.45 V.

The adsorption of anions leads to a modification of the electronic structure at the surface and hence to a surface reconstruction [10], that is, a rearranging of the surface atomic structure. The structural and electronic changes at the surface give rise to a new potential distribution at the interface, thus altering the electron transfer rate of electrochemical reactions, for instance the hydrogen evolution [11] and the reduction of CO_2 [12].

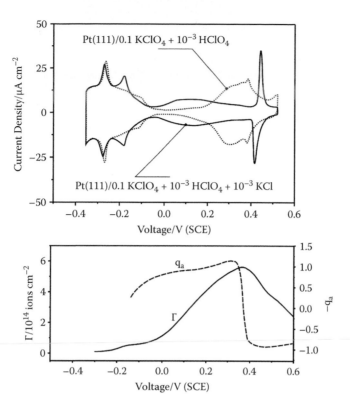

FIGURE 4.4 Adsorption of chloride on Pt(111) surface studied by cyclic voltammetry in perchlorate solutions [after Li and Lipkowski [8]].

Initial attempts to describe the metal–electrolyte interface quantum mechanically use the Jellium model to account for the metal surface [13]. This model basically considers the replacement of positive charge of the atom cores by a homogeneous distribution of charge density, which is represented by a step function as follows:

$$n^+(z) = \begin{cases} \bar{n}, & z \le 0 \\ 0 & z > 0 \end{cases} \tag{4.16}$$

where $\bar{n}$ is the background charge density, normally expressed as a function of the dimensionless electron distance r_s (Wigner–Seitz–Abstand) and the Bohr radius

$$r_0 = \frac{\hbar^2}{m_e e^2} = 0.529 \text{Å}$$

$$\bar{n} = \frac{3}{4\pi} \frac{1}{(r_0 r_s)^3} \tag{4.17}$$

The profile of the electron density was calculated by Lang and Kohn [14] by applying the density functional theory with the assumption of local density approximation. The density functional theory is based on the theorem of Hohenberg and Kohn, after which the system energy is determined by the electron density $n(\vec{r})$, so that the energy becomes a function of the electron density $E = E[n(\vec{r})]$. Hence, this function reaches a minimum as the $n(\vec{r})$ is in its ground state, which can be found by solving a system consisting of a series of one-electron Schrödinger equations, also known as Kohn–Sham equations [15]:

$$-\frac{\hbar}{2m_e}\nabla^2\psi_i + v_{eff}(\vec{r})\psi_i(\vec{r}) = E_i\psi_i(\vec{r}), \qquad (4.18)$$

where $v_{eff}(\vec{r})$ is the effective one-electron potential, given by the expression

$$v_{eff}(\vec{r}) = -e^2\sum_{\vec{R}}\frac{Z}{|\vec{r}-\vec{R}|} + e^2\int\frac{n(\vec{r}')}{|\vec{r}-\vec{r}'|}d\vec{r}' + v_{ex}[n(\vec{r})] \qquad (4.19)$$

where $\vec{R}$ gives the position of atom cores and $v_{ex}(\vec{r})$ is the exchange correlation potential. The electron density results from the sum of the probabilities of each electron given by the square one-electron wave functions:

$$n(\vec{r}) = \sum_i|\psi_i(\vec{r})|^2$$

The local density approximation provides a simple method for calculating $v_{ex}(\vec{r})$ as the derivative of the correlation energy $\partial E_{ex}/\partial n(\vec{r})$. In this approximation, E_{ex} for a local electron density $n(\vec{r})$ in an inhomogeneous electron distribution is equal to the energy of a homogeneous gas of electrons with the same density, and thus

$$E_{ex}[n(\vec{r})] = \int n(\vec{r})\varepsilon_{ex}[n(\vec{r})]dr' \qquad (4.20)$$

where $\varepsilon_{ex}[n(\vec{r})]$ is the exchange correlation energy per electron. The result of these complex calculations is shown graphically in Figure 4.5. It should be noted, that the electrons leak out from the positive core lattice 0.1 to 0.3 nm into the vacuum, so that an electrostatic dipole layer arises at the metal surface because of the charge imbalance. Moreover, damped oscillations toward the metal bulk with a period of $\pi/(3\pi\bar{n})^{1/3}$ appear. This phenomenon, known as Friedel oscillations [17], can be experimentally resolved by scanning tunneling microscopy (STM) at low temperatures. The effect arises from an abrupt change of the positive charge background and can be also observed around atomic defects in metals [18], as for instance after the deposition of metal nanoclusters onto single crystal surfaces. The calculated effective potential course at the metal–vacuum interface is depicted in Figure 4.5b. It

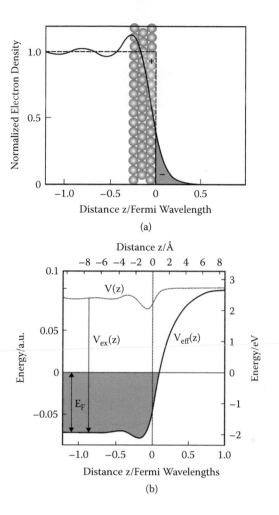

FIGURE 4.5 (a) Electron density profile as calculated with the Jellium model, taking $r_s = 2$. The distance is given in the Fermi wavelength $2\pi/(3\pi\bar{n})^{1/3}$, equal to 0.345 nm for $r_s = 2$; (b) spatial distribution of the electrostatic potential $V(z)$ calculated for $r_s = 5$ (16).

is composed of an electrostatic contribution $v(z)$ and an interaction potential $v_{ex}(z)$. These potentials define the metal work function as follows:

$$\Phi = v_{eff}(+\infty) - v_{eff}(-\infty) - E_F \tag{4.21}$$

where the Fermi energy is calculated by the expression

$$E_F = \frac{\hbar}{2m_e}(3\pi^2\bar{n})^{2/3}.$$

The Jellium model constitutes the basis for quantum mechanical calculations of the metal–electrolyte interface. For instance, molecular dynamics calculations on

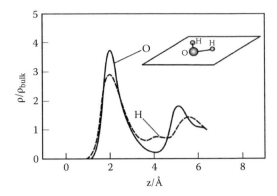

FIGURE 4.6 Normalized O and H atomic density as a function of the distance to the Pt(100) surface as calculated by using molecular dynamics methods (after Spohr et al., 1996 [23]).

the system of Pt(111)- and Ag(100)-water interfaces [19,20] indicate a preferential orientation of the water molecule with the negative pole toward metal for a zero charge at the metal surface [21,22]. The coincidence of the normalized atomic density of O and H at 2 nm from the Pt(100) surface, as shown in Figure 4.6, indicates that water molecules are oriented practically horizontally. Further oscillations with a period of 0.3 nm give evidence for an ordered water structure up to a distance of 1 nm away from the interface. According to these calculations, the ordered structure extends further to 2 nm for an electric field of 4×10^8 V cm^{-2}.

The preferential water orientation at the metal–electrolyte interface was indirectly indicated from the appearance of a maximum of the interfacial entropy at a negative metal excess surface charge [24] (see Figure 4.7). The interfacial entropy can be experimentally determined by the temperature dependence of the surface tension:

$$\Gamma_S = -\left(\frac{\partial \gamma}{\partial T}\right)_{P,\mu,V} \tag{4.22}$$

The water contribution to the entropy is given by

$$\Gamma_S^W = \Gamma_S - \sum_i \Gamma_i \bar{S}_i \tag{4.23}$$

where Γ_i is the excess surface concentration of the ion i and $\bar{S}_i$ is the corresponding partial entropy. The interaction between the metal surface and water dipoles leads to a change of the water structure, and thus entropy changes, with contributions arising from the molecular configurations and the rotation, libration, and vibration modes. The experimental results can be modeled by means of the statistical thermodynamics by the formulation of an appropriate distribution function for each entropy contribution. The configuration entropy (Eq. 4.24) has the major contribution to the total water surface entropy, and it is determined by the number of possible configuration

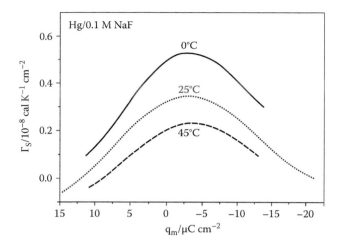

FIGURE 4.7 Surface excess entropy of the metal–electrolyte interface serves as a function of the surface charge in absence of specific adsorption. The presence of an entropy maximum at slight negative surface charge indicates a preferential orientation of water dipoles with the negative pole (oxygen) toward surface (Harrison et al., 1973 [25]).

combinations of oriented dipoles $N_\downarrow$ and $N_\uparrow$. An additional contribution is given by the number of combinations of monomers N_m and dimers N_d, assuming the presence of associated water molecules.

$$S_{konf} = k \ln \Omega(N_\downarrow, N_\uparrow) + k \ln \Omega(N_m, N_d) \qquad (4.24)$$

The first term of this equation is a strong function of the metal surface charge and provides the parabolic form to the whole entropy. The maximum of the configuration entropy is reached at a zero total orientation, that is, $N_\downarrow = N_\uparrow$. It can be demonstrated that at the maximum, the following relation holds:

$$q_{M|S=S_m} = (\Delta G_\uparrow - \Delta G_\downarrow)\frac{\varepsilon_0}{8\pi\bar{\mu}} \qquad (4.25)$$

where $\Delta G_\uparrow - \Delta G_\downarrow$ [eV] represents the change of the adsorption Gibbs free energy for the inversion of the orientation, $\bar{\mu}$ [C cm] is the dipole moment of the water molecule, and ε_0 [F cm^{-1}] is the permittivity of a vacuum.

The preferential orientation of water molecules at the metal surface requires a breakdown of hydrogen bonds in the water structure, leading to a change of the profile of the dielectric constant. Calculations based on a correlation of the number of broken bonds with the dielectric constant have shown that the dielectric constant increases linearly from 5 at the surface to 55 at the Helmholtz plane. It decreases further asymptotically to 80 toward the water bulk [26,27].

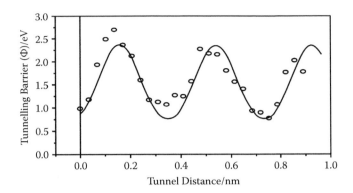

FIGURE 4.8 Tunnel barrier as a function of the distance away from the Ag(111)/HClO₄ interface [after Hugelmann et al. [28]].

The structure of the metal–electrolyte interface was also investigated by Hugelmann and Schindler by means of scanning tunnel spectroscopy experiments [28]. The measured tunnel current at the Au(111)/0.01 M HClO₄ interface shows a modulation with a periodicity of 0.32 nm to 0.35 nm with an increasing distance of the scanning tip from the surface. This seems to agree with the width of water layers in an oriented structure. This result indicates a modulation of the potential barrier calculated with Eq. (4.26) with the tunnel distance, where each minimum corresponds to an entire number of water layer (see Figure 4.8). This would mean that the tunnel process occurs upon the energy states of hydrogen in the water molecule.

$$\Phi = \frac{\hbar^2}{8m_e}\left(\frac{\partial\ln J_T}{\partial z}\right)^2 \tag{4.26}$$

A similar experiment was carried out in sulfate solutions on Au surfaces [29], where an oscillatory course of the tunnel current with a period of 0.32 nm was also observed. These results were correlated with theoretical calculations of the interface structure by applying the density functional theory. Here, it was assumed that sulfate ions absorb

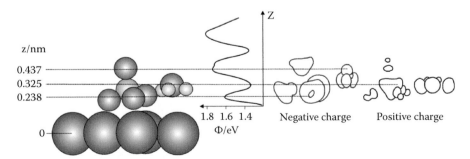

FIGURE 4.9 Correlation of the charge distribution (calculated by applying the density functional theory) and the measured effective tunnel barrier at the Ag(111)/H₂SO₄ interface at V = 0.8 V (vs. SCE) [after Simeone et al. [29]].

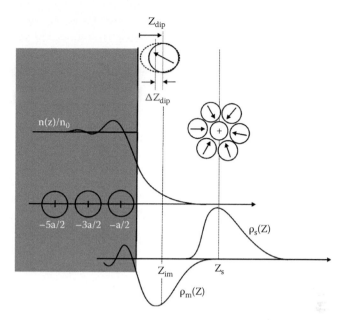

FIGURE 4.10 Schematic of the Amokrane and Dadiale model for describing the metal–electrolyte interface based on quantum mechanics concepts.

covalently on the metal surface. On the other hand, the hydroniums, which are associated to the oxygen atom of the sulfate ion by a hydrogen bond, are oriented preferentially with their hydrogen atoms toward the surface. The comparison of the results of the tunnel current spectroscopy with those of the density functional theory calculations offers the searched for relation between the tunnel barrier and the atomic structure (see Figure 4.9). The absolute distance between the scanning tip and the substrate is determined by assigning the first barrier maximum to the center of the negative charge of the sulfate ion. This procedure is justified if one takes into account that the effective tunnel barrier reflects the distribution of the negative charge.

Now, we will explore the interpretation of the interfacial capacity from a quantum mechanical point of view (see Figure 4.10). The analysis of the interfacial capacitance can be separated into the contribution of the metal and that of the solution as two capacitance elements connected in series, C_m and C_s, respectively. Hence, the capacitance of the Helmholtz layer is given by

$$\frac{1}{C_H} = \frac{1}{C_m} + \frac{1}{C_s} \tag{4.27}$$

According to the Jellium model, a dipole layer is formed at the metal surface. Thus, a surface potential χ_m arises, which can be calculated by integrating the net density of excess charge ρ_m [C cm^{-3}] [30]:

$$\chi_m(\sigma_m) = -\frac{1}{\varepsilon\varepsilon_0} \int_{-\infty}^{+\infty} \rho_m(z)z dz = \chi_{m(\rho_M=0)} + \delta\chi_{m(\rho_M)} \tag{4.28}$$

The surface potential can be considered the sum of the surface potential of an uncharged metal plus a term accounting for the change introduced by image charges. Following this idea, Lang and Kohn [31] introduced the concept of an image plane, which is interpreted as the gravity center of charge changes induced by an electric field. The position of this plane is given by

$$z_{im} = \int_{-\infty}^{+\infty} \frac{\rho_m(z, \sigma_m + \delta\sigma_m) - \rho_m(z, \sigma_m)}{\delta\sigma_m} z \, dz \qquad (4.29)$$

where σ_m [C cm^{-2}] is the charge induced at the metal surface. In a similar way, a plane for the charge excess in the solution can be defined, which position is given by

$$z_s = \int_{-\infty}^{+\infty} \frac{\rho_s(z, \sigma_s + \delta\sigma_s) - \rho_s(z, \sigma_s)}{\delta\sigma_s} z \, dz \qquad (4.30)$$

According to this model, one can calculate the interfacial capacity considering a planar capacitor between planes at z_{im} and z_s. Hence,

$$\frac{1}{C_H} = \frac{z_s - z_{im}}{\varepsilon\varepsilon_0} \qquad (4.31)$$

The distance $(z_s - z_m)$ is additionally changed because of the deformation of water molecules caused by electrostriction and polarization effects, which lead to a modification of the charge distribution at the interface. Assuming that the water molecules behave as hard spheres, a closest position for them can be defined: z_{dip}. Due to electrostriction effects, this distance varies with the surface charge according to an inverse parabolic function $\Delta z_{dip} \propto (\sigma - \sigma^0)^{-n}$, where $\sigma = \sigma_m = -\sigma_s$ is the surface charge density at the interface and σ^0 is the charge density at the function minimum.

On the other hand, polarization effects can be included in the capacity calculation by the following expression:

$$\Delta z_s = \frac{\partial}{\partial\sigma} \int_{-\infty}^{+\infty} P^{dis}(z) + P^{or}(z) dz \qquad (4.32)$$

where P^{dis} and P^{or} [C cm^{-2}] represent the deformation and the orientation polarization of adsorbed water per unit length, respectively. Thus, the inclusion of these effects in the formulation of the double-layer capacity results in the expression

$$\frac{1}{C_H} = \frac{z_s + \Delta z_{dip} - z_{im}}{\varepsilon\varepsilon_0} - \frac{\partial}{\partial\sigma} \int_{-\infty}^{+\infty} \frac{P^{dis}(z) + P^{or}(z)}{\varepsilon\varepsilon_0} dz \qquad (4.33)$$

In this equation one can see a contribution from the metal, given by

$$\frac{1}{C_m} = \frac{\Delta z_{dip} - z_{im}}{\varepsilon\varepsilon_0} \tag{4.34}$$

and a contribution from the solution:

$$\frac{1}{C_s} = \frac{z_s}{\varepsilon\varepsilon_0} - \frac{\partial}{\partial\sigma} \int_{-\infty}^{+\infty} \frac{P^{dis}(z) + P^{or}(z)}{\varepsilon\varepsilon_0} \, dz \tag{4.35}$$

This model represents an important advance in the understanding of the metal-electrolyte interface, since it includes the influence of metal properties in the expression of the capacity. For instance, the experimentally determined capacity for different metals in concentrated solutions, show parabolic curves with a maximum at positive surface charges. In fact, these results cannot be described by the Gouy–Chapman model, since it loses its validity under nondiluted solutions.

This model is a semi-empirical method of calculus; the total capacity is an experimentally accessible quantity whereas the metal contribution is calculated by applying the Jellium model. Hence, the contribution of the solution is calculated as $1/C_s = 1/C_H - 1/C_m$. Figure 4.11 depicts the solution contribution to the capacity. It should be noted that a maximum of C_s appears at slight negative charged surface, in agreement with the picture shown by the surface excess entropy. The bell shape of the solution capacity is related to the reorientation of the water dipoles, the surface charge, and the polarization changes of the water layer, $P(z)$. All these effects can be better analyzed thanks to the separation of the metal and solution contributions possible with this model.

4.2 THE ELECTRON TRANSFER AT METAL–ELECTROLYTE INTERFACES: FROM BUTLER–VOLMER TO QUANTUM MECHANICAL CONCEPTS

The electron transfer rate in electrochemical processes can be described in a phenomenological way by the Butler–Volmer equation. This expression is derived from the general activation state theory for chemical reactions and is also based on thermodynamic concepts. Let us consider the general electrochemical equation:

$$Ox + e^- \underset{k_{ox}}{\overset{k_{red}}{\rightleftarrows}} Red \tag{4.36}$$

The total current is thus given by

$$j = ev \left\{ c_{ox} \exp\left[-\frac{\Delta\mu_{red}^{\#}(\phi)}{kT}\right] \exp\left[-e\frac{\alpha\Delta\phi}{kT}\right] - \right. \tag{4.37}$$

$$\left. - e c_{red} k_0 \exp\left[-\frac{\Delta\mu_{ox}^{\#}(\phi)}{kT}\right] \exp\left[e\frac{(1-\alpha)\Delta\phi}{kT}\right] \right\}$$

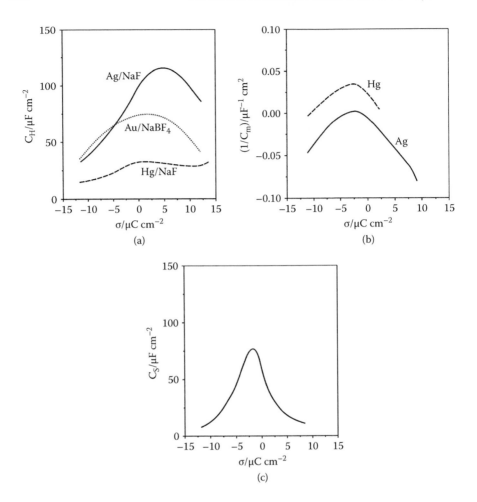

FIGURE 4.11   (a) Interfacial capacity as a function of the surface charge for different metals without contact adsorption in concentrated electrolytes (compact double layer); (b) inverse of the metal capacity contribution $1/C_m$ as a function of the surface charge; (c) dependence of C_s on the surface charge [after Amokrane and Badiali [30]].

where $\Delta\mu_{red}^{\#}(\phi)$ and $\Delta\mu_{ox}^{\#}(\phi)$ are the activation energies for the reduction and the oxidation process at a potential ϕ, respectively. $\Delta\phi = \phi - \phi_0$ is the deviation of the potential from that in equilibrium, and v is the pre-exponential factor accounting for the attempt frequency of the reacting species. The transfer coefficient α, also known as the symmetry factor, can be considered as the fraction of interfacial potential drop at the point where the activation complex is formed (maximum of the potential energy profile). The potential energy profiles are constructed by assuming Morse potential profiles for the oxidizing and reducing species. The success of this interpretation is due to the reproduction of the experimentally observed linear $\log j \propto \Delta\phi$ relation, referred to as Tafel behavior. The asymmetry coefficient can be calculated from the slope of the measured "Tafel lines" at potentials $|\Delta\phi| > kT/e = 25\,\text{mV}$.

In the case of multielectron reactions, the experimentally obtained transfer coefficient is a combination of the individual transfer coefficients for each one-electron step. Hence, different transfer coefficients for the cathodic and the anodic branch of the Tafel diagram, $\log j$ vs. $\Delta\varphi$, are expected; that is, $\alpha_{red} \neq (1 - \alpha_{ox})$. The values of the apparent transfer coefficients serve as the first criterion for elucidating the reaction mechanism, specially by the question what of the different reaction steps is the rate determining.

The Butler–Volmer equation is usually presented in the simplified form:

$$j = j_0 \left\{ \exp\left[-e \frac{\alpha\eta}{kT} \right] - \exp\left[e \frac{(1-\alpha)\eta}{kT} \right] \right\} \tag{4.38}$$

with $\eta = \phi - \phi_0$ as the applied overtension, and j_0 as the exchange current density. This latter quantity represents the rate of electron transfer at reaction equilibrium (i.e., for $j = 0$). Under these conditions, it follows $\Delta\mu_{red}^{\#}(\phi_0) = \Delta\mu_{ox}^{\#}(\phi_0) = \Delta\mu_0^{\#}$, and hence

$$j_0 = ev \exp\left[-\frac{\Delta\mu_0^{\#}}{kT} \right] c_{red}^{1-\alpha} c_{ox}^{\alpha} \tag{4.39}$$

The exchange current density serves as a guide for comparing the catalytic properties of different materials for a determined electrochemical reaction. We will return to this quantity in the next subsection.

Quantum mechanics describes in a general way the electron transfer at the interface separating two phases with the so-called golden law. According to this law, the rate of electron exchange, k_{et}, from an initial state in a solid to a final state in the solution is given by

$$k_{et} = \frac{2\pi}{\hbar} \int f(E_{\vec{k}}) \sum_{\vec{k}} \left| V_{\vec{k}f} \right|^2 \delta\left[E_{\vec{k}} - E_f \right] dE \tag{4.40}$$

where the wave vector $\vec{k}$ represents the electronic state in the solid, and f that in the solution. In this equation, $f(E_{\vec{k}})$ represents the Fermi–Dirac distribution function, $\left| V_{\vec{k}f} \right|$ the coupling matrix between the initial and final electronic states, and $\delta\left[E_{\vec{k}} - E_f \right]$ the delta function, which accounts for the condition of an isoenergetic transfer. The coupling matrix can be interpreted as the probability that the electron exchange occurs and is the result of a mixing of the wave functions of the initial and final states, $\Psi_{\vec{k}}$ and Ψ_f, by the Hamiltonian $H_{\vec{k}f}$:

$$\left| V_{\vec{k}f} \right| = \left\langle \psi_{\vec{k}} \middle| H_{\vec{k}f} \middle| \psi_f \right\rangle \tag{4.41}$$

Since we are dealing with molecules, the wave functions must include the quantum state of the nucleus. In the so called Born–Oppenheimer approximation [32] the electron and the nucleus wave functions are considered as uncoupled, and hence the molecular wave function can be simply expressed as the product of the electron and the nucleus wave function: $\psi_M(\vec{r}, \vec{R}) = \psi_e(\vec{r}, \vec{R}) \times \psi_n(\vec{R})$. As a result of this principle, the coupling matrix $\left|V_{kf}\right|$ can be written as the product of a perturbation Hamiltonian $\left|H_{kf}\right|$ and the vibration Franck–Condon factor $\left\langle \psi_{nk} | \psi_{nf} \right\rangle$, which accounts for the mixing of the nuclear wave functions:

$$\left|V_{kf}\right| = \left|H_{kf}\right| \left\langle \psi_{nk} | \psi_{nf} \right\rangle \tag{4.42}$$

The quantum mechanical description of the electron exchange revolves around the formulation of an adequate Hamiltonian. The Hamiltonian is normally presented as the sum of a series of terms for each of the subsystems constituting the analyzed transfer system: the electron acceptor in the solution H_e [33], the electron in the metal H_m, the transfer Hamiltonian H_t [33], and the Hamiltonians for the energetic fluctuations (librational and oscillatory modes) of the electron acceptor in the solution H_s. Thus,

$$H_T = \sum_i H_i$$

Electron acceptors in the solution processes are solvated ions or metal complexes. The ion solvation shell can be considered as formed by a first and a second hydration layer. The first solvation layer is characterized by completely oriented water dipoles strongly bound to the central ion by electrostatic forces. The secondary solvation layer embraces the transition zone from the primary layer to the undisturbed bulk water structure. The number of bounded water dipoles is referred to as the hydration number, which can be determined indirectly by their influence on different solution properties, such as ion mobility [34], solution entropy [35], and solution compressibility [36], as well as by means of diffraction [37] and spectroscopic methods [38]. As a rule of thumb, this number is proportional to the inverse of the ion radius. In the case of transition metals, the hydration number is instead given by the coordination number, which depends on the particular orbital structure.

The energy fluctuations caused by molecular librational motion and vibration, as well as by electronic polarization of the solvent molecules, lead to an energy distribution of the electronic states in the solution. Librational motion and vibration fluctuation modes are in general much slower compared to the electron transfer rate. Polarization effects, however, show a lower time constant than that for the electron transfer.

Slow fluctuation modes can be described as classical harmonic oscillators that are subdivided into two categories: those occurring inside the solvation layer (inner sphere) and those appearing beyond the first solvation layer. The association of vibration modes to the molecular structure around the ion species is schematically represented in Figure 4.12. Water molecules bounded to the central ion by electrostatic forces connect that ion to the surrounding water structure by hydrogen bonds. In the case of metal complexes, some of the coordinated water molecules are replaced by ligands such as Cl⁻, NH_3, or CN⁻. The libration and vibration modes of the internal ion structure and those of the solvation shell (shown in Figure 4.12 as a shaded

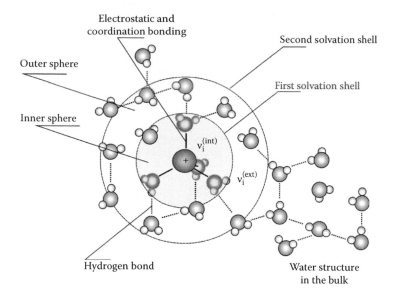

Electrostatic and
coordination bonding

Second solvation shell

Outer sphere

First solvation shell

Inner sphere

$v_i^{(int)}$

$v_i^{(ext)}$

Hydrogen bond

Water structure
in the bulk

FIGURE 4.12 (**See color insert.**) Schematic depicting the inner and outer solvation shells of a cation. The internal frequency modes are represented by v_i. The extent of coupling of ion fluctuations with the surrounding water structure is given by the interaction constant g_i.

sphere) can be described as a phonon bad, that is, as harmonic oscillators. Hence, the Hamiltonian adopts the form

$$H_S = \frac{1}{2} \sum_i \hbar v_i (\bar{p}_i^2 + \bar{r}_i^2) \qquad (4.43)$$

where $\bar{p}_i$ and $\bar{r}_i$ are the dimensionless momenta and coordinates, respectively: $\bar{p}_i = p_i / (\hbar m_i v_i)^{1/2}$, $\bar{r}_i = r_i (m_i v_i / \hbar)^{1/2}$. The magnitude of influence of fluctuation modes on the electron energy level in the ion is introduced by a further Hamiltonian:

$$H_{int} = (z_{ion} - n) \sum_i \hbar v_i g_i r_i \qquad (4.44)$$

where z_{ion} is the valence of the ion and g_i is the coupling constant corresponding to the vibration mode i. n accounts for the number operator, which indicates whether the electron is present on the reactant.

A particular electron transfer process can be analyzed in terms of two limit cases: the diabatic and the adiabatic case. The former considers that the time constant of the faster phonons is smaller than that of the electron exchange. Thus, there is here no chance that electronic equilibrium between metal and redox systems can be set during the transfer process. This situation is usually found in systems for which the interaction of solution species and substrate, quantified by the matrix element V_k, is very weak. Hence, the transfer Hamiltonian is small and can be calculated according to the perturbation method. Applying a first order perturbation form for the transfer Hamiltonian, the following expression for the transfer rate can be obtained [39,40]:

$$j = \frac{4e}{\hbar} \operatorname{Re}\left\{ \int_{-\infty}^{0} dt \sum_{k} |V_k|^2 \exp\left[i \frac{e\eta - \left(E(\vec{k}) - E_F\right)}{\hbar} \right] f\left[E(\vec{k})\right] W(z) \right\} \qquad (4.45)$$

where $W(z)$ is the so-called generation function accounting for the reorganization of the inner and outer solvation shells during the transfer step. Formally, this function is given by

$$W(z) = \frac{Tr\left[\exp\left(-\frac{H_S^*}{kT} \right) \exp\left(-i \frac{zH_S^*}{\hbar} \right) \right]}{Tr\left[\exp\left(-\frac{H_S^*}{kT} \right) \right]} (41) \qquad (4.46)$$

η accounts for the reaction overtension, given by the difference between the energy at which the transfer occurs and the energy of the electron acceptor. Regarding the molecular fluctuations as classical harmonic oscillators, the function $W(z)$ reduces to

$$W(z) = \exp\left[-\frac{iz\lambda}{\hbar} - \frac{z^2 \lambda kT}{\hbar^2} \right] \qquad (4.47)$$

where λ represents the energy for the reorganization of the solvation shell. This parameter reflects the degree of interaction between the acceptor species in the solution and the surrounding solvent. Each fluctuation mode is characterized by its eigenfrequency and the coupling constant g_i:

$$\lambda = \frac{1}{2} \sum_{i} \hbar v_i g_i^2 \qquad (4.48)$$

Equation (4.45) can be further simplified by replacing the sum upon the moment $\vec{k}$ by an integral upon energy by introducing the electron density of state $\rho(E)$. Moreover, the coupling matrix $|V_k|$ is replaced by the mean value $|\overline{V}|$ obtained after the integration of impulse and energy. This approximation is justified by the fact that the strongest electronic coupling is found perpendicular to the surface; on the other hand, the major contribution to the integral upon energy is provided by the energy states closest to the Fermi level. Hence, we have

$$j = \frac{2e}{\hbar} \sqrt{\frac{\pi}{\lambda kT}} |\overline{V}|^2 \rho(E_F) \int f(E) \exp\left[-\frac{(\lambda + E - E_F - e\eta)^2}{4\lambda kT} \right] dE \qquad (4.49)$$

This equation is very similar in its structure to that derived considering a classical treatment of molecular fluctuations as harmonic oscillators:

$$j = ev \int \rho(E) f(E) \exp\left[-\frac{(\lambda + E - E_F - e\eta)^2}{4\lambda kT} \right] dE \qquad (4.50)$$

Equation (4.50) was formulated by Marcus [42] and improved upon by Gerischer (see refs. 20, 21, and 147 in Chapter 1) in terms of probabilistic concepts. Marcus related the exponential term to the state density of the acceptor in the solution by comparing with the classical formulation of the rate constant of a thermal-activated reaction:

$$k_{et} = v \exp\left(-\frac{\Delta \tilde{\mu}_\alpha}{kT} \right).$$

The pre-exponential factor v remained, however, without explicit interpretation and was associated with the frequency of the oscillations of the electron acceptor perpendicular to the surface, also known as attempt frequency. The Marcus–Gerischer concept is limited by attempting to describe some quantum mechanical phenomena of the electron transfer. However, this expression constitutes an invaluable tool for the qualitative elucidation and prediction of electrochemical processes at metal- and semiconductor-electrolyte interfaces.

A quantum mechanical interpretation of the pre-exponential factor v of Eq. (4.50) can be derived by comparing it with Eq. (4.49). Thus, we have

$$v = \frac{2}{\hbar} \sqrt{\frac{\pi}{\lambda kT}} \left| \overline{V} \right|^2 \rho(E_F) \qquad (4.51)$$

This expression offers a functionality of the attempt frequency with the electronic properties of the metal and the electronic coupling of the electron acceptor with the substrate.

In the case of an electron transfer in the adiabatic limit, the time constant of the solvent fluctuations is much longer than that of the electron exchange. Thus, electronic equilibrium is established during the reorganization of the solvation shell. Accordingly, the electron transfer may be described in terms of the activation energy theory. Adopting parabolic potential surfaces, the following expression results:

$$v = \frac{1}{2\pi} \sqrt{\frac{\sum_i v_i^2 \lambda_i}{\lambda}}; \quad \lambda_i = \frac{\hbar v_i g_i^2}{2} \qquad (4.52)$$

Although the quantum mechanical treatment represents a large progress for the description of elemental transfer reactions, its application in the description of complex reactions involving intermediate adsorption steps, for instance, the hydrogen evolution or the CO_2 reduction may become extremely cumbersome.

4.3 REACTION MECHANISMS IN SELECTED SYSTEMS: HYDROGEN EVOLUTION AND CO_2 REDUCTION

Hydrogen evolution and CO_2 reduction are the most intensive investigated electrochemical processes for the conversion of sunlight into chemical energy [43–46]. In both cases, the reaction path involves a series of one-electron steps with reactant and product adsorption. Hydrogen evolution is undoubtedly the most investigated electrochemical reaction. The first reports of cathodic proton reduction can be found in the works of Tafel from 1905 [47]. Since this time, a large number of contributions to clarify the reaction mechanism have been published. From the long list, we can mention the pioneering works of Gurney in 1932 [48], Horiuti and Polanyi in 1935 [49], Butler in 1936 [50], Conway and Bockris in 1957 [51], Marcus in 1965 [52], Conway in 1964 [53], Trasatti in 1972 [54], and Enyo in 1983 [55].

The reaction path consists basically of two consecutive steps: the electrosorption of atomic hydrogen (step I), followed by desorption of molecular hydrogen by path II or path II*.

$$H_3O + e^-_{metal} \rightarrow H_{ads\ metal}\ (I) \tag{4.53}$$

$$H_{ads\ metal} + H_3O^+ + e^-_{metal} \rightarrow H_2O + H_2\ (II) \tag{4.54}$$

$$H_{ads\ metal} + H_{ads\ metal} \rightarrow H_2\ (II^*) \tag{4.55}$$

The outstanding electrochemical activity of metals of the platinum group (Pt, Rh, Ir, Os) can basically be ascribed to the strong adsorption of atomic hydrogen in the first step of the reaction path. Intuitively, one understands that a strong enough metal–hydrogen bond should reduce the activation energy. Hence, the first reaction step (I) accelerates. A too strong bonding, however, should slow down the subsequent desorption step (II or II*). This qualitative interpretation was developed after discovery of the so-called volcano-type curve obtained by graphing the exchange current density on different metals as a function of the M-H binding energy (Figure 4.13) [56]. The values of M-H binding energies were obtained from electrochemical experiments.

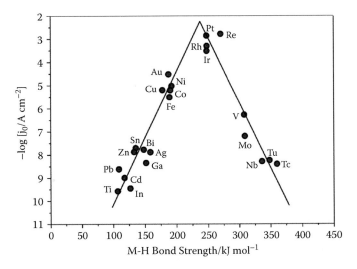

FIGURE 4.13 Exchange current density of the hydrogen evolution reaction for different metals as a function of M-H binding energy.

The influence of the metal on the reaction rate pointed out by this type of diagram gives birth to the concept of *electrocatalysis*.

Considerable progress in the elucidation of the proton reduction process has been made by applying quantum mechanical concepts. A first approach is depicted in Figure 4.14. The electronic energy levels are quantized within the potential hole given by the Morse curve of the hydronium group. Initially, the reducing proton is situated close to the surface at its ground state characterized by a frequency ν_0. Thermal fluctuations drive the system energy to higher excitation states. If the system overcomes the activation barrier, it passes to the final states as adsorbed H [57]. After the interpretation of Conway [57], an increase of the M-H binding strength

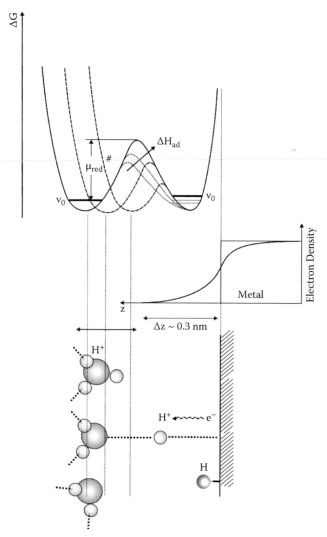

FIGURE 4.14 Schematic depicting the proton reduction (hydronium) using quantum mechanical concepts.

leads to an increment of the steepness of the Morse curve for the adsorbed state as shown in Figure 4.14 by the dotted line. Therefore, it is expected that an increase of the adsorption strength results in an increase of the activation energy barrier μ_{red}^* and a shift of the transition point toward the metal surface. This picture suggests that the accelerating effect caused by a higher adsorption Gibbs free energy should be thought of in terms of an increase of the probability of transfer of one electron at the transition point. This in turn is caused on approaching the tail of the electron density profile of the metal (Jellium model).

The molecular fluctuations of the inner sphere of a hydrated proton (hydronium) consist basically of stretch and deformation oscillation modes, which can be experimentally analyzed by Fourier transform infrared spectroscopy (FTIR). There is experimental evidence for the formation of coordination groups of hydroniums with several water molecules by hydrogen bridges. Two fundamental structures can be detected, which are known by the names of Eigen ($H_5O_2^+$) and Zündel ($H_9O_4^+$) [58]. Typical FTIR spectra arise from characteristic oscillation modes of these groups. Figure 4.15 shows an example of such spectra obtained by subtracting the spectra of water from that of a HCl solution [59]. The absorption maxima arise from coupled libration and vibration modes of the O-H bonds in the local structures (Eigen and Zündel) (see schemes in Figure 4.15c).

According to the Jellium model, the electron density spreads out from the metal lattice up to 0.3 nm into vacuum. Therefore, it is expected that a similar distance is to be found in the metal–electrolyte solution.

General relations between physical properties of metals and proton reduction were first analyzed by Trasatti in the 1970s [54]. The relation between the logarithm of the exchange current density, j_0, and the work function of the metallic substrate, Φ_m, is particularly astonishing. Figure 4.16 shows a reproduction of the graphic presented by Trasatti. It can be seen that the data locate on two lines with the same slope. The separation of metals in two groups was ascribed to the different orientations of water molecules at the hydrogen standard potential. The energy difference is about 0.4 eV, it being in agreement with the change of surface dipole χ_{H2O} due to the rotation of the adsorbed water molecules. After the interpretation of Trasatti [54], the lower group belongs to metals on which water dipoles are oriented perpendicularly to the surface. In the other case, water molecules lie parallel to the surface. The linear relation between $\log j_0$ vs. Φ_m points out the role of the metal surface charge on the hydrogen evolution rate.

The quantum mechanical description of proton reduction involves a coupling matrix, which accounts for the interaction of the electronic states of the metal surface and the molecular orbitals of the species in the solution. In most cases, theoretical calculations are made on the basis of an oversimplification of the electronic description of the metal. The influence of the band structure, for instance is not taken into account because of the extreme complexity of the calculus. The high electrocatalytic activity of d-metals in comparison with sp-metals can be primarily ascribed to the strong electronic coupling of the spatially localized d-bands with the reducing ions. There is an interesting relation between the electrochemical activity of the metal surface and the distance of the center of the d-band from the Fermi level. Figure 4.17 shows the density of states of different metals projected at the

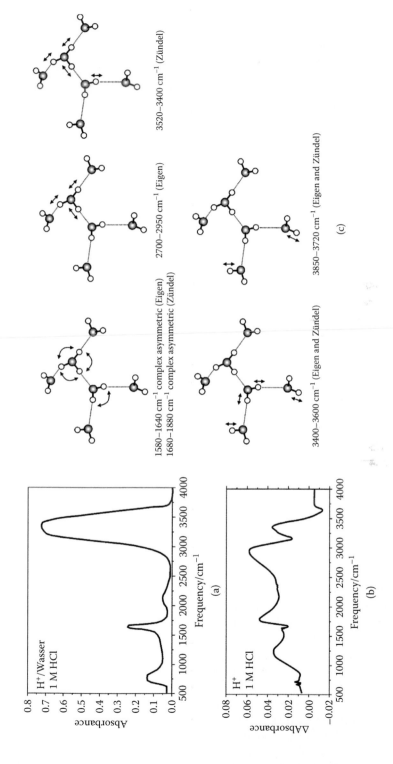

FIGURE 4.15 (a) FTIR spectrum of a 1 M HCl solution; (b) spectrum obtained after subtracting the water spectrum from (a); (c) main vibration modes for its frequency interval [after Kima et al. [59]].

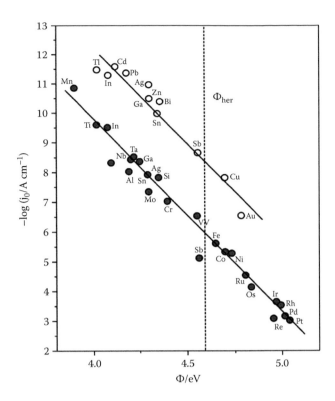

FIGURE 4.16 Exchange current density for the hydrogen evolution reaction on different metals as a function of their work functions.

(100) surface and the corresponding density of surface states as calculated by the self-consistent localized orbital method (SCLO) (see Ref. 60 for more details). According to this representation one can see that the catalytic metals for the hydrogen evolution such as Rh ($-\log j_0 = 3.5$), Pd ($-\log j_0 = 3.2$), and Ni ($-\log j_0 = 5.3$) have a d-band close to the Fermi level (in a vacuum). However, the d-band center of known bad catalytic metals such as Ag ($-\log j_0 = 7.9$) and Cu ($-\log j_0 = 7.8$) is found by 1.5–3.0 eV from the Fermi level. It should be noted, that in spite of the comparable electronic properties of Pd and Ni, the latter shows low catalytic activity. Moreover, it seems that there is no correlation between the position of the work function respective to the redox level of the hydrogen evolution reaction, Φ_{NHE}, and the catalytic activity. Here, it is opportune to mention that the linear relationship between the exchange current density and work function discussed by Trasatti have to be revised in virtue of the disparity of reported work functions.

  Another seldom-treated point in the discussion of the electrochemical activity of metal surfaces for the proton reduction is the role played by localized surface states. Localized states in d-metals arise as a consequence of a d-band splitting caused by a potential change at the surface. Figure 4.18 depicts a schematic representation of this type of surface states, also known as Tamm states. These are different from the Shockley states, which arise by adding the weak potential in a gas of free electrons, a

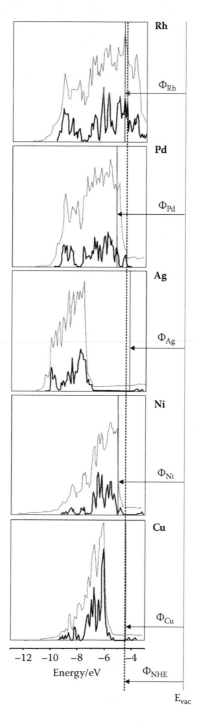

FIGURE 4.17 Projected density of bulk electronic states on the surface (100) (····) and corresponding surface state (——) calculated by Arlinghaus et al. [58] for different metals.

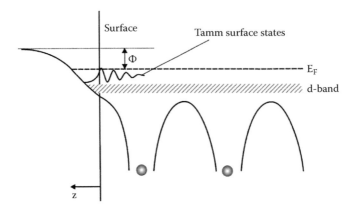

FIGURE 4.18 Schematic depicting the formation of Tamm states.

formalism that is used to describe sp-metals. Tamm states result from the description of d-bands by applying the Kronig–Penney formalism with infinite potential walls, which is an acceptable approximation for the tight-binding states (d- and f-valence band states). This difference is, however, only artificial; both states can be regarded as Bloch waves vanishing toward the metal bulk.

The energy position of surface states can, for instance, be determined by first-principle calculations and the use of the surface embedded Green function for a semi-infinite geometry (consult Ref. 61 for details). The result of such calculations is shown in Figure 4.19 for the case of a Pt(100) surface at the wave vector Γ. The spectrum shows peaks at binding energies of 1.2 eV and 0.3 eV, as well as at −0.5 eV

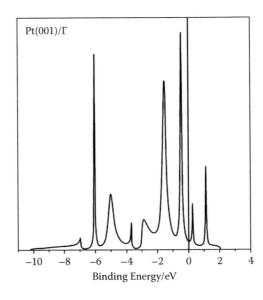

FIGURE 4.19 Density of states (DOS) on Pt(001) surface along the Γ wave vector [after Benesh et al. [61]].

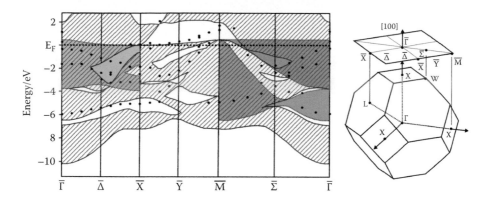

FIGURE 4.20 Projected bands, surface states (ss), and resonance states (rs)* along the principal symmetry directions of the first the Brillouin zone of the Pt(001) surface. Dashed: projected , and bands. Shadowed: and bands. ss and sr are represented by points [after Benesh et al. [61]]. *When the surface states are localized within the gap of projected bulk bands, one speaks of true surface states, to distinguish them from those which overlap the projections, referred to as surface resonances. Symmetry points and directions are indicated using BSW (Bouckaert-Smoluchowski–Wigner) notation [62].

and −1.5 eV. The first two peaks correspond to Tamm states; the other two peaks are classified as Shockley states. The strong localized density of states appears as points at the projected band structure along selected directions of the Brillouin zone on the Pt(100) surface (see Figure 4.20).

The presence of localized electronic states presupposes a preferential electron transfer upon them. According to Heisenberg uncertainty principle, the localization of the electron density in the band structure leads to an extension of their spatial outreach, that is, $\Delta k \times \Delta z \geq 1/2$. Thus, this can be interpreted as a stronger coupling of these states with the molecular orbitals of the solvated proton with the metals. This would mean according to the quantum mechanical treatment of the electron transfer a larger matrix element $V_{\bar{k}}$ and hence, a shorter transfer time, as given by the Heisenberg relation [63]:

$$\frac{\hbar}{\tau_r} = 2\pi \sum_k \left|V_{\bar{k}}\right|^2 \delta(E - E_{\bar{k}}) \,(64) \tag{4.56}$$

where τ_r represents the transfer time constant.

Arlinghaus et al. [60] carried out self-consistent surface electronic-structure calculations which determine the fraction of electron in surface states at the surface of a pile of several atomic planes. Comparison of the energy distribution of the electron density of the atomic plane at the surface with those of the crystal bulk revealed that the fraction of electrons in surface states at the different surfaces is Cu 36%, Rh(23%), Ni(23%), Ag(22%), and Pd (19%). It was found [60] that there is a correlation between the surface state fraction and the shifts of core s- and p-levels, so that the density of surface states is essentially determined by an upward shift of the surface potential relative to the bulk.

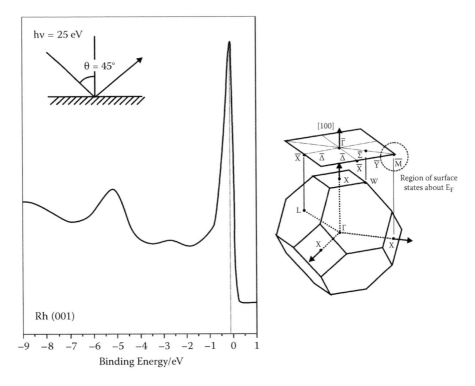

FIGURE 4.21   Photoemission spectrum of Rh(001) surface obtained at an excitation energy $h\nu = 25$ eV with the geometry shown in the figure (after Morra et al. [65]).

Surface states are experimentally investigated by means of a strong emission signal in the photoelectron spectra (PES) and inverse PES above and below the Fermi level, respectively. For instance, the spectrum shown in Figure 4.21 presents a strong emission at 0.15 eV below the Fermi level, which can be assigned to the presence of surface states [65].

In an attempt to investigate further the correlation between the electronic properties of metal surfaces and the corresponding electrocatalytic activities, one finds that several factors are involved: the density of occupied electronic surface states, the work function and its position with respect to the Fermi level of the hydrogen evolution reaction, and the energy difference between the d-band center and the Fermi level. For instance, the binding energy of metal–hydrogen shows an interesting correlation with the distance of the d-center ε_d, from the Fermi level, as can be seen in Figure 4.22. It is worth noting, that experimentally proved catalysts such as Pt, Rh, and Ir are situated very close to other d-metals such as Cu, Ni, Wo, Mo, and Co. Surprisingly, Cu, a very bad catalyst for the *her*, is placed near Pt! In principle, this finding suggests that additional electronic and chemical factors are decisive for the catalytic activity of metals in solution, which cannot be ascertained by vacuum techniques. It should be noted though that the adsorption strength is directly linked with the proximity of the d-band center to the Fermi level [66]. The binding strength of M-H, on the other hand, shows a direct proportional relation with the metal work function as shown in Figure 4.23.

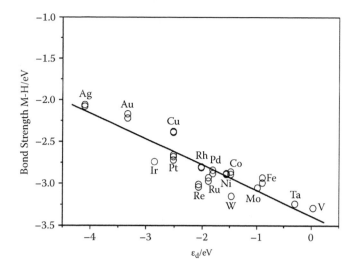

FIGURE 4.22 Bond strength of M-H as a function of the distance of d-band center from Fermi level for various metals. (Data were taken from the work of Greeley and Mavrikakis, 2005 [66].).

Out of this tendency are the metals W, Au, Ag, and Cu. This relation can be explained in terms of a larger surface electron density with localized character as result of the occupation of d-orbitals with larger work functions.

A look at the physical adsorption of molecular H_2 on metal surfaces helps to understand the role played by surface localized electronic states in the catalytic process. In the adsorption process, an equilibrium occurs between attractive van der Waals forces and repulsive forces arising close to the surface. The former are a

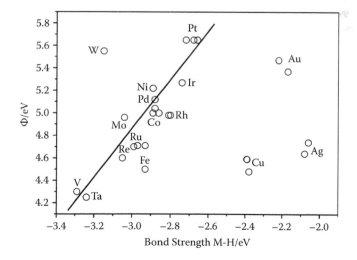

FIGURE 4.23 Bond strength of M-H as a function of the work function for various metals. (Data were taken from the work of Greeley and Mavrikakis, 2005 [66].).

consequence of the interaction of fluctuating molecular dipoles with image charges at the metal surface. The latter are determined by Coulomb repulsion forces and the Pauling principle by overlapping of electron clouds. The Pauli principle requires the orthogonalization of the wave function of substrate and adsorbate. Since H_2 is a closed system with a large energy difference between the highest occupied (HOMO) and lowest unoccupied molecular orbitals (LUMO), the orthogonalization needs such a large energy input that a steep increase of the physical sorption potential near the surface results. The sum of attractive and repulsive potential courses gives a minimum that characterizes the adsorption strength. The magnitude of repulsion forces depends on the electron density at the surface. Occupied surface states leads to a considerable increase of the physical adsorption potential because of their long decaying length (see dashed line in Figure 4.24).

Adsorbed H_2 may split into adsorbed H. This process must overcome an energy barrier determined by the crossing point of the physical adsorption curve and that for chemisorption. Since individual H-atoms do not have close valence shells, only the binding states of the orthogonalized atom-substrate wave function will be occupied, whilst the antibinding states remain empty. Thus, a chemical bond is formed, characterized by a minimum of the potential curve.

The crossing point of the physical adsorption and chemisorption curves defines the energy barrier for H_2 splitting. Figure 4.24 shows that increasing density of occupied surface electronic states leads to an increase of the adsorption barrier. This fact constitutes a starting point for the discussion of the influence of Shockly states on the catalytic activity of metals for the hydrogen evolution reaction. For instance, it was observed that the occupation of surface states by adsorption of alkali metals or oxygen results in a large increment of the H_2 adsorption barrier on Pd, Pt, and Ni [67–69]. Even if this type of experiment were performed under conditions that are far

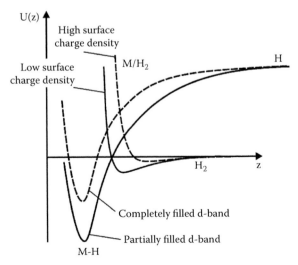

FIGURE 4.24 Physical adsorption potential curve of H_2 and chemisorption potential curve of atomic H for high (— —) and low (----) surface charge density.

away from the conditions found in a metal–solution interface (ultra-high vacuum), some conclusions can be extrapolated to electrolytic reactions.

Surface steps and defects, on the other hand, bring about a dispersion of electrons of surface states and hence a local drop of the electron density, thus leading to stronger adsorption bonds.

The occupation degree of d-bands depends on their energy distribution and their position with respect to the Fermi level. A change of the occupation degree and the center of the band is expected by reduction of the dimensions of the metal phase to the nanoscale, which is known to cause a narrowing of band [70]. This, in turn, leads to a modification of catalytic properties. There is experimental evidence of an enhanced electron transfer rate of the proton reduction [71,72], as well as of an enhanced catalytic activity for reactions in the gas phase [73,74] of small metal clusters. Here, the electronic structure of the surface plays a fundamental role because of an increase of the number of steps and kink sites.

The metal–electrolyte interface presents a more complex scenario than does the metal–vacuum system. The electrosorption of anions such as Cl^- or SO_4^{2-} involves, for instance, surface reconstruction [75,76] and the blocking of adsorption sites, with consequences for the catalytic activity. Of large significance is the formation of surface oxides upon contact of metal electrodes with an aqueous solution. The driving force for surface oxidation is given by the Gibbs free energy, ΔG_{ox}, or the oxide formation potential, $V_{ox}^0 = -\Delta G_{ox}/zF$. An interesting relationship between the exchange current density for the *her* reported in the literature and the oxide formation potentials

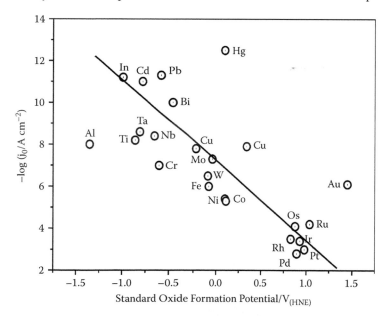

FIGURE 4.25 Exchange current density of the hydrogen evolution reaction as a function of standard potentials of oxide formation for different metals. (Values of j_0 were taken from the work of Trasatti [54]; oxide formation potentials were extracted from *CRC Handbook*.) Values are compared on the basis of the reaction $M + n\,H_2O = M(OH)_n + n\,H^+ + n\,e^-$.

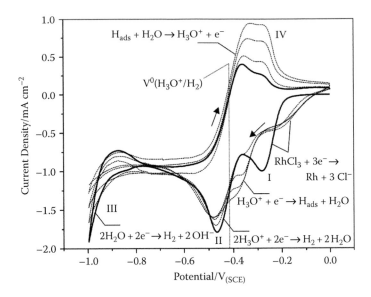

FIGURE 4.26 Cyclovoltammetry performed on p-type Si(111) in 1 mM RhCl₃ + 0.5 M NaCl of pH 2.2 under illumination. Full line: First cycle. Scan rate: 10 mV s⁻¹. Illumination by a W-I lamp at 100 mW cm⁻².

for the metal substrate can be observed (see Figure 4.25). It should be noted that typical electroactive materials such as Pt, Rh, Ir, Os, and Pd fall at potentials between 0.5 V and 1.0 V. In spite of their similar electronic properties, Ni, W, and Mo seem to lose their activity by surface oxidation. In the case of Ni, for instance, the adsorption of oxygen in the form of surface oxides or hydroxide monolayers hinders H adsorption and, hence, slows down the hydrogen evolution rate. Ni belongs together with Pt and Rh to the group of so-called d-metals, which have similar work function, position of the d-band center, and M-H bond strength (see Figs. 4.17, 4.22, and 4.23). Regarding its oxygen affinity, Ni can be grouped together with the iron-metals, with an oxide formation potential near the standard hydrogen potential (NHE). The adsorption of hydroxyl groups as the first step in the oxide formation mechanism deactivates adsorption sites for H.

$$Ni_s + H_2O \rightarrow Ni\text{-}OH_s + e^- + H^+ \qquad (4.57)$$

Also under high concentrations of acid solutions, often HClO₄ or H₂SO₄, the adsorption of –OH cannot be avoided. The adsorption of –OH is proposed as an intermediate step in the reduction of Ni²⁺ from acid solutions [77–79], allowing a thin oxide film to form. This is a convenient example that shows that, apart from the electronic properties of the metal surface, the chemical reactivity to the electrolyte determines the surface electrocatalysis. The potential of zero charge can be regarded as an indicator of the reactivity of a metal at a given potential. If the *pcz* is more negative than 0 V (NHE), it is expected that water molecules orient with the oxygen atom toward the surface, hence favoring the formation of M-OH bonds.

In the above discussion, we have analyzed the hydrogen evolution reaction in an acid solution (i.e., the proton reduction). If we apply a cathodic overtension higher than ~1 V, we will observe that hydrogen evolution occurs by direct reduction of water. To exemplify these processes, a photoelectrochemical system is conveniently selected: p-type Si/Rh/electrolyte (mildly acidified: pH 2 to 3). A voltammetry performed in a solution containing $RhCl_3$ on illuminated p-type Si(111) allows identification of the metal phase formation on p-Si(111) and the different electrochemical processes occurring on the metal surface (see Figure 4.26). The first cycle shows cathodic peaks at −0.28 V (I) and −0.47 V (II), which can be ascribed to the reduction of Rh(III) complexes and the reduction of protons on the surface of formed metal islands. We will return to the discussion of the deposition of noble metals on silicon surface in Chapter 5. The increase of the cathodic current at $V < -0.9$ V indicates the onset of water reduction, which should occur by a different mechanism than the proton reduction. From the thermodynamic point of view, the standard potential of these processes differ by 0.692 V (at pH = 1) (see reactions 4.58 and 4.59).

$$2\,H_3O^+ + 2\,e^- \rightarrow 2\,H_2O + H_2 \qquad V^0 = -0.24\,V - 0.059\,pH\ (SCE) \qquad (4.58)$$

$$2\,H_2O + 2\,e^- \rightarrow H_2 + 2\,OH^- \qquad V^0 = -1.067\,V\ (SCE) \qquad (4.59)$$

The appearance of a current peak in voltammetric experiments is the result of limited mass transport at the electrode surface. This is namely the case of proton reduction in mildly acidified solutions. It should be noted that (I) decreases markedly after the first cycle and shifts to −0.2 V (lower overtension for metal deposition), a natural result of the consumption of ions at the surface and the lower activation energy for nucleation on the metal islands.

Because of the strong H-adsorption on Rh, the proton reduction peak shows a shoulder attributed to the electrosorption reaction Eq. (4.53). The reduction of adsorbed H_2 and H is indicated by the related double anodic peak IV in the reverse potential scan. Note that this peak increases each cycle as a consequence of the increasing surface of growing metal islands. The wide anodic peak at −0.88 V can be ascribed to the reduction of intercalated hydrogen, as can be inferred by the similar electrochemical behavior found in TiO_2 (see Chapter 2).

Now, it is convenient to analyze how water reduction occurs. In contrast to the large volume of literature dedicated to the elucidation of proton reduction, water reduction has not been sufficiently explored. A reliable mechanism can be formulated in terms of the reduction of protons arising from water dissociation. Consider the number of dissociated water molecules in the Helmholtz layer of $d \sim 0.3$ nm. A value of 1.8×10^6 molecules cm^{-2} is calculated. Supposing that the O-H bond in the H_3O^+ stretches toward the electrode surface with a frequency of 8.8×10^6 Hz (taken from an adsorption maximum at 3400 cm^{-1}), then a current density of 2.55 $\mu A\ cm^{-2}$ should be attained, regarding an instantaneous charge transfer at the transition point. This implies delivery of additional energy for the dissociation, provided by the electric field at the double layer. Water reduction typically occurs at potentials 1 V more negative than the potential of zero charge, so that an electric field between 1×10^7 V cm^{-1} and $\times 10^8$ V cm^{-1} is established. The energy necessary for a complete orientation of the water molecule is given by $E = \vec{p} \cdot \vec{E}$, where $\vec{p}$ is the dipole moment of water, assumed to be 1.85 Debye. This energy should be sufficient to break down the hydrogen bridges of a binding energy of 0.24 eV, which hinders water rotation. The strong orientation of water molecules near the electrode surface brings about a drop in the dielectric constant from 80 for the bulk water to 8.5 near the metal surface. This in turn causes a tenfold increase of the electric field, and hence of the density of ionized molecules.

Theoretical investigations of the water dynamics have shown that water ionization assisted by high electric fields is a common process in bulk water. Calculations made by applying the method of Car–Parrinello [80,81] and the so-called way probe indicate that water dissociation takes place following energy fluctuations arising from particular transitory structures of solvated water which leads to destabilization of O-H bonds [82]. The coordinated motion of various water molecules generates a localized transitory electric field of a lifetime of some tens of fs, which stabilizes an ion pair H_3O^+-OH^- by a Grottus-like mechanism. The energy barrier for the formation of this ion pair is about 0.43 eV. But since electron transfer steps occur within a time span of 100 fs to 200 fs, the reduction probability of dissociated water is rather low.

The influence of the interfacial electric field on the water ionization was analyzed by Brüesch and Christen [83] by applying a modified Poisson–Boltzmann equation (4.60), which relates the concentration of generated protons $n(z)$ and the potential distribution $\varphi(z)$ in pure water in contact with a metal.

$$n(z) = \gamma(z) n_0 \exp\left(-\frac{ze\varphi(z)}{kT} \right) \qquad (4.60)$$

The correction factor $\gamma(z)$ accounts for field-assisted deviations arising from chemical activity of protons, image charges at the metal surface, ionic polarization, and water dissociation. The relation between the electric potential $\varphi(z)$ and the charge density $\rho(z)$ is given by the Poisson equation:

$$-\rho(z) = n(z) \varepsilon_w \varepsilon_0 \frac{\partial^2 \varphi}{\partial z^2} \qquad (4.61)$$

In particular, field-assisted water dissociation is accounted for by the introduction of a local degree of dissociation $\alpha(z)$ into the correction factor $\gamma(z)$. Figure 4.27 shows a concentration profile of protons as calculated by Brüesch and Christen [83], taking into account the electric field dependency of the correction factor $\gamma(z)$ for an interfacial potential drop of 0.4 V. Note that this analysis predicts an increase of water dissociation over one order of magnitude for a distance of 2–3 nm from the negatively charged electrode surface. The water reduction process is thus completed by transfer of one electron to the generated proton. Although the above considerations offer a theoretical frame for further analysis of the water reduction process, they cannot be directly extrapolated for the case when water is reduced from an electrolyte solution, where the interfacial potential should basically be compensated by solvated salt cations. Polarization effects on water dissociation are also known in the

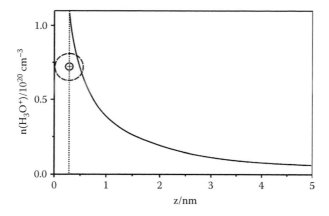

FIGURE 4.27 Proton concentration and potential distribution at the metal–water interface as calculated by the corrected Poisson–Boltzmann equation for a potential drop of 0.4 V (after Brüesch and Christen [83]).

anodic growth of oxides [84]. In this case, the condition of continuity of the density of dielectric shift $\vec{D} = \varepsilon\varepsilon_0\vec{E}$ at the oxide–electrolyte interface must hold. Hence, the electric field in the double layer is given by $\left|\vec{E}\right| = (\Delta\varphi / d_{ox})(\varepsilon_{ox} / \varepsilon_H)$, where ε_H stands for the dielectric constant of the double layer. On the other hand, the orientation of water dipoles lowers the local dielectric constant with a consequent increase of the electric field. The dielectric constant of the double layer increases from a minimum value for the water layer in contact with the metal to the water bulk as [85]

$$\varepsilon(z) = \frac{\varepsilon_\infty}{\left[1 + (\dfrac{\varepsilon_\infty}{\varepsilon_{min}} - 1)\exp\left(-\dfrac{z}{\kappa}\right)\right]} \quad ; \quad \kappa^{-1} = \left(\frac{\varepsilon_0\varepsilon_\infty kT}{e^2\sum_i^\infty n_i^\infty Z_i^2}\right) \tag{4.62}$$

where n_i is the concentration of species i and Z_i is the ion valence.

In the particular case of photoelectrodes based on semiconductor–metal interfaces, the electric field at the metal–electrolyte interface depends on the partition of the electrode potential drop between the semiconductor and the double layer. At a high concentration of surface states, for instance, Fermi level pinning occurs as the Fermi level reaches the neutrality level of the semiconductor. Thus, any change of electrode potential ends in a change of the electric field at the double layer. Let us consider the water reduction of n-Si(111)/oxide/Rh interfaces. Figure 4.28 shows the current-voltage curves obtained in 0.5 M KCl with a natural pH of 5.5 on n-type Si(111):H and n-Si(111)/oxide/Rh electrodes. This latter was prepared by potentiostatic electrodeposition of Rh islands at −0.6 V_{SCE} from 1 mM $RhCl_3$ + 0.5 M KCl.

Capacitance measurements can be used to obtain additional information about changes of the interfacial structure.

The first feature one can observe is a significant drop of the water reduction overtension, of ~0.7 V on the Rh-plated electrode with respect to that on n-type Si(111):H.

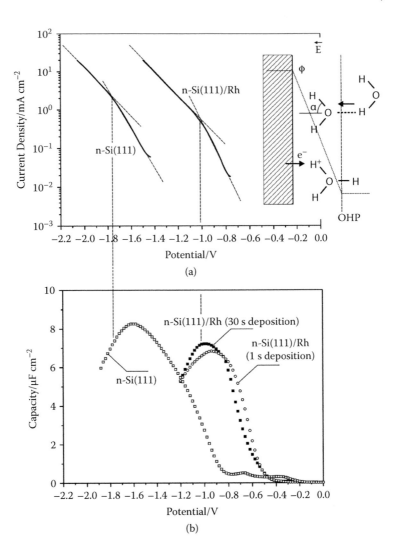

FIGURE 4.28 Current density- (a) and capacity-voltage (b) of n-type Si(111):H before and after the deposition of Rh in 0.5 M KCl. Rh deposition was performed potentiostatically at −0.6 V from mM RhCl₃ + 0.5 M KCl. Frequency: 100 Hz.

A reduction of the overtension due to the better catalytic activity of Rh for the proton reduction than n-type silicon can be quantified by means of the Butler-Volmer equation (4.38):

$$\Delta\eta = \frac{kT}{e}\frac{1}{\alpha}\ln\left(\frac{i_{0(Rh)}}{i_{0(si)}}\right) \tag{4.63}$$

Taking $i_{0(Rh)} = 3.16 \times 10^{-4}$ A cm^{-2} and $i_{0(Si)} = 1.26 \times 10^{-8}$ A cm^{-2}, as reported in the literature for 0.5 M H₂SO₄ [54], we have $\Delta\eta = 0.506$ V assuming $\alpha \approx 0.5$. The surface chemical changes arising during the metal deposition, however, modify the partition of the potential drop between semiconductor and double layer, and hence the driving electric field at the metal–electrolyte interface. The rise of the capacitance-voltage curves at −0.85 V and −0.45 V for hydrogenated and Rh-plated n-type Si(111) indicates that the semiconductor enters the accumulation potential region. This means the flat band potential shifts 0.4 V toward more positive potentials after Rh-deposition. The reduction of the space charge layer under the nanodimensioned metal islands manifested as an increase of the semiconductor donor concentration might be regarded as a possible argument. This, however, is insufficient considering a shift larger than the $E_{CB} - E_F \sim 0.25$ eV and the fact that no further shift is found after more prolonged plating time (1 s and 30 s deposition). In this case, we have to consider the change of surface chemistry induced by deposition. As explained in detail in Chapter 5, deposition of noble metals with standard reduction potentials more positive than the flat band potential leads to simultaneous surface oxidation. As a result, a surface oxide is exposed to the electrolyte on the uncovered parts of the plated electrodes and surface acid–base reactions charge the double layer depending on solution pH. So, the measured potential at the flat band condition is

$$\equiv\text{Si-OH} \leftrightarrow \equiv\text{Si-O}^- + \text{H}^+ \tag{4.64}$$

$$\equiv\text{Si-OH} + \text{H}^+ \leftrightarrow \equiv\text{Si-OH}_2{}^+ \tag{4.65}$$

Assuming that the surface excess charge is compensated by solvated ions in the solutions at the outer Helmholtz plane, the standard electrochemical potential for these reactions in equilibrium is given by

$$\Delta\tilde{\mu}^0_{a,b} = \Delta\mu^0_{a,b} \pm e\Delta\varphi_H + \beta_{a,b}\theta_{a,b} \tag{4.66}$$

where the subscript refers to acid and base reactions (4.64 and 4.65 respectively), $\Delta\mu^0_{a,b}$ is the chemical potential, and $\Delta\varphi_H$ is the interfacial potential drop. $\beta_{a,b}$ refers to the Frumkin parameter for interaction of adsorbed acid–base species [86], and $\theta_{a,b}$ is the corresponding coverage. Considering the concentration dependence of the electrochemical potential, we have

$$\exp\left(-\frac{\Delta\tilde{\mu}^0_a}{kT}\right) = \frac{N_{Si-OH}}{a_{H^+}N_{Si-O^-}}; \exp\left(-\frac{\Delta\tilde{\mu}^0_b}{kT}\right) = \frac{N_{Si-OH}a_{H^+}}{N_{Si-OH_2^+}} \tag{4.67}$$

where N_i represents the density of surface species. From Eqs. (4.66) and (4.67), we have

$$\Delta\varphi_H = -kT(\Delta\mu^0_{a,b} + \beta_{a,b}\theta_{a,b}) - 2.303kT \times pH \tag{4.68}$$

Figure 4.29 shows the change of measured flat band potential, $V_{fb} = \Delta\varphi_H - \Delta\varphi_{ref}$, with the solution pH for an oxidized p-type Si(100), where experimental values are found in good agreement with the predicted relation (4.68).

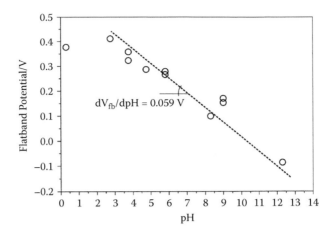

FIGURE 4.29   pH-dependency of the flatband potential (vs. NHE) of p-type Si(100) after Muñoz et al. (Ref. 108, Chapter 2).

Now, considering that the isoelectric point of SiO_2 is about 1.5 to 3, it is evident that the oxidized silicon surface is negatively charged at neutral pHs with respect to the H-terminated one, a fact that is reflected in the positive potential shift of the flat band potential.

Another feature to observe in Figure 4.28 is the change of the Tafel slope of the water reduction at −1.75 V and −1.0 V for the hydrogenated and plated n-type Si(111) electrode: from 160 mV dec^{-1} ($\alpha = 0.37$) to 313 mV dec^{-1} ($\alpha = 0.19$). Here, it is interesting to note that the break of the Tafel line appears after the capacitance curve starts to decrease with overpotential. In accumulation, the semiconductor capacitance increases with the interfacial potential drop. This occurs until the Fermi level reaches the conduction band edge: a typical band Fermi level pinning. Further increase of interfacial drop will be taken up by the double layer with a consequent increase of the interfacial electric field. Note that the onset of water reduction is observed near the capacitance maximum. The capacitance decrease is related to the drop of dielectric constant in the double layer because of the orientation of water molecules with the increasing electric field. The change of the slope of the Tafel line may be related to a change of water orientation with the electric field. Su and Bühl [84] modeled the dissociation of water on Al_2O_3 by calculating the energy for the displacement of an H atom from one water molecule to the neighboring one by elongation of the O-H bond beyond 0.2 nm. In this case, the presence of an electric field stabilizes the dissociation configuration, given by: $\Delta E = C |\vec{E}| \cos\alpha$, where C is a constant, and α is the angle of the breaking bond with respect to the direction of the electric field. Thus, the dissociation rate constant increase with the electric field as

$$k_{diss} = k_{diss}(E = 0) \times \exp[C |\vec{E}| \cos\alpha] \tag{4.69}$$

The Tafel-like behavior of water reduction might be ascribed to the dependence of dissociation rate with the electric field.

Let us now to take a look at the energy of molecular orbitals of water and hydronium. The energy levels were calculated by Chaplin [87] by applying the restricted Hartree–Fock method [88] and are shown in Figure 4.30, together with the spatial distribution of the residence probability of electrons. It can be seen that the lowest unoccupied molecular orbital (LUMO) of hydronium (H_3O^+) presents an s-type character. The discrete level of isolated molecules transforms into a quasi-continuous Boltzmann-type energy distribution due to the interaction with the surrounding solvent. It is not totally clear at what energy level the electron exchange for the reduction of these cations takes place. Considering the hypothetical concept of work function for the proton reduction at −4.9 eV, it is probably that the electron transfer occurs, inasmuch as the energy of unoccupied levels of protons reach this threshold. The calculated energy level of hydronium corresponds to an isolated entity. If we consider a hydration heat of $\Delta H_h = -3.90$ eV for H_3O^+ [7], the energy level of the

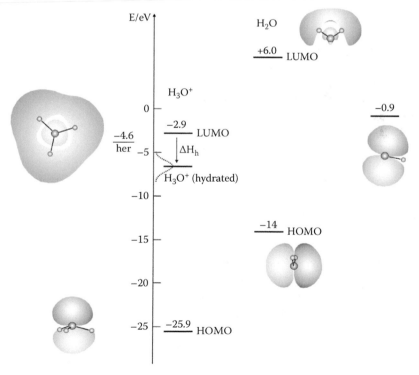

FIGURE 4.30   Calculated molecular orbitals for the hydronium ion, water, and hydroxyl ion (after M.F. Chaplin [87]).

LUMO-level is expected to fall to −6.80 eV. But, the direct reduction of H_2O is energetically restricted.

The reduction of CO_2 on sp-metals involves an initial one-electron transfer step for the formation of the radical CO_2^- [89]:

$$CO_{2\,(ads)} + e^- \rightarrow CO_2^- \qquad (4.70)$$

This process demands a high activation energy, which is reached by applying large cathodic potentials; for instance the standard potential for this reaction is −2.21 V vs. SCE in DMF (dimethylformamide) and −1.90 V vs. NHE in aqueous electrolytes. Further reduction of radical $CO_2^{\bullet-}$ follows different routes depending on the substrate and electrolyte as depicted in Figure 4.31. It can be seen that the radical CO_2^- adsorbs on some metals, thus leading to the formation of CO and CO_3^{2-} as end products. Otherwise, the reduction of CO_2^- continues to formic acid or oxalic acid in aqueous and nonaqueous solutions respectively.

Formic acid, CO, CH_4, C_2H_6, and C_2H_4 are the main products of the CO_2 reduction in aqueous electrolytes, which competes with the strong hydrogen evolution reaction at high cathodic overpotentials. The efficiency of each reduction product depends on the electrode material and the electrolysis operation conditions: potential, current density, temperature [90–92], partial CO_2 pressure [93,94], mixed supercritical fluids [94], and mixed electrolytes [90,92]. The use of aprotic organic solvents was regarded as a way to increase the solubility of CO_2 [95,96] and to suppress the hydrogen evolution reaction. CO, oxalic acid, and formic acid were observed as main products on different electrode materials. However, no hydrocarbons were obtained in solvents such as N,N-dimethyl formamide, propylene carbonate, and acetonitrile [97]. A considerable increment of the efficiency of hydrocarbon formations was found by using methanol as a solvent. The solubility of CO_2 in methanol is five times that in water at room temperature and increases up to 15 times at temperatures below 273 K [98]. Kaneco et al. [99,100] have investigated the reduction of CO_2 in methanol-based electrolytes, LiOH, CsOH, RbOH, and KOH at 243 K, and in the potential region from −3 V to 5 V using Cu as the cathode material. They have found that the Faradaic efficiency for the production of ethylene increased 20% with respect to that found under similar temperature and pressure conditions in hydrogen carbonate solutions (see Table 4.1). Particularly, a considerable efficiency for the production of methane is observed in lithium-supporting electrolytes—for instance, 71.8% by using $LiClO_4$ as the supporting salt [101] at 243 K and a Cu cathode at −3.0 V vs. Ag/AgCl (sat. KCl) [102]. Figure 4.32 exemplifies the potential dependency of Faradaic efficiencies in a LiOH/methanol solution for different products. It can be seen that the hydrogen evolution is suppressed to efficiencies less than 2% at potentials more negative than −3.0 V. The following reaction pathway is inferred from the reported experimental evidence in methanol:

$$CO_2^-{}_{ads} + H^+ + e^- \rightarrow CO_{ads} + OH^- \qquad (4.71a)$$

$$CO_{ads} + 4\,H^+ + 4\,e^- \rightarrow CH_{2\,ads} + H_2O \qquad (4.71b)$$

$$CH_{2\,ads} + 2\,H^+ + 2\,e^- \rightarrow CH_4 \qquad (4.71c)$$

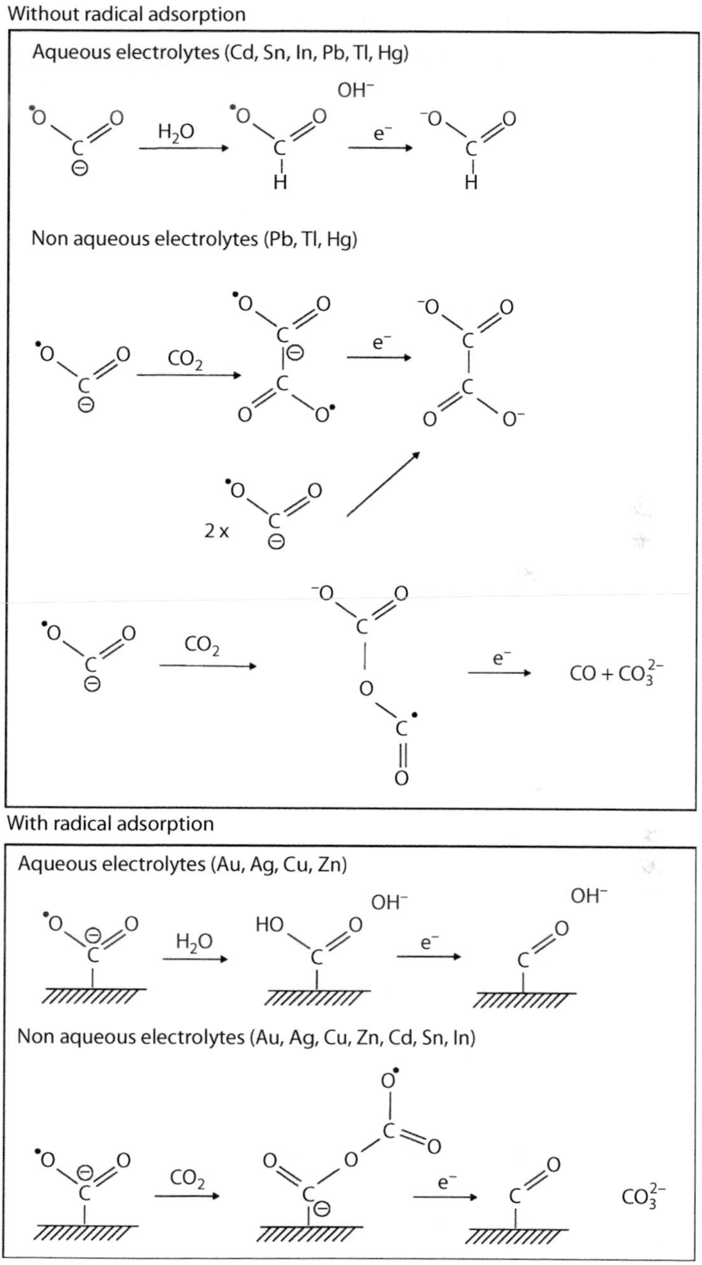

FIGURE 4.31 First steps of CO_2 reduction on sp-metals in aqueous and nonaqueous electrolytes (after Y. Hori [89]).

TABLE 4.1

Faradaic Efficiency of Ethylene and Methane in the Electrochemical Reduction of CO_2 in Aqueous and Methanol-Based Electrolytes on Cu Cathodes at 243 K

	Faradaic Efficiency/%			
	Methanol		Water	
Cation of Supporting Salt	**Ethylene**	**Methane**	**Ethylene**	**Methane**
Li	14.7	63.0	4	26
K	37.5	16.0	14	16
Rb	31.0	4.6	—	—
Cs	32.7	4.1	13	15

Note: Methanol: supporting salts: MOH; water: supporting salts: $MHCO_3$

Source: Kaneco, S., Katsumata, H., Suzuki, T. and K. Ohta. 2006. Electrochemical reduction of carbon dioxide to ethylene at a copper electrode in methanol using potassium hydroxide and rubidium hydroxide supporting electrolytes. *Electrochim. Acta* 51:3316.

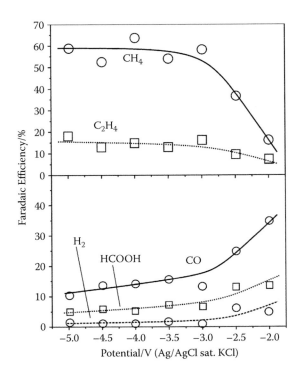

FIGURE 4.32 Potential dependence of the Faradaic efficiencies for the products of electrochemical reduction of CO_2 on Cu cathodes at 243 K in LiOH/methanol (after S. Kaneco et al. [92]).

$$2 \, CH_{2 \, ads} \rightarrow C_2H_4 \tag{4.71d}$$

$$CO_2^{\cdot-}{}_{ads} + H^+ + e^- \rightarrow HCOO^-{}_{ads} \rightarrow HCOO^- \, (aq) \tag{4.71e}$$

$$CO_2^{\cdot-}{}_{ads} + CO_2 + e^- \rightarrow CO_3^{2-} + CO \tag{4.71f}$$

The enhanced solubility of CO_2 in aqueous electrolytes at high pressures leads to higher Faradaic efficiencies. Hara et al. [93] have found that Cu, Fe, and Co show the higher efficiencies for the production of CH_4 in 0.1 M $KHCO_3$ at 30 atm. In this case, however, CO and formic acid constitute the main products of CO_2 reduction.

The reduction of CO_2 on metals of the platinum group is particularly interesting, since the large energy-demanding first step for the formation of the radical $CO_2^{\cdot-}$ is replaced by another mechanism involving the adsorbed hydrogen at potentials close to that for the hydrogen evolution reaction [103,104]:

$$H^+ + e^- \rightarrow H_{ads} \tag{4.72a}$$

$$CO_2 + 2 \, H_{ads} \rightarrow CO_{ads} + H_2O \tag{4.72b}$$

On Pt surfaces, the reaction course is retarded by a strong adsorption of CO and small amounts of products such as COH and HCOOH. Thus, hydrogen evolution becomes the predominant reaction. As expected for reactions involving adsorbed species, the reduction rate of Eq. (4.72) depends on the surface density of kink and edge sites. The activity of Pt, Rh, and Ir single crystal surfaces decreases as follows: (110) > (111) > (100). Hoshi et al. [104] have explained the superior reactivity of Pt(110) surface in terms of an energetically more favorable H adsorption on fourfold hollow sites. The formation of a poisoning-reduced CO_2 layer on Pt electrodes was circumvented in a device invented by Inuzuka [105] based on a porous Pt anode connected with a porous catalytic metal cathode by a proton-conductive solid.

The use of Pd was particularly effective for the reduction of CO_2 in aqueous $KHCO_3$ solutions. Ohkawa et al. [106] have shown that the previous absorption of hydrogen increases the production of CO and HCOOH. Ayers and Farley [107] proposed a method for the reduction of CO_2 in HCO_3^- solutions based on bipolar Pd electrodes. Hydrogen penetrates from the negative of the bipolar electrode and reduced the neutral CO_2 once it arrived at the slightly positively polarized side of the electrode according to the following reactions:

$$(HCO_3^-/CO_2)_{ads} + 2 \, Pd\text{-}H \rightarrow HCOOH + 2 \, Pd \tag{4.73}$$

$$(HCO_3^-/CO_2)_{ads} + 4 \, Pd\text{-}H \rightarrow HCHO + H_2O + 4 \, Pd \tag{4.74}$$

$$(HCO_3^-/CO_2)_{ads} + 6 \, Pd\text{-}H \rightarrow CH_3OH + H_2O + 6 \, Pd \tag{4.75}$$

A series of other proposed approaches have been listed in Ref. 108. Recently, Varghese et al. [109] have reported a photocatalytic device based on annealed N-doped TiO_2 nanotubes covered with ultrathin layers of sputtered Pt or Cu, which reduces CO_2 to methane when exposed to solar light. It was shown that CO_2 can be reduced by photocatalysis in the presence of vapor or water under UV illumination, where the presence of Cu seems to be crucial for the reduction process [110–112]. For instance, methane and ethylene were found to be the main products in a suspension of Cu(<5wt%)/TiO_2 in water when irradiated with a Xe lamp and pressured with CO_2 at 27.09 atm [111]. On the other hand, the presence of Pt on TiO_2 leads to dissociative adsorption of CO_2 with the production of CO and oxygen [113]. The mechanism for the photocatalytic formation of methane is not totally elucidated. It is believed that it involves the oxidation of water by photoinduced holes at the TiO_2 valence band and the reduction of CO_2 by electron injection:

$$H_2O + H^+ \rightarrow OH^\bullet + H^+ \tag{4.76a}$$

$$H^+ + e^- \rightarrow H_{ads} \tag{4.76b}$$

$$2\,H_{ads} \rightarrow H_2 \tag{4.76c}$$

$$2\,CO_2 + 4\,e^- \rightarrow 2\,CO_{ads} + O_2 \tag{4.76d}$$

$$CO_{ads} + 6\,e^- + 6\,H^+ \rightarrow CH_4 + H_2O \tag{4.76e}$$

The wide band gap of TiO_2, however, limits the conversion efficiency upon solar light irradiation. The use of more adequate p-type semiconductors, such as Si, GaAs, and InP, as photocathodes was evaluated by Hirota et al. [114] in a methanol solution of CO_2 at 40 atm. Hydrogen, CO, and methyl formate were found as main products of the photo-induced reduction of CO_2. Figure 4.33a compares the photocurrent-voltage curves for each semiconductor. It can be seen that p-type InP shows a better performance in spite of the more unfavorable position of the energy bands as shown in Figure 4.33b. It should be noted, that the standard potential for the formation of the initial CO_2^- is energetically quite distant from the affinity value of the three analyzed semiconductors. Based on previous studies of the reduction mechanism of CO_2 in organic solvents [95,115,116] Hirota et al. [114] suggested the formation of a complex radical anion [117]:

$$O = C^\bullet - O - C \overset{\displaystyle O^-}{\underset{\displaystyle O}{<}} \tag{I}$$

to explain a substantial decrease of the energy barrier for the first electron transfer step:

$$2\,CO_{2\,ads} + e^- \rightarrow (CO_2)_2^{\bullet-}{}_{ads} \tag{4.77}$$

The electron affinity of I was found to be +0.9 eV [118], 1.5 eV more positive than that for the radical CO_2^- (−0.58 eV) [117]. Thus, further chemical reduction of $(CO_2)_2^{\bullet-}{}_{ads}$ via CO_2^- leads to the formation of CO by decomposition of the resulting dianion:

$$(CO_2)_2^{2-}{}_{ads} \rightarrow CO + CO_3^{2-} \tag{4.78}$$

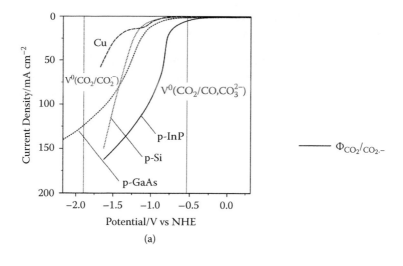

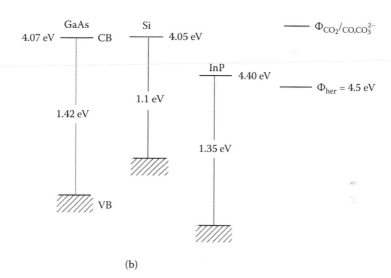

FIGURE 4.33 (a) Current-voltage curves for p-type semiconductors in 0.3 M TBAP (tetra-butylammonium perchlorate) in methanol obtained under illumination with a Xe lamp: 480 mW cm^{-2}. Spectrum was cut at wavelengths shorter than 370 nm [taken from Ref. 114]; (b) energy band diagrams depicting the relative position of bands respective of the work function for the reactions: $CO_2 + e^- \rightarrow CO_2^-$ and $2CO_2 + 2\,e^- \rightarrow CO + CO_3^{2-}$.

It was observed that the introduction of copper particles suspended in the methanol-based catholyte makes it possible to obtain methane and ethylene with a maximal efficiency of 1.56% at –10°C on illuminated p-type InP [119].

The introduction of nitrogen containing ions such as NH_4^+ [120], NR_4^+ [121], and pyridinium [122] has been shown to reduce the reaction overpotential acting as intermediate in the photoreduction process to form the CO_2^- on p-type semiconductors.

4.4 DEVELOPMENT OF NEW ELECTROCHEMICAL CATALYSTS: FUTURE DIRECTIONS

The modification of the electronic structure of metal surface and hence of its electro-catalytic properties by adding alloying elements makes it possible to use more abundant and inexpensive metals such as Ni and Fe for the hydrogen evolution reaction. This does not restrict the use of noble metals such as Pt, Rh, or Ir, regarding their costs. In fact, inorganic photoelectrodes designs based on thin film absorber materials covered by small metal catalyst particles in the size range below 50 nm requires an extremely small amount of material to collect the light-induced carriers. For instance, assuming medium-size cubic metal islands with 50 nm dimensions, 4×10^{10} nanoemitter islands are needed to efficiently collect carriers from 1 m^2. Hence, with a gram of Pt at presently 48 USD [123], the noble metal costs for 1 km^2 amount to 5150 USD. This is not a cost relevant factor of such a solar cell field. For materials such as crystalline Si, the distance among nanoemitters can be increased substantially because of the larger minority carrier diffusion length; this gain in cost would be partly compensated by the necessity to form "fingers" into the material for better spectral sensitivity in the long wavelength regime of this indirect absorber. If one uses, however, less noble metals to form the semiconductor–nanoemitter junction and subsequently "caps" these metals with a noble metal for photoelectrocatalytic operation, an additional potential for pronounced further cost reduction will result (see scheme in Figure 4.34).

By now, it is known that the electrocatalytic properties of binary alloys are closely related to the modification of the work function and d-band character respective to the individual constituents [124]. Following this concept, the activation of close-packed surfaces of Au, Pt, PtRu, Rh, Ir, Ru, and Re by deposition of pseudomorphic

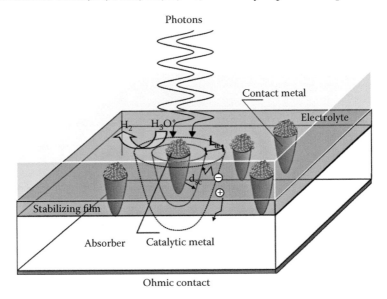

FIGURE 4.34 **(See color insert.)** Schematic of a nanoemitter photoelectrode based on non-catalytic metals covered by a thin film of a metal of the platinum group.

Pd overlayers was investigated by Greeley et al. [125]. According to their results (see Figure 4.35a), the measured exchange current density presents a volcano-type functionality with the calculated Gibbs free energy of hydrogen adsorption, ΔG_H, with a maximum at $\Delta G_H = 0$. This analysis was performed in terms of a theoretical model introduced by Nørskov et al. [126,127] and is based on density functional theory (DFT) calculations of the free energy of reaction intermediates, which includes the solvation energy. According to this model, the exchange current density for very exothermic hydrogen adsorption (i.e., $\Delta G_H < 0$) is given by

$$j_0 = -ek_0 \frac{1}{1+\exp(-\Delta G_H / kT)} \tag{4.79}$$

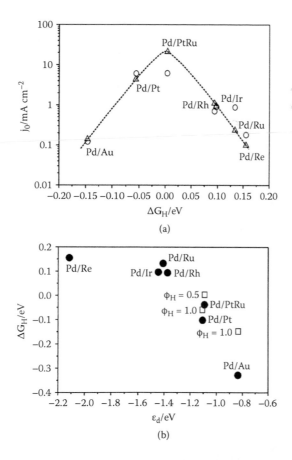

(a)

(b)

FIGURE 4.35 (a) Exchange current density for the hydrogen evolution reaction as a function of Gibbs free energy of adsorption at 293 K on a Pd overlayer on different close-packed surfaces. Circles: experimental values; triangles: predicted by the model of Nørskov et al. [127]. Dashed line: theoretical predicted exchange current density; (b) Free energy of hydrogen adsorption as a function of the d-band center for $\theta_H = 0.25$ (circles). Squares represent the adsorption energies corresponding to the system coverage [after Greeley et al. [125]].

whereas for endothermic adsorption (i.e., $\Delta G_H > 0$), the model predicts

$$j_0 = -ek_0 \frac{\exp(-\Delta G_H / kT)}{1 + \exp(-\Delta G_H / kT)} \tag{4.80}$$

where k_0 is an adjustable preexponential factor that includes all effects related to the reorganization of the solvent during the proton transfer to the surface, and which are assumed to be independent of the metal at the equilibrium potential. The adsorption energy at a given coverage θ_H is calculated by

$$\Delta G_H = \Delta H_H(\theta_H) + \Delta E_{zpe} - T\Delta S_H \tag{4.81}$$

where ΔH_H is the coverage dependent adsorption enthalpy, calculated as $H[\text{surf} + n\, H_{ads}] - [\text{surf} + (n-1)\, H_{ads})] - \frac{1}{2}H[H_2]$. ΔE_{zpe} is the difference in zero-point energy between adsorbed and gas phase hydrogen, and ΔS_H is the adsorption entropy: $\Delta E_{zpe} - T\Delta S_H \cong 0.24$ eV [127]. Moreover, the coverage is calculated by convergence of DFT calculations with the isotherm:

$$\exp(-\Delta G_H / kT) = \frac{\theta_H}{1 - \theta_H} \tag{4.82}$$

The plot in Figure 4.35a shows the experimentally measured exchange current densities and those calculated by Eqs. (4.79 and 4.80) as a function of calculated free energy of hydrogen adsorption. The parameter k_0 was obtained by fitting experimental data of j_0 of individual metal surfaces [127]. It should be noted that a monolayer of Pd on a closely packed Pt surface increases the exchange current density from 0.47 mA cm^{-2} to 6.32 mA cm^{-2}. Particularly interesting is the activation of the otherwise bad catalyst Au(111): compare $j_0 = 2.51 \times 10^{-4}$ mA cm^{-2} for Au(111) and 1.3×10^{-1} mA cm^{-2} for Pd$_{ML}$/Au(111). The changes of the surface electronic structure by deposition of a Pd monolayer can be quantified by the shift of the d-band center, defined as

$$\varepsilon_d = \frac{\int N(\varepsilon)\varepsilon\, d\varepsilon}{\int N(\varepsilon)\, d\varepsilon} \tag{4.83}$$

where $N(\varepsilon)$ is the density of states. As stated above, the energy position and density of d-states projected onto the reactive surface determine the adsorption properties of transition metals, as is clearly shown in Figure 4.35b. This is also visualized in the volcano-shaped relation between the rate of oxygen reduction and the position of the d-band center in nanoparticles of bimetallic alloys Pt$_3$M (M: Ti, V, Fe, Co, Ni) [128].

The experimentally observed link between the d-band center and the electrocatalytic activity was explored by Hammer and Nørskov [129,130] with regard to the dissociative adsorption of molecular hydrogen onto close-packed surfaces of Al, Cu, Pt, and Cu$_3$Pt. The authors refer to the energetic changes related with the modification of the

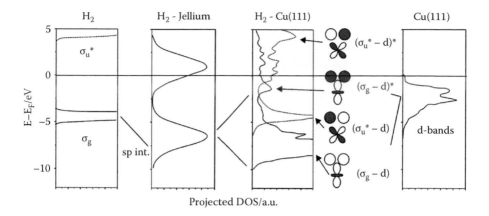

FIGURE 4.36 Calculated evolution of the molecular orbital energy of hydrogen upon adsorption onto Cu(111) surfaces (after Nørskov [130]).

electronic structure of the metal surface upon approaching the adsorbate. A qualitative picture of the theoretical background is given in Figure 4.36 for the case of H_2 adsorption onto Cu(111). Firstly, the interaction of the bonding and antibonding states of H_2 with the s and p electrons (Al or Jellium) gives rise to the formation of down-shifted broad resonance states. The interaction with the d-states of metal further splits the σ_g and σ_u^* projected density of states in molecule–surface bonding, $\sigma_u^* - d$ and $\sigma_g - d$, and antibonding, $(\sigma_u^* - d)^*$ and $(\sigma_g - d)^*$, contributions. The downward shift of the lowest of the two adsorbate states and the upward shift of the highest state upon interaction with the d-band of the metal can be quantified by the perturbation expression:

$$\Delta\varepsilon \sim \frac{|V_{ad}|^2}{|\varepsilon_d - \varepsilon_\sigma|} \tag{4.84}$$

where ε_d refers to the d-band center and $|V_{ad}|$ is the coupling matrix element between the adsorbate states and a state representative of the metal d-states. The use of the second-order perturbation theory is justified by the fact that the valence bands for simple gas atom adsorbates like hydrogen, oxygen, and nitrogen are shifted well below the d-bands due to the interaction with the metal sp-electrons. Three factors are thus considered for the formulation of reactivity criterion: (i) the position of the molecular bonding and antibonding states relative to the d-bands, (ii) the coupling matrix element between the molecular orbitals and the metal d-states, and (iii) the filling f of the molecule–surface antibonding states given by the position of the Fermi level, E_F. Accordingly, a measure of the metal reactivity can be estimated by the energy difference arising from the coupling to the d-bands as

$$\delta E = -2\frac{V_{ad}^2}{\varepsilon_{\sigma^*} - \varepsilon_d} - 2(1-f)\frac{V_{ad}^2}{\varepsilon_d - \varepsilon_\sigma} + \alpha V_{ad}^2 \tag{4.85}$$

where the first term represents the energy gain due to the hybridization between σ_u^* and d-states; the factor 2 is for spin. The second term corresponds to the $\sigma_g - d$ interaction. The factor $(1 - f)$ represents the fraction of unoccupied antibonding states. The last term accounts for the repulsion due to the orthogonalization of both σ_u^* and σ_g states to the metal d-states, where α is a proportionality factor. Equation (4.85) offers a consistent picture of the reactivity of metals and alloys. For instance, it helps explain the different reactivity of Pt and Cu atoms in Cu_3Pt alloys, for which the absence of an activation barrier and the presence of a positive value for the dissociative hydrogen adsorption, respectively, can be predicted. Such a higher reactivity cannot be predicted by large Fermi level density of states of localized d-holes just above the Fermi level.

The theory of uncompleted occupied d-bands was invoked, however, to correlate the observed activity of surface Ni-alloys. According to the Brewer–Engel theory [131,132], the introduction of atoms of an sp-metal into metals with a strong d-character leads to an increase of the vacant d-states with a consequent shift of the d-band center toward the Fermi level. The Brewer–Engel theory predicts that the s- and p-electrons, the long-range order effects of which go beyond nearest neighbors, determine the crystal structure. There is a given correlation between the ratio of s- to p-electrons and the resulting structure. The d-electrons, on the other hand, interact with the nearest neighbors to indirectly fix the crystal structure by determining the number of available p-electrons. Thus, the bcc structure is to be correlated with an electron configuration approximating the d^{n-1} one, with $n = d + s + p$-electrons. When the valence state corresponds to $d^{n-2}sp$ or $d^{n-3}sp^2$ the hcp and fcc structures are expected. On the other hand, the increasing overlap of d-orbitals from the 3d to the 5d element series increases the cohesive force of materials. The application of the Brewer–Engel theory in electrocatalysis is based on the parallelism between the alloy stability and the observed electrocatalytic activity for hydrogen evolution. Following these concepts, the basic idea arises that a synergetic effect appears by combining metals of the left half of the transition series that have empty or half-filled vacant d-orbitals with metals of the right half of the transition series with internally paired d-electrons not available for bonding in the pure metal. The Brewer multicomponent alloying approach provides a criterion for planning and prediction to select appropriate combinations leading to optimal electrocatalytic activity, such us Mo-Co, Mo-Ni and W-Ni alloys. In those cases, the synergetic effect is explained in terms of (i) the extension of the d-orbitals and (ii) the creation of semi-vacant d-pairs due to the sharing effect, which leads to a larger availability of d-orbitals for the hydrogen adsorption and related electron transfer. The use of this qualitative predicting correlation can be considered as a first orientation for the investigation of alternative alloys with outstanding electrocatalytic properties.

Other factors than the electronic structure of the surface may play a decisive role for the reactivity in an aqueous electrolyte. The electronic properties of Ni, for instance, let envisage good electrocatalytic properties. Experimentally, however, an exchange current density of just $-\log j_0$ (A cm^{-2}) = 5.2–5.9 is observed. Electrochemical studies of Ni-Mo alloys [133] have shown an exchange current of $-\log j_0 = 4.3$ for a rich Ni alloy ($Ni_{16}Mo, Ni_{24}Mo$), an order of magnitude larger than that for Ni, and two larger than that for Mo. Numerous studies of the hydrogen evolution on Ni surfaces show that it occurs by a Volmer–Heyrovsky-type mechanism where the electrochemical desorption of H_{ads} is the rate-determining step [134]. Highfield et al. [133] invoked a *spillover* effect to explain the synergetic effect upon alloying Ni and Mo. According to these authors, adsorption of H on Ni leads to a depletion of *d*-electron density at the Fermi level, and hence to a slowdown of the electrochemical desorption step (H $^+$-H_{ads} recombination) [135]. Mo is characterized by a strong H-adsorption (compare Ni and Mo in the volcano plot in Figure 4.13). According to Nørskov et al. [136], Mo should be the best promoter of H_2 dissociated adsorption due to the efficient filling of incoming adsorbate antibonding molecular orbitals. Thus, the synergy mechanism should involve the ready migration of H_{ads} created on Ni to neighboring Mo trap centers. This explanation, based practically in a mechanistic point of view, leaves more questions than answers, especially if one takes into account the energy demanding desorption of hydrogen from Mo sites.

In spite of their large influence, dielectric effects and surface reactions with water at the metal–electrolyte interface were ignored in most reported studies attempting to find a trend of metal and alloys respective to their electrocatalytic properties for hydrogen evolution. Ni presents a potential of zero charge of −0.30 V vs. NHE. This implies an orientation of the water molecules with the O-atom toward the surface at the hydrogen reversible potential. Moreover, Scherer et al. [137] have shown by STM and x-ray reflectivity experiments that dissolution and the formation of an oxide monolayer by adsorption with H_2O takes place on Ni(111) between −0.03 and 0.15 V vs. NHE. This leads to an increase of the surface roughness and the formation of $Ni(OH)_2$. It is not totally clear, however, by what mechanism Mo exerts its acceleration of the hydrogen evolution reaction. The preferential orientation of water molecules with the negative pole toward the surface implies an energy-demanding reorientation of the inner Helmholtz layer during the proton discharge [138].

If we look at the *pzc* of Mo, estimated by Trasatti [54] to be about −0.71 V, we can infer a stronger interaction of the Mo surface with water, which is related to a partial occupancy and the overlapping extent of the 4d orbitals. Thus, the synergetic effect found in NiMo alloys may be due to an electron transfer from Ni to Mo, in a similar way as predicted by the Brewer–Engel theory. This, in turn, decreases the water polarization at the Ni–electrolyte interface, and hence the energy required for reorientation during proton discharge.

REFERENCES

1. Grahame, D.C. 1947. The electrical double layer and the theory of electrocapilarity. *Chem. Rev.* 41:441.
2. The name was introduced in reference to the theory of electrical interface by Helmholtz. von Helmholtz, H. L. 1879. *Wied. Ann.* 7:337.
3. Gouy, G. 1909. *Comt. Rend. Acad. Sci.* 149:654.
4. Gouy, G. 1910. *J. Phys. (France)* 4:457.
5. Chapman, D.L. 1913. A contribution to the theory of electrocapillarity. *Philos. Mag.* 25:475.
6. Stern, O. 1924. *Z. Electrochem.* 30:508.

7. Bockris, J. O'M. and A.K.N. Reddy. 1970. *Modern electrochemistry*. New York: Plenum Publishing; Bockris, J. O'M. and S.U.M. Khan. 1993. *Surface Electrochemistry: A Molecular Level Approach*. New York: Plenum.

8. Li, N. and J. Lipkowski. 2000. Chronocoulometric studies of chloride adsorption at the Pt(111) electrode surface. *J. Electroanal. Chem.* 491:95.

9. Kolb, D.M. 1978. *Advances in Electrochemistry and Electrochemical Engineering*. Gerischer, H., and C.W. Tobias (Eds.), Vol. 11. New York: John Wiley & Sons.

10. Lipkoski,. J. and R.N. Ross (Eds.). 1993. *Structure of Electrified Interfaces*. Weinheim: VCH Verlagsgesellschaft GmbH.

11. Kibler, L.A. 2006. Hydrogen electrocatalysis. *ChemPhysChem* 7:985.

12. Hoshi, N., Mizumura, T. and Y. Hori. 1995. Significant difference of the reduction rates of carbon dioxide between Pt(111) and Pt(110) single crystal electrodes. *Electrochim. Acta* 40:883.

13. Zangwill, A. 1998. *Physics at surfaces*. Cambridge: Cambridge University Press.

14. Lang, N.D. and W. Kohn. 1970. Theory of metal surfaces: charge density and surface energy. *Phys. Rev. B* 1:4555.

15. Levine, I.N. 2000. *Quantum chemistry,* 5th ed. New Jersey: Prentice Hall.

16. Oura, K., Lifshits, V.G., Saranin, A.A., Zotov, A.V. and M. Katayama. 2003. *Surface Science: An Introduction*. Berlin: Springer Verlag.

17. Crommie, M.F., Lutz, C.P. and D.M. Eigler. 1993. Imaging standing waves in a two-dimensional electron gas. *Nature* 363:524.

18. Suzuki, T. and Y. Hasegawa. 2001. Electron standing-wave observation in the Pd over-layer on Au(111) and Cu(111) surfaces by scanning tunneling microscopy. *Phys. Rev. B* 64:081403(R).

19. Xia, X. and M.L. Berkowitz. 1995. Electric-field induced restructuring of Water at a Platinum-Water interface: a molecular dynamics computer simulation. *Phys. Rev. Lett.* 74:3193.

20. Schweighofer, K.J., Xia, X. and M.L. Berkowitz. 1996. Molecular dynamics study of water next to electrified Ag(111) surfaces. *Langmuir* 12:3747.

21. Spohr, E. 1999. Molecular simulation of the electrochemical double layer. *Electrochim. Acta* 44:1697.

22. Price, D.L. and J.W. Halley. 1995. Molecular dynamics, density functional theory of the metal–electrolyte interface. *J. Chem. Phys.* 102:6603.

23. Spohr, E., Tóth, G. and K. Heinzinger. 1996. Structure and dynamics of water and hydrated anions near platinum and mercury surfaces as studied by MD simulations. *Electrochim. Acta* 41:2131.

24. Bockris, J.O'M. and S.U.M. Khan 1993. *Surface electrochemistry. A molecular level approach*. New York: Plenum Press.

25. Harrison, J.A., Randles, J.E.B. and D.J. Schiffrin. 1973. The entropy of formation of the mercury-aqueous solution interface and the structure of the inner layer. *J. Electroanal. Chem.* 48:359.

26. Stenschke, H. 1985. Polarization of water in the metal/electrolyte interface. *J. Electroanal. Chem.* 196:261.

27. Teschke, O., Ceotto, G. and E.F. de Souza. 2000. Interfacial aqueous solutions dielectric constant measurements using atomic force microscopy. *Chem. Phys. Lett.* 326:328.

28. Hugelmann, M. and W. Schindler. 2004. in situ distance tunneling spectroscopy at Au(111)/0.02 M HClO4. *J. Electrochem. Soc.* 151:E97.

29. Simeone, F.C., Kolb, D.M., Venkatachalam, S. and T. Jacob. 2007. The Au(111)/electrolyte interface: a tunnel-spectroscopic and DFT investigation. *Angew. Chem. Int. Ed.* 46:8903.

30. Amokrane, S. and J.P. Badiali. 1992. Analysis of the capacitance of the metal-solution interface: Role of the metal and the metal-solvent coupling. *Modern Aspects of Electrochemistry*, Vol. 22, Bockris, J.O'M. et al. (Eds.). New York: Plenum Press.
31. Lang, N.D. and W. Kohn. 1973. Surface-dipole barriers in simple metals. *Phys. Rev. B* 8:6010.
32. Born, M. and R. Oppenheimer. 1927. Zur Quantentheorie der Molekeln. *Ann. Phys.* 389:457.
33. Schmickler, W. 1986. A theory of adiabatic electron-transfer reactions. *J. Electroanal. Chem.* 204:31.
34. Koneshan, S., Rasaiah, J.C., Lynden-Bell, R.M. and S.H. Lee. 1998. Solvent structure, dynamics, and ion mobility in aqueous solutions at 25°C. *J. Phys. Chem. B* 102:4193.
35. Ohtaki, H. and T. Radnai. 1993. Structure and dynamics of hydrated ions. *Chem. Rev.* 93:1157.
36. Bockris, J. O'M and P.P.S. Saluja. 1972. Ionic solvation numbers from compressibilities and ionic vibration potentials measurements. *J. Phys. Chem.* 76:2140.
37. Caminiti, R., Nacheri, G., Paschina, G., Piccaluga, G. and G. Pinna. 1980. Interactions and structure in aqueous $NaNO_3$ solutions. *J. Chem. Phys.* 72:4522.
38. Michaellian, K. H. and M. Moskovits. 1978. Tetrahedral hydration of ions in solution. *Nature* 273:135.
39. Dwayne Miller, R.J., McLendon, G.L., Nozik, A.J., Schmickler, W. and F. Willig. 1995. *Surface electron transfer processes.* New York: VCH Publishers.
40. Details about the derivation of this expression can be read in the works of Levich and Dogonadze (*Dokl. Akad. Nauk SSSR* 1959,124, 123), Schmickler and Ulstrup (*Chem. Phys.* 1977, 19, 217), and Fischer (*J. Chem. Phys.* 1970, 53, 3195).
41. Tr is *trace*, which is defined as

$$Tr|A| = \sum_{i=1}^{n} \lambda_i,$$

where λ_i are the eigenvalues of the matrix A.
42. Marcus, A. 1956. On the theory of oxidation-reduction reactions involving electron transfer I. *J. Chem. Phys.* 24:966.
43. Heller, A., Aharon-Shalom, E., Bonner, W.A. and B. Miller. 1982. Hydrogen-evolving semiconductor photocathodes: Nature of the junction and function of the platinum group metal catalyst. *J. Am. Chem. Soc.* 104:6942.
44. Chandra, N., Wheeler, B.L. and A.J. Bard. 1985. Semiconductor electrodes. 59. Photocurrent efficiencies at p-indium phosphide electrodes in aqueous solutions. *J. Phys. Chem.* 89:5037.
45. Lewerenz, H.J., Heine, C., Skorupska, K., Szabo, N., Hannappel, T., Vo-Dinh, T., Campbell, S.A., Klemm, H.W. and A.G. Muñoz. 2010. Photoelectrocatalysis: Principles, nanoemitter applications and routes to bio-inspired systems. *Energy & Environmental Science* 3:748.
46. Muñoz, A.G. and H.J. Lewerenz. 2010. Advances in Photoelectrocatalysis with Nanotopological Photoelectrodes. *ChemPhyChem* 11:1603.
47. Tafel, J. 1905. Über die Polarisation bei kathodischer Wasserstoffentwicklung. *Z. Phys. Chem. Stöchiom. Verwand.* 50:641.
48. Gurney, R.W. 1932. The quantum mechanics of electrolysis. *Proc. Roy. Soc.* A134:137.
49. Horiuti, J. and M. Polanyi. 1935. Outlines of a theory of proton transfer. *Acta Physicochim. URSS* 2:505 (translation from Russian in *J. Mol. Cat. A: Chem.* 2003, 199, 185).
50. Butler, J.A.V. 1936. Hydrogen overvoltage and the reversible hydrogen electrode. *Proc. Roy. Soc.* A157:423.

51. B.E. Conway and J.O´M. Bockris. 1957. Electrolytic hydrogen evolution kinetics and its relation to the electronic and adsorptive properties of the metal. *J. Chem. Phys.* 26:532.
52. Marcus, R.A. 1965. On the theory of electron-transfer reactions. VI. Unified treatment for homogeneous and electrode reactions. *J. Chem. Phys.* 43:679.
53. Conway, B.E. and M. Salomon. 1964. Electrochemical reaction orders: applications to the hydrogen- and oxygen evolution reactions. *Electrochim. Acta* 9:1599; Studies on the hydrogen evolution reaction down to $-150°C$ and the role of proton tunnelling. *J. Chem. Phys.* 41:3169.
54. Trasatti, S. 1971. Work function, electronegativity and electrochemical behaviour of metals (I). *J. Electroanal. Chem.* 33:351. (II); 1972, 39:163 (III); 1973, 44:367 (IV).
55. Enyo, M., Conway, B.E., Bockris, J.O'M, Yeager, E., Khan, S.U.M. and R.E. White (Eds.). 1983. *Comprehensive Treatise of Electrochemistry*. New York: Plenum Press.
56. Kibler, L. 2006. Hydrogen electrocatalysis. *ChemPhysChem* 7:985; Nørskov, J.K., Bligaard, T., Logadottir, A., Kitchin, J.R., Chen, J.G., Pandelov, S. and U. Stimming. 2005. Trends in the exchange current for hydrogen evolution. *J. Electrochem. Soc.* 152:J23.
57. Conway, B.E. 1999. Electrochemical processes involving H adsorbed at metal electrode surfaces, in A. Wieckowski (Ed.), *Interfacial Electrochemistry, Theory, Experiment and Applications*. New York: Marcel Dekker.
58. Marx, D., Tuckerman, M.E., Hutter, J. and M. Parrinello. 1999. The nature of the hydrated excess proton in water. *Nature* 397:601; Eigen, M. 1964. Proton transfer, acid-base catalysis and enzymatic hydrolysis. *Angew. Chem. Int. Edn* 3:1; Zündel, G. 1976. *The hydrogen bond: Recent developments in theory and experiments. II. Structure and spectroscopy*, Schuster, P., Zündel, G., Sandorfy, C. (Eds.). Amsterdam: North-Holland.
59. Kima, J., Schmitt, U.W., Gruetzmacher, J.A., Voth, G.A. and N.E. Scherer. 2002. The vibrational spectrum of the hydrated proton: Comparison of experiment, simulation, and normal mode analysis. *J. Chem. Phys.* 116:737.
60. Arlinghaus, F.J., Gay, J.G. and J.R. Smith. 1981. Surface states on d-band metals. *Phys. Rev. B* 23:5152.
61. Benesh, G.A., Liyanage, L.S.G. and J.C. Pingel. 1990. The surface electronic structure of (1×1) Pt(001). *J. Phys.: Condens. Matter* 2:9065.
62. Bouckaert, L.P., Smoluchowski, R. and E. Wigner. 1936. Theory of Brillouin zones and symmetry properties of wave functions in crystals. *Phys. Rev.* 50:58.
63. Muscat, J.P. and D.N. Newns. 1978. Chemisorption on metals. *Prog. Surf. Sci.* 9:1.
64. According to the scaling property of the delta: $\delta(xa) = (1/|\alpha|)\,\delta(x)$. Hence, delta-function $\delta(E)$ is given in eV^{-1}.
65. Morra, R.M., Almeida, F.J.D. and R.F. Willis. 1990. ARUPS observation of a strong surface state on Rh(001). *Physica Scripta* 41:594.
66. Greeley, J. and M. Mavrikakis. 2005. Surface and subsurface hydrogen: Adsorption properties on transition metals and near-surface alloys. *J. Phys. Chem. B* 109:3460.
67. Lindgren, S.A. and L. Walldén. 1987. Discrete valence-electron states in thin metal overlayers on a metal. *Phys. Rev. Lett.* 59:3003.
68. Memmel, N., Rangelov, G. and E. Bertel. 1993. Influence of the substrate on the band structure of alkali metal monolayers. *Surf. Sci.* 285:109.
69. Memmel, N., Rangelov, G., Bertel, E. and V. Dose. 1991. Modification of surface states by alkali-metal adsorption and surface reconstruction. An inverse-photoemission study of Na/Ni(110). *Phys. Rev. B* 43:6938.
70. Wertheim, G.K. 1989. Electronic structure of metal clusters. *Z. Phys. D—Atoms, Molecules and Clusters* 12:319.
71. Sánchez, C.G., Leiva, E.P.M. and W. Schmickler. 2003. On the catalytic activity of palladium clusters generated with the electrochemical scanning tunnelling microscope. *Electrochem. Commun.* 5:584.

72. Mukerjee, S. 1990. Particle size and structural effects in platinum electrocatalysis. *J. Appl. Electrochem.* 20:537.

73. Heiz, U., Sanchez, A., Abbet, S. and W.-D. Schneider. 1999. The reactivity of gold and platinum metals in their cluster phase. *Eur. Phys. J. D* 9:35.

74. Somorjai, G.A., Contreras, A.M., Montano, M. and R.M. Rioux. 2006. Clusters, surfaces, and catalysis. *PNAS* 103:10577.

75. Wang, J., Ocko, B.M., Davenport, A.J. and H.S. Isaacs. 1992. In situ x-ray-diffraction and -reflectivity studies of the Au(111)/electrolyte interface: Reconstruction and anion adsorption. *Phys. Rev. B* 46:10321.

76. Schneider, J. and D.M. Kolb. 1988. Potential induced surface reconstruction of Au(100). *Surf. Sci.* 193:579.

77. Chassaing, E., Joussellin, M. and R. Wiart. 1983. The kinetics of nickel electrodeposition inhibition by adsorbed hydrogen and anions. *J. Electroanal. Chem.* 157:75.

78. Muñoz, A.G. and J.W. Schultze. 2004. Effects of NO_2^- on the corrosion of Ni in phosphate solutions. *Electrochim. Acta* 49:293.

79. Wiart, R. 1990. Elementary steps of electrodeposition analysed by means of impedance spectroscopy. *Electrochim. Acta* 35:1587.

80. The Car–Parrinello method combines molecular dynamics with density functional theory and allows making *ab initio* quantum mechanical calculations. The validity of classical mechanics for the ionic motion and the Born–Oppenheimer approximation to separate the nuclear and electron coordinates are assumed.

81. Car, R. and M. Parrinello. 1985. Unified approach for molecular dynamics and density-functional theory. *Phys. Rev. Lett.* 55:2471.

82. Geissler, P.L., Dellago, C., Chandler, D., Hutter, J. and M. Parrinello. 2001. Autoionization in liquid water. *Science* 291:2121.

83. Brüesch, P. and Th. Christen. 2004. The electric double layer at a metal electrode in pure water. *J. Appl. Phys.* 95:2846.

84. Su, Z., Bühl, M. and W. Zhou. 2009. Dissociation of water during formation of anodic aluminium oxide. *J. Am. Chem. Soc.* 131:8697.

85. Teschke, O., Ceotto, G. and E.F. de Souza. 2000. *Chem. Phys. Lett.* 326:328; Teschke, O., Ceotto, G. and E.F. de Souza. 2001, *Phys. Rev. E* 64:011605.

86. Frumkin, A.N. 1925. *Z. Phys. Chem.* 166:466.

87. Chaplin, M.F. 2011. Water structure and behavior, http//www.lsbu.ac.uk/water.

88. The Hartree–Fock method is used to calculate the orbital energy and wave function of quantum mechanical many-particle systems. It consists of an *ab initio* procedure, whereby no empiric parameters and only natural constants are required. This method neglects interactions among electrons so that the Hamiltonian is given by

$$H(r_1, r_2 \ldots r_N) = \sum_{i=1}^{N} H_i(r_i); \quad H_i = -\frac{1}{2m_e}\nabla_i^2 - \frac{e^2}{4\pi\varepsilon_0}\sum_i\sum_k \frac{Z_k}{r_{ik}}$$

where $Z_k\,e$ represents the charge of nucleus k. The molecular wave function will be expressed as the product of one-particle functions for electrons $1,2,\ldots,N$. The Schrödinger equation is then solved as a system of equations, $H_i(i)\psi_i(i) = \varepsilon_i\psi_i$. Each electron will be described by a given wave function $\psi_i(\vec{r}_i, \vec{R})$, which for a fixed configuration of atom cores depends only on the electron coordinates $\vec{r}_i$. To include the electron permutation among the one-particle wave functions, the total wave function is written in the form of a Slater determinant, where the molecular orbitals represent the product of a spatial and a spin function, $\psi_i(i)\chi_i(i)$. Thus, we have

$$\Psi(1,2,....N) = \frac{1}{\sqrt{N!}} \begin{vmatrix} \psi_1(1) & \psi_1(2)... & \psi_1(N) \\ \psi_2(1) & & \\ \vdots & & \vdots \\ \psi_N(1) & & \psi_N(N) \end{vmatrix}$$

wherein the Pauli principle is fulfilled, since the determinant is zero when two rows or columns are equal. The molecular orbitals can be regarded as a linear combination of atomic wave functions:

$$\psi = \sum_i c_i \psi_i^A$$

This approach is known as *linear combination of atomic orbitals* (LCAO). The energy of each molecular orbital is given by $E_i = \langle \psi_i | H | \psi_i \rangle / \langle \psi_i | \psi_i \rangle$. The effective potential for a given electron is constituted by the Coulomb potential of the cores and those resulting from the distribution of all other $N-1$ electrons (which can be determined only after resolving the Schrödinger equation). Thus, the problem must be resolved iteratively by starting with one-particle wave function $\psi_i^0(i)$ obtained by linear combination of atomic orbitals. This method is known as *self-consistent field Hartree–Fock* (SCF-HF). The iteration must be continued until the calculated wave function in step k does not differ largely from that in step $k-1$.

89. Hori, Y. 2008. Electrochemical CO_2 reduction on metal electrodes, in *Modern Aspects of Electrochemistry*, Vol. 42, C. Vayenas et al. (Eds.). New York: Springer.
90. Mizuno, T., Naitoh, A. and K. Ohta. 1995. Electrochemical reduction of CO_2 in methanol at $-30°C$. *J. Electroanal. Chem.* 391:199.
91. Kaneco, S., Hiei, N., Xing, Y., Katsumata, H., Ohnishi, H., Suzuki, T. and K. Ohta. 2002. Electrochemical conversion of carbon dioxide to methane in aqueous $NaHCO_3$ solution at less than 273 K. *Electrochim. Acta* 48:51.
92. Kaneco, S., Iiba, K., Suzuki, S., Ohta, K. and T. Mizuno. 1999. Electrochemical reduction of carbon dioxide to hydrocarbons with high Faradaic efficiency in LiOH/methanol. *J. Phys. Chem. B* 103:7456.
93. Hara, K., Kudo, A. and T. Sakata. 1995. Electrochemical reduction of carbon dioxide under high pressure on various electrodes in an aqueous electrolyte. *J. Electroanal. Chem.* 391:141.
94. Abbott, A.P., and C.A. Eardley. 2000. Electrochemical reduction of CO_2 in a mixed supercritical fluid. *J. Phys. Chem. B* 104:775.
95. Amatore, C. and J.M. Sevéant. 1981. Mechanism and kinetic characteristics of the electrochemical reduction of carbon dioxide in media of low proton availability. *J. Am. Chem. Soc.* 103:5021.
96. Kaneco, S., Katsumata, H., Suzuki, T. and K. Ohta. 2006. Electrochemical reduction of carbon dioxide to ethylene at a copper electrode in methanol using potassium hydroxide and rubidium hydroxide supporting electrolytes. *Electrochim. Acta* 51:3316.
97. Ikeda, S., Takagi, T. and K. Ito. 1987. Selective formation of formic acid, oxalic acid, and carbon monoxide by electrochemical reduction of carbon dioxide. *Bull. Chem. Soc. Jpn.* 60:2517.
98. Hochgesand, G. 1970. Rectisol and purisol. *Ind. Eng. Chem.* 62:37.
99. Kaneco, S., Iiba, K., Hiei, N., Ohta, K., Mizuno, T. and T. Suzuki. 1999. Electrochemical reduction of carbon dioxide to ethylene with high Faradaic efficiency at a Cu electrode in CsOH/methanol. *Electrochim. Acta* 44:4701.

100. Kaneco, S., Iiba, K., Yabuuchi, M., Nishio, N., Ohnishi, H., Katsumata, H., Suzuki, T. and K. Ohta. 2002. High efficiency electrochemical CO_2-to-methane conversion method using methanol with lithium supporting electrolytes. *Ind. Eng. Chem. Res.* 41:5165.

101. Hoshi, N., Noma, M., Suzuki, T. and Y. Hori. 1997. Structural effect on the rate of CO_2 reduction on single crystal electrodes on palladium. *J. Electroanal. Chem.* 421:15.

102. Nikolic, B.Z., Huang, H., Gervasio, D., Lin, A., Fierro, C., Adzic, R.R. and E.B. Yeager. 1990. Electroreduction of carbon dioxide on platinum single crystal electrodes: electrochemical and in situ FTIR studies. *J. Electroanal. Chem.* 295:415.

103. Hoshi, N., Ito, H., Suzuki, T. and Y. Hori. 1995. CO_2 reduction on Rh single crystal electrodes and the structural effect. *J. Electroanal. Chem.* 395: 309.

104. Hoshi, N., Mizumura, T. and Y. Hori. 1995. Significant difference of the reduction of carbon dioxide between Pt(111) and Pt(110) single crystal electrodes. *Electrochim. Acta* 40:883.

105. Inuzuka, N. 1992. Jpn. Kokai Tokkyo Koho JP 04 314 881, 6 Nov. *Chem. Abstr.* 120 (1994) 333692g.

106. Ohkawa, K., Hashimoto, K., Fujishima, A., Noguchi, Y. and S. Nakayama. 1993. Electrochemical reduction of carbon dioxide on hydrogen storing materials: Part 1. The effect of hydrogen absorption on the electrochemical behavior on palladium electrodes. *J. Electroanal. Chem.* 345:445.

107. Ayers, W.M. and M. Farley. 1988. *Carbon dioxide reduction with an electric field assisted hydrogen insertion reaction. Catalytic activation of carbon dioxide,* Ch. 11, pp. 147–154. ACS Symposium Series, Vol. 363. American Chemical Society.

108. Jitaru, M., Lowy, D.A., Toma, M., Toma B.C. and L. Oniciu. 1997. Electrochemical reduction of carbon dioxide on flat metallic cathodes. *J. Appl. Electrochem.* 27:875.

109. Varghese, O.K., Paulose, M., LaTempa, T.J. and C.A. Grimes. 2009. High-rate solar photocatalytic conversion of CO_2 and water vapor to hydrocarbon fuels. *Nano Lett.* 9:731.

110. Hirano, K., Inoue, K. and T. Yatsu. 1992. Photocatalysed reduction of CO_2 in aqueous TiO_2 suspension mixed with copper powder. *J. Photochem. Photobiol. A: Chem.* 64:255.

111. Adachi, K., Ohta, K. and T. Mizuna. 1994. Photocatalytic reduction of carbon dioxide to hydrocarbon using copper-loaded titanium dioxide. *Sol. Energy* 53:187.

112. Tan, S.S., Zou, L. and E. Hu. 2007. Photosynthesis of hydrogen and methane as key components for clean energy systems. *Sci. Technol. Adv. Mater.* 8:89.

113. Tanaka, K., Miyahara, K. and I. Toyoshima. 1984. Adsorption of CO_2 on TiO_2 and Pt/TiO_2 studied by X-ray photoelectron spectroscopy and Auger electron spectroscopy. *J. Phys. Chem.* 88:3504.

114. Hirota, K., Tryk, D.A., Yamamoto, T., Hashimoto, K., Okawa, M. and A. Fujishima. 1998. Photoelectrochemical reduction of CO_2 in a high-pressure CO_2+methanol medium at p-type semiconductor electrodes. *J. Phys. Chem. B* 102:9834.

115. Aylmer-Kelly, A.W.B., Bewick, A., Cantrill, P.R. and A.M. Tuxford. 1973. Studies of electrochemically generated reaction intermediates using modulated specular reflectance spectroscopy. *Faraday Discuss. Chem. Soc.* 56:96.

116. Vassiliev, Y.B., Bagotzky, V.S., Osetrova, N.V., Khazova, O.A. and N.A. Mayorova. 1985. Electroreduction of carbon dioxide: Part III. Adsorption and reduction of CO_2 on platinum *J. Electroanal. Chem.* 189:311.

117. Rossi, A.R. and K.D. Jordan. 1979. Comment on the structure and stability of $(CO_2)_2^-$. *J. Chem. Phys.* 70:4422.

118. Bard, A.J. and L.R. Faulkner. 1980. *Electrochemical methods: Fundamentals and applications.* New York: John Wiley & Sons.

119. Kaneco, S., Ueno, Y., Katsumata, H., Suzuki, T. and K. Ohta. 2009. Photoelectrochemical reduction of CO_2 at p-InP electrode in copper particle-suspended methanol. *Chem. Eng. J.* 148:57.

120. Aurian-Blajeni, B., Habib, M.A., Taniguchi, I. and J.O´M. Bockris. 1983. The study of adsorbed species during the photoassisted reduction of carbon dioxide at a p-CdTe electrode. *J. Electroanal. Chem.* 157:399.

121. Bockris, J.O´M. and J. Wass. 1989. The photoelectrocatalytic reduction of carbon dioxide. *J. Electrochem. Soc.* 136:2521.

122. Barton, E.E., Rampulla, D.M. and A.B. Bocarsly. 2008. Selective solar-driven reduction of CO_2 to methanol using a catalyzed p-GaP based photoelectrochemical cell. *J. Am. Chem. Soc.* 130:6342.

123. Kitco Metals Inc., Montreal, Canada. http://www.kitco.com.

124. Brooman, E.W. and A.T. Kuhn. 1974. Correlations between the rate of the hydrogen electrode reaction and the properties of alloys. *Electroanal. Chem. Int. Electrochem.* 49:325.

125. Greeley, J., Nørskov, J.K., Kibler, L.A., El-Aziz, A.M. and D.M. Kolb. 2006. Hydrogen evolution over bimetallic systems: understanding the trends. *ChemPhysChem* 7:1032.

126. Nørskov, J. K., Rossmeisl, J., Logadottir, A., Lindqvist, L., Kitchin, J. R., Bligaard, T. and H. Jonsson. 2004. Origin of the overpotential for oxygen reduction at a fuel-cell cathode. *J. Phys. Chem. B* 108:17886.

127. Nørskov, J. K., Bligaard, T., Logadottir, A., Kitchin, J. R., Chen, J. G., Pandelov, S. and U. Stimming. 2005. Trends in the exchange current for hydrogen evolution. *J. Electrochem. Soc.* 152:J23.

128. Stamenkovic, V.R., Mun, B.S., Arenz, M., Mayrhofer, K.J.J., Lucas, C.A., Wang, G., Ross, P.N. and N.M. Markovic. 2007. Trends in electrocatalysis on extended and nanoscale Pt-bimetallic alloy surfaces. *Nature Mater.* 6:241.

129. Hammer, B. and J.K. Nørskov. 1995. Electronic factors determining the reactivity of metal surfaces. *Surf. Sci.* 343:211.

130. Nørskov, J.K. 1990. Chemisorption on metal surfaces. *Rep. Prog. Phys.* 53:1253.

131. Jakšić, M.M. 1984. Electrocatalysis of hydrogen evolution in the light of the Brewer-Engel theory for bonding in metals and intermetallic phases. *Electrochim. Acta* 29:1539.

132. Ezaki, H., Morinaga, M. and S. Watanabe. 1993. Hydrogen overpotential for transition metals and alloys, and its interpretation using an electronic model. *Electrochim. Acta* 38:557.

133. Highfield, J.G., Claude, E. and K. Oguro. 1999. Electrocatalytic synergism in Ni/Mo cathodes for hydrogen evolution in acid medium: a new model. *Electrochim. Acta* 44:2805.

134. Appleby, A.J., Kita, H., Chemla, M. and G. Bronoel. 1982. *Encyclopaedia of electrochemistry of the elements*, A.J. Bard (Ed.), Vol. 9, Ch. 3. New York: Marcel Dekker.

135. Christmann, K.R. 1988. *Hydrogen Effects in Catalysis*, Z. Paal, P.G. Menon (Eds.), Ch. 1. New York: Marcel Dekker.

136. Nørskov, J.K. and F. Besenbacher. 1987. Theory of hydrogen interaction with metals. *J. Less-Common Met.* 130:475.

137. Scherer, J., Ocko, B.M. and O.M. Magnussen. 2003. Structure, dissolution, and passivation of Ni(111) electrodes in sulfuric acid solution: an in situ STM, X-ray scattering, and electrochemical study. *Electrochim. Acta* 48:1169.

138. Schultz, Z.D., Shaw, S.K. and A.A. Gewirth. 2005. Potential dependent organization of water at the electrified metal-liquid interface. *J. Am. Chem. Soc.* 127:15916.

5 Electronics and Chemistry of Electrolyte–Metal–Oxide–Semiconductor Contacts

Apart from the electrocatalytic activity of the metal–electrolyte contact, the conversion efficiency of photoelectrochemical cells depends on the extent of the recombination rate of photogenerated charge carriers at interface states with respect to the rate of charge transfer from the semiconductor across the interfacial layers to the electrolyte. The formation of metal–semiconductor junctions by electrochemical deposition is accompanied by often-unavoidable lateral surface chemical and electrochemical reactions that lead to the formation of localized energy states closely related to the chemical transformation of the surface structure of the semiconductor. Thus, the selection of plating conditions for controlling the level of arising pernicious interface states requires a better knowledge of chemical and electronic behavior of the semiconductor–electrolyte interface. In this chapter, combined electrochemical and spectroscopic analysis of surface states will be presented with emphasis on the analysis of the Si-SiO$_x$ interface as a model system.

5.1 METHODS: COMBINATION OF ELECTROCHEMICAL TECHNIQUES WITH SURFACE-SENSITIVE SPECTROSCOPY

Photoelectron spectroscopy is the most commonly used technique to investigate the chemical structure of surfaces and elucidate the course of chemical and (photo) electrochemical reactions. This contributes to the interpretation of electrochemical investigations of the surface electronic properties by means of electrochemical impedance spectroscopy (EIS) and scanning electrochemical microscopy (SECM), which will be briefly reviewed in this chapter.

Firstly, let us review the fundamentals of the technique, which often appear a bit foreign to electrochemists.

Photoelectron spectroscopy is a technique based on the analysis of emitted electrons upon irradiation with x-rays. Figure 5.1 shows a schematic of the photoemission process. Electrons with a binding energy E_B with respect to the Fermi level are excited by photons of an energy $h\nu$ leaving the solid with a kinetic energy given by [1]

$$E_{kin} = h\nu - E_B - \Phi \tag{5.1}$$

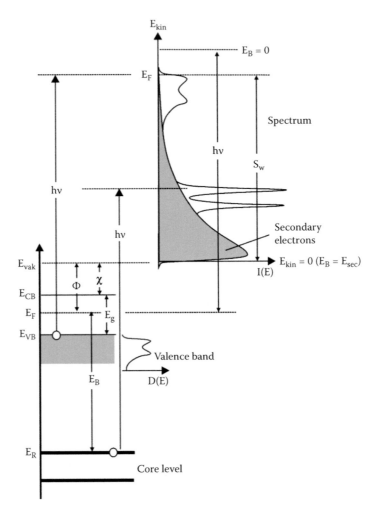

FIGURE 5.1　Schematic showing the electron excitation process by an incident x-ray beam.

As schematized in Figure 5.1, the incident photons can excite electrons from the valence band and the deep core level of an atom. The irradiation of the surface with photons of constant energy causes the excitation of electrons, which acquire different kinetic energies resulting in an energy distribution of the transduced signal of the detector in the form of a spectrum $I(E_{kin})$. The intensity is proportional to the density of occupied electronic states $N(E)$. Electrons emitted from deep core levels appears as sharp lines, the broadening of which arises from statistical processes inherent to the lifetime of photoemitted electrons and instrumental resolution. In addition to the peaks due to elastic photoelectrons, a number of additional features appear in the spectrum, such as a continuous background of inelastic secondary electrons, Auger peaks, and peaks arising from plasmon losses. The secondary electron edge defines the position for $E_{kin} = 0$. Thus, the affinity of the material can be calculated from the spectrum width S_W as

$$S_w = E_{sec} - E_{VB} = h\nu - E_g - \chi \tag{5.2}$$

This feature is used to quantify the formation of surface dipoles by adsorption or formation of surface bonds. The change of the binding energy of a given core level with the chemical environment is the core of this technique, which allows identification of elements and their oxidation or chemical states. The binding energy measured for an element in a solid deviates from that corresponding to an isolated atom. This deviation can be regarded as arising from three fundamental contributions: (i) the formation of chemical bonds, which involves redistributions of the charge density around the atom (ΔE_B^{chem}); (ii) an electrostatic term related with the influence of long range interactions on the local potential (ΔE_B^{el}), especially important in the case of ionic crystals; and, (iii) a relaxation term representing the behavior of valence electrons during the photoemission process (E_B^{relax}). Therefore, we can write

$$E_B = E_B^0 + \Delta E_B^{chem} + \Delta E_B^{el} - E_B^{relax} \tag{5.3}$$

ΔE_B^{chem} is proportional to the change of the charge distribution around the atom, and hence to the electronegativity difference between the atom and the nearest neighbors. ΔE_B^{el} is quantified by the Madelung equation, which for an atom of charge z_i is given by

$$\Delta E_B^{el} = z_i \frac{e^2}{4\pi\varepsilon_0} \sum_{j \neq i} \frac{z_j}{r_{ij}} \tag{5.4}$$

where r_{ij} is the distance from nearest neighbors of charge z_j. The Madelung term is defined for ionic solids as $\Delta E_B^{el} = \alpha_M e^2 / 4\pi\varepsilon_0 r_0$, where α_M is the Madelung constant and r_0 is the lattice distance. The first three terms in Eq. (5.3) describe the photoionization process as transition from a state with N electrons to a system with $N-1$ electrons, where it is assumed that the orbitals of the ion are identical to those of the neutral atom. This description is known as the frozen orbital approximation or Koopmans theorem [2]. The additional relaxation term represents the gain of kinetic energy due to secondary processes, which leads to a screening of the remaining hole.

The core level of elements with an angular momentum quantum number $l > 0$ shows a line splitting because spin-orbit coupling according to the total angular momentum $j = l + s$. The relation between the line intensities is given by their multiplicities $m = 2j+1$ as

$$\frac{I(j_+)}{I(j_-)} = \frac{2(l + \frac{1}{2})+1}{2(l - \frac{1}{2})+1} \tag{5.5}$$

The excitation process can be theoretically treated by applying the perturbation theory of first order and the Fermi golden rule. In the photoemission process, the excitation of an electron in its ground state, characterized by the wave function $|\psi_0\rangle$, goes to the final state $|\psi_f\rangle$. This leaves a photoelectron with an impulse $\bar{k}$ and a kinetic energy $E_{kin} = \hbar^2 k^2 / 2m_e$ and a system of $N-1$ electrons with a total energy $E_{f,s}$. According to the golden rule, the probability of the excitation process is given by [3]

$$\omega(h\nu) = \frac{2\pi}{\hbar} \sum_s \left|\langle \psi_{f,s} | H_{PE} | \psi_0 \rangle\right|^2 \delta(E_{kin} + E_{f,s} - E_i - h\nu) \tag{5.6}$$

where the subscript s refers to the different excited states of the left $(N-1)$ electron system. The perturbation operator H_{PE} accounts for the interaction of electrons with the vector potential, $\vec{A}$, of the photon field. Because the electromagnetic field consists of two vector fields, the electric $\vec{E}(r,t)$ and the magnetic field, $\vec{B}(r,t)$, a third vector $\vec{A}(r,t)$ connecting both is used to represent the electromagnetic interaction. The electric and magnetic fields are given by $\vec{E}(r,t) = -\partial \vec{A}(r,t)/\partial t$ and $\vec{B}(r,t) = \vec{\nabla} \times \vec{A}(r,t)$, respectively. The perturbation operator is obtained by replacing the impulse operator $\vec{p} = -(\hbar/i)\vec{\nabla}$ in the absence of an electromagnetic field by $\vec{p} - (e/c)\vec{A}$ in the unperturbed Hamiltonian $H = \vec{p}^2 / 2m_e + V(\vec{r})$. Hence, we have

$$H = H_0 + H_{PE} = \underbrace{\frac{p^2}{2m_e} + V(\vec{r})}_{H_0} + \underbrace{\frac{e}{2m_e c}(\vec{A}\cdot\vec{p} + \vec{p}\cdot\vec{A}) + \frac{e^2}{2m_e c^2}A^2}_{H_{PE}} \tag{5.7}$$

Some simplifications are introduced in the above description to make it manageable for the interpretation of most photoemission experiments. The term A^2 represents two-photon processes. Thus, for one-photon processes, $A^2 = 0$. Local field corrections due to inhomogeneous distribution of the electron density are not considered, so that $\vec{\nabla}\cdot\vec{A} = 0$ and

$$H_{PE} = \frac{e}{m_e c}\vec{A}\cdot\vec{p} \tag{5.8}$$

In the so-called sudden approximation, it is assumed that the excitation of the electron is fast enough to consider that the orbitals of the left system of $N-1$ electron remain unchanged during the emission. Accordingly, the final state $|\psi_{f,s}\rangle$ of the system with N electrons can be decoupled in the photoexcited electron and the left $N-1$ system by introducing a creation operator in the secondary quantization notation as follows:

$$|\psi_{f,s}\rangle = |f, N-1, s\rangle \rightarrow c_{f,s}^+ |N-1, s\rangle \tag{5.9}$$

Substituting Eq. (5.9) into (5.6) yields

$$I(h\nu) \propto \frac{2\pi}{\hbar} \sum_{f,i,k} \left|\langle \psi_f | H_{PE} | \psi_{i,k} \rangle\right|^2 \underbrace{\sum_s \left|\langle N-1, s | c_k | N \rangle\right|^2 \delta(E_{kin} + E_{s(N-1)} - E_{i(N)} - h\nu)}_{A_k(E)} \tag{5.10}$$

where c_k is an electron annihilation operator.

$A_k(E)$ is often called the spectral function. The photocurrent detected in a photoemission process consists of lines arising from the photoionization of various orbitals k. Each line appears together with its satellites according to the number of excited states s created in the photoexcitation of the particular orbital k.

Because of the *fast* infinite degrees of freedom, the core level photoemission spectrum in metals deserves a different treatment. The spectral function $A_k(E)$, equal to the sum of the overlap matrix elements, is written as

$$A_k(E) = \sum_s |\langle i | f_s \rangle| \tag{5.11}$$

The time variation of the outer electron system due to its interaction with the core hole $|f(t)\rangle$ is given by

$$|f(t)\rangle = \exp\left[-\frac{i}{\hbar}(H - E_0)t\right]|i\rangle = g(t)|i\rangle \tag{5.12}$$

where $H = H_0 + H_{int}$ is the final state Hamiltonian $E_0 = H_0|i\rangle$ and $g(t)$ as the dynamical response of the outer electron system to the hole core, which accounts for many body effects. The spectral function is calculated by the Fourier transform of $g(t)$ as

$$A(E) = \frac{1}{\pi}\text{Re}\int_0^{+\infty} dt \exp\left[i\left(\frac{E - E_B}{\hbar}\right)t\right]g(t) \tag{5.13}$$

In the absence of interaction between the core hole and the valence electrons, $H_{int} = 0$, $g(t) = 1$ and hence

$$A(E) = \frac{1}{\pi}\delta(E - E_B) \tag{5.14}$$

In cases where conducting electrons screen the core hole, the $g(t)$ function adopts the form

$$g(t) = \frac{1}{t^\alpha}$$

and then the spectral function is given by [4,5]

$$A(E) = \frac{1}{(E - E_B)^{1-\alpha}} \tag{5.15}$$

where

$$\alpha = 2\sum_l (2l + 2)(\delta_l / \pi)^2$$

The sum extends over the angular moment of the conducting electrons and δ_l is the phase shift of a conducting electron with an angular momentum l. The spectral function (5.15) represents an asymmetrical line, the origin of which is the creation of an electron–hole pair.

Photoemitted core level electrons appear with a continuous energy distribution—a peak. The width of the emission line is idealized by a convolution of a Gaussian and Lorentzian function. The former accounts for the measurements factors, such as the instrumental response, x-ray line shape, Doppler and thermal broadening. The latter accounts for the lifetime broadening because of the uncertainty principle relating lifetime and energy of ejected electrons. The experimental observed asymmetry of the peak, appearing as a consequence of the perturbation of the final state of a metal due to the hole, is taken into account by the expression proposed by Doniach and Sunjic [6]:

$$I(E) \propto \frac{\Gamma(1-\alpha)}{[(E-E_B)^2 + \gamma^2]^{\frac{(1-\alpha)}{2}}} \cos\left\{ \tfrac{1}{2}\pi\alpha + (1-\alpha)\tan^{-1}\left[\tfrac{(E-E_B)}{\gamma} \right] \right\} \tag{5.16}$$

where Γ is the gamma function and γ represents the natural line width or full width at half maximum (FWHM) related to the lifetime of the hole state. Experimental values of the asymmetry factor α are obtained by fitting Eq. (5.16) to experimental spectra. The largest α-values are observed for magnetic materials such as those of the 4d and 5d series. These metals have the largest density of states close to the Fermi level and hence a strong electron–electron interaction as well as a large screening effect by d-electrons.

The photon sources in current XPS equipment are x-ray tubes in which the x-ray flux is obtained by bombarding a target with high-energy electrons. Common targets are Mg and Al, for which the emission is dominated by the unresolved doublet $K_{\alpha1,2}$ ($2p_{1/2} \rightarrow 1s$ and $2p_{3/2} \rightarrow 1s$ transitions) at 1253.6 eV for Mg and 1486.6 eV for Al. The line width of the $K_{\alpha1,2}$ doublet is about 700–800 meV for both materials. The emission of satellites is eliminated by using monochromators. For UPS (ultraviolet photoelectron spectroscopy) experiments, a gas discharge lamp is used. The filling gas is often He. Depending on the gas pressure and discharge current, one of the two intense lines with photon energies of 21.2 eV (He I) or 40.8 eV (He II) can be produced.

The synchrotron radiation emitted from the accelerated beam of charged particles offers a broad continuous spectrum from 6 eV to 6000 eV. By using appropriate monochromators, photons with the required energy can be selected. This photon source is characterized by its high brightness and intensity as well as a complete polarization. High-resolution photoemission spectra are obtained by means of an undulator and a monochromator of high resolution, which allow setting the incident radiation with a precision of 1 meV. The small incident spot and the possibility of performing time-resolved experiments make this source optimal for investigations of many-body effects at the valence band and core level.

The resolution of detectors depends on the kinetic energy of emitted electrons as shown in Figure 5.2a. Thus, the possibility of performing measurements with incident radiation energy slightly above the binding energy of the investigated element allows obtaining an improved signal resolution. In addition, the precise adjustment of the incident ray energy offers a way to control the mean free path of emitted

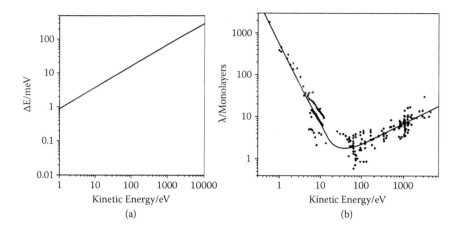

FIGURE 5.2 (a) typical relation between maximal achievable resolution and the kinetic energy of photoelectrons; (b) inelastic mean free path as a function of the kinetic energy of photoexcited electrons expressed in units of monolayer.

electrons and hence, the surface sensitivity. The inelastic mean free path shows a typical V-like course when represented as a function of the kinetic energy. Figure 5.2b shows the compilation of a series of measured values on different elements and the universal approximation found empirically by Seah [7]:

$$\lambda(E_{kin}) = \frac{C_1}{E_{kin}^2} + C_2\sqrt{d_{ml}E_{kin}} \qquad (5.17)$$

where d_{ml} corresponds to the thickness of a monolayer. For elements, $C_1 = 538$ and $C_2 = 0.4$. For inorganic compounds, they are $C_1 = 2170$ and $C_2 = 0.72$. In another approach, Tanuma et al. [8,9] calculated the mean free path of photoemitted electrons for different elements and compounds for kinetic energies larger than 50 eV by applying the modified Bethe equation:

$$\lambda(E_{kin}) = \frac{E_{kin}}{E_p^2[\beta \ln(\gamma E_{kin}) - C/E_{kin} + D/E_{kin}^2]} \text{ [Å]} \qquad (5.18)$$

where $E_p = 28.8\sqrt{n_v\rho/M}$ [eV] is the plasmon energy determined by the number of valence electrons (per atom or per molecule), n_v, the specific density ρ [g cm^{-3}], and the atomic or molecular mass M.

The parameters β, γ, C, and D were determined empirically as a function of different physical properties:

$$\beta = -0.10 + \frac{0.944}{\sqrt{E_p^2 + E_g^2}} + 0.069\rho^{0.1}; \ \gamma = \frac{0.191}{\sqrt{\rho}};$$

$$C = 1.97 - 0.91\frac{n_v\rho}{M}; \ D = 53.4 - 20\frac{n_v\rho}{M}$$

For sensitive surface analysis, the excitation energy must be selected so that a minimum of the mean free path is achieved.

The emission intensity of a given core level line of an element i in the spectrum can be quantitatively expressed by

$$I_i(h\nu) = I_0 \sigma_j(h\nu) T \int_{\gamma=0}^{\pi} \int_{\phi=0}^{2\pi} L_i(\gamma) \int_{z=0}^{\infty} n_i(z) \exp\left[-\frac{z}{\lambda_i \cos\theta} \right] dz \, d\gamma \, d\phi \qquad (5.19)$$

where I_0 is the intensity of the incident photons and $\sigma_j(h\nu)$ is the photoionization cross section for absorption of a beam of photons of energy $h\nu$. $L_i(\gamma)$ represents the angular dependency of the photoemission process, where γ is the angle between the incident x-ray beam and the direction of the ejected photoelectron. $n_i(z)$ is the concentration of atoms of the species i as a function of the depth z, λ_i is the inelastic mean free path, and θ is the angle of emission with respect to a surface perpendicular to the surface. T is an instrumental factor depending upon, among other things, the transmission function of the spectrometer.

The photoionization cross section was calculated for the first time by Cooper [10], who applied the dipole approximation formula:

$$\sigma_j(h\nu) = \frac{4\pi\alpha_f a_0^2}{3} h\nu \left| \int \psi_i(\vec{r}_0...\vec{r}_N) \sum_i \vec{r}_i \psi_f(\vec{r}_0...\vec{r}_N) d\tau \right|^2 \qquad (5.20)$$

where $\psi_{i,f}$ are the wave functions of the initial/final state of the excitation process, α_f is a fine structure constant, and $\vec{r}_i$ represents the electron coordinate. The integration is performed over the entire electron configuration space. Yeh and Lindau [11] carried out a comprehensive calculation of photoionization cross sections of all elements from $Z = 1$ to $Z = 103$ by applying numerical methods and a Hartree–Fock–Slater approximation for the calculation of the wave functions. The photoionization cross sections show, in general, a maximum at photon energies lower than 100 eV, decreasing exponentially with the photon energy. The Cooper minimum effect is a minimum in the photoionization cross section for particular orbitals and excitation energies, which was predicted in 4d and 5d orbitals on the basis of the presence of a node in the radial part of the 4d and 5d atomic wave functions.

Particularly in the case of a uniform layer of a material A covering a substrate B, Eq. (5.19) can be used to determine the layer thickness by measuring the intensity of the core level signals of A and B and comparing them with the corresponding signal intensities for the pure components:

$$I_A = I_A^{\infty} \left[1 - \exp\left(-\frac{d_A}{\lambda_A(E_A)\cos\theta} \right) \right] \qquad (5.21)$$

$$I_B = I_B^{\infty} \exp\left(-\frac{d_A}{\lambda_A(E_B)\cos\theta} \right)$$

where λ_A is the attenuation length in the film A.

Surface chemical and electronic changes occurring during (photo)electrochemical treatments can be followed by SRPES using a specially designed cell connected to the vacuum system by an entry lock system maintained with an overpressure of nitrogen. The general procedure is schematized in Figure 5.3 [12]. Sample surfaces are firstly cleaned and chemically prepared. In the case of Si, hydrogen termination can be achieved by immersing the sample in a deaerated 40% NH_4F solution for 10 min in an oblique position and washed with pure water saturated with nitrogen. Samples are glued onto the sample support with an ultrahigh vacuum conducting carbon paste under protected atmosphere. The sample is then introduced in the vacuum chamber by a fast entry lock and transported through the buffer chamber to the measure cells with the help of a vertical glass tube, the lower end of which is wrapped with a Pt-sheet that acts as a counter electrode. A reference electrode is inserted in the glass cell by a Luggin capillary. Electrolyte is conducted to the capillary by a siphon system driven with nitrogen. A drop is formed on the sample connecting counter and reference electrodes (see scheme of the cell in Figure 5.4). After the electrochemical experiment has been performed, the sample is rinsed with deaerated water and dried, before it is again introduced in the analysis chamber.

SRPES experiments shown in this chapter were generated at BESSY II, Berlin. This synchrotron is one of the third generation, which works with electrons with 1.7 GeV energy. The storage ring of a dimension of about 240 m delivers with the help of an undulator polarized light in the plane of the ring of energy in the range of 86 eV to 1890 eV. The energy incident beam was calibrated by the $4f_{7/2}$ line at 84.0 eV of Au samples.

Photoelectrochemical experiments are performed in a special designed three-electrode arrangement with a quartz window, so that it does not absorb UV-radiation as schematized in Figure 5.5. The semiconductor wafer is glued at the end of a rod of

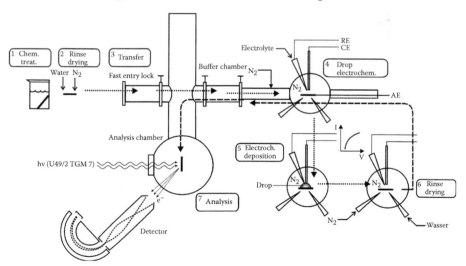

FIGURE 5.3 Schematic of measuring system and procedure for surface chemical analysis at the synchrotron.

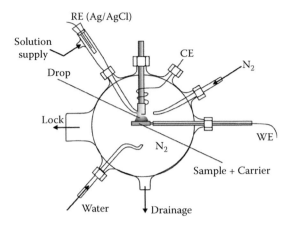

FIGURE 5.4 Schematic of the electrochemical cell used in *in-system* experiments.

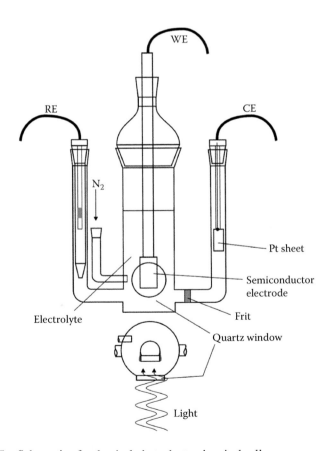

FIGURE 5.5 Schematic of a classical photoelectrochemical cell.

insulating material with an internal connection. Work, reference, and counter electrodes (generally a Pt sheet or a graphite rod) are connected to a potentiostat to control the potential of the electrode surface respective to the reference. The reference electrode is connected to the work electrode by a high impedance element. The desired electrode potential is set by means of a control loop, so that the current flowing between work and counter electrode drives the potential of the former to adopt the desired potential set by its electrode kinetics: I (V). More details about equipment and interpretation of electrochemical measurements can be found by consulting Refs. [13,14]. In this type of cell, electrochemical impedance (EIS) experiments are performed. The usual form of this technique consists of applying an additional alternating voltage of small amplitude (5 to 25 mV) to the DC applied voltage:

$$V(t) = V_0 + \Delta V \sin(\omega t) \tag{5.22}$$

Small input alternating voltages should ensure a linear response of the system. Thus, the resulting current has the form

$$J(t) = J_0 + \Delta J \sin(\omega t + \phi) \tag{5.23}$$

where ϕ is the phase difference between the voltage and current (0 for pure resistive behavior, $\pi/2$ for pure capacitive behavior). One can define the conventional impedance as

$$Z(\omega) = \frac{\Delta V \sin(\omega t)}{\Delta J \sin(\omega t + \phi)} \tag{5.24}$$

which can be expressed as a vector of modulus $|Z(\omega)|$ and phase ϕ. Applying the Euler identity, we have $Z(\omega) = |Z(\omega)| \exp(-i\phi) = Z'(\omega) + iZ''(\omega)$, where Z' and Z'' are the real and imaginary part of the vector in the complex plane. The impedance behavior of the metal–electrolyte interface, for instance, is represented by a Faradaic impedance element connected in parallel with a capacitance. The former is derived from the electrochemical reactions taking place at the electrode surface. The later represents the capacity of the double layer. To derive an impedance expression that reproduces the experimental results, it is a common practice to propose an electrochemical mechanism from which a temporal behavior is derived. The current density at the electrode is a complex function of potential, concentration of the reactive species, and surface concentration of adsorbed species. Thus, we can solve the differential equation:

$$dJ = \left(\frac{\partial J}{\partial V}\right)_{c,\theta} dV + \sum_i \left(\frac{\partial J}{\partial c_i}\right)_{V,\theta,c_{j\neq i}} dc_i + \sum_i \left(\frac{\partial J}{\partial \theta_i}\right)_{V,c_i,\theta_{j\neq i}} d\theta_i \tag{5.25}$$

Because c_i and θ_i are time dependent, one can convert Eq. (5.25) in a time-independent function by applying Laplace transform. Thus, we have

$$\Delta \overline{J} = \left(\frac{\partial J}{\partial V}\right)_{c,\theta} \Delta \overline{V} + \sum_i \left(\frac{\partial J}{\partial c_i}\right)_{V,\theta,c_{j\neq i}} \Delta c_i(s) + \sum_i \left(\frac{\partial J}{\partial \theta_i}\right)_{V,c_i,\theta_{j\neq i}} \Delta\theta(s)_i \quad (5.26)$$

where s is the complex argument of Laplace functions.

After finding the different s-functions in Eq. (5.26), we arrive at

$$\frac{\Delta \overline{V}}{\Delta \overline{J}} = Z_F(s) \quad (5.27)$$

from which we can obtain the complex impedance $Z' + i\, Z''$ by substituting s by $i\omega$. The capacitance of the double layer is expressed by the term $Z_c(s) = 1/Cs$. In some cases, the capacitive response is better represented by a constant phase element (CPE), the impedance of which is given by $Z_{cpe}(s) = 1/s^\alpha(\overline{R}^{\alpha-1}\overline{C}^\alpha)$, where α is usually known as the dispersion coefficient. $\overline{R}$ and $\overline{C}$ are mean values of the double layer resistance and capacitance, respectively. This impedance function often finds application in rough systems, the surfaces of which are characterized by a lateral distribution of RC domains. The total electrode capacitance will be given by

$$Z_T(s) = R_\Omega + \left(\frac{1}{Z_c(s)} + \frac{1}{Z_F(s)}\right)^{-1} \quad (5.28)$$

where R_Ω represents the electrolyte resistance. In the case of semiconductors, a classical representation is given by

$$Z_{sc}(s) = \left(\frac{1}{Z_{c,sc}(s)} + \frac{1}{Z_{ss}(s)} + \frac{1}{R_{sc}}\right)^{-1} \quad (5.29)$$

where $Z_{c,sc}$, Z_{ss}, and R_{ss} are the capacitive impedance of the space charge layer, the impedance of the surface states, and the resistance for the electron transfer in the space charge layer, respectively. The schematic in Figure 5.6 provides a visual representation of electrodic processes in different impedance elements.

Another electrochemical technique implemented for the study of electrode kinetics is the scanning electrochemical microscope [15,16]. The foundation of this technique is basically the change of current observed upon approaching a micro-dimensioned Pt electrode to the substrate being investigated in the presence of a redox couple. The position of the microelectrode (often named UME: ultra-microelectrode) is positioned by means of a computer-controlled piezoelectric positioner with linear encoders (100-nm resolution, for instance in Model M370 of Princeton Applied Research) on all axes, which independently read the true position of the probe (Figure 5.7a).

The UME is generally constructed with a wire of a noble metal (Pt, Ir, Au) wrapped with glass and cut horizontally to form a circular active surface surrounded by an inactive surface. This allows control of the diffusion boundaries.

Let as assume a generalized redox reaction: Ox + e⁻ → red. After applying a potential pulse at the UME for which the redox reaction is under diffusional control, the time dependence of the current will be given by the Cottrell equation for small electrodes:

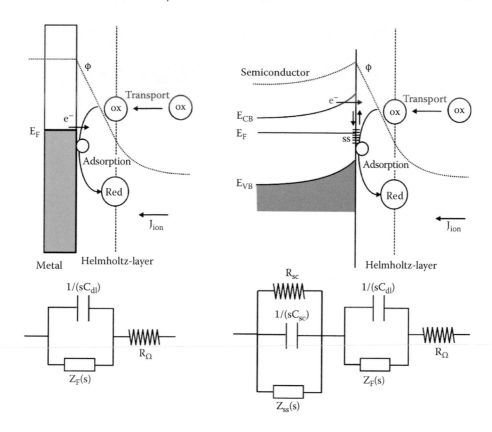

FIGURE 5.6 Schematic representation of an electron transfer reaction at metals (left) and semiconductors (right) and corresponding equivalent circuits.

$$j(t) = \frac{nF}{\sqrt{\pi}}\sqrt{D_{ox}}\,c_{ox}\frac{1}{\sqrt{t}} + nFD_{ox}c_{ox}\frac{1}{r_e} \qquad (5.30)$$

where r_e is the radius of the active surface. It must be noted that at $t \to \infty$ the current tends toward a constant value $j_\infty = nFD_{ox}c_{ox}1/r_e$. Upon approaching the UME to the substrate, the diffusion field is distorted and the current increase or decrease depending on the rate of the backward reaction at the substrate. A current-distance curve is obtained. The potentials of the UME and the substrate can be independently controlled with respect to a reference electrode by a bi-potentiostat. Two limit cases can be distinguished: an insulating substrate (negative feedback) and a conducting substrate (positive feedback). In the former case, the current decreases by approaching the UME according to the nondimensional equation:

$$J_{ins} = \frac{j}{j_\infty} = \frac{1}{0.15 + 1.5385/L + 0.58\exp(-1.14/L) + 0.0908\exp[(L-6.3)/1.017L]} \qquad (5.31)$$

where $L = d/r_e$. For a conductive substrate, the current is given by

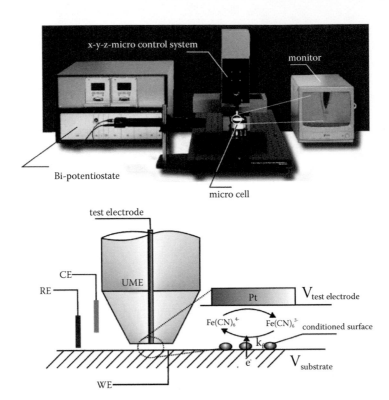

FIGURE 5.7 (Top) Photograph showing a typical SECM experimental set up; (bottom) schematic depicting the working principle of the SECM.

$$J_c = \frac{j}{j_\infty} = 0.68 + 0.78377/L + 0.3315\exp(-1.067/L) \tag{5.32}$$

For the case of a substrate with a finite kinetics, the nondimensional current-distance curve is given by

$$J_T = J_s\left(1 - \frac{J_{ins}}{J_c}\right) + J_{ins} \tag{5.33}$$

where J_s is

$$J_s = \frac{0.78377}{L(1+1/\Lambda)} + \frac{0.68 + 0.3315\exp(-1.0672/L)}{1 + F(L,\Lambda)} \tag{5.34}$$

where

$$\Lambda = k_f \frac{d}{D_{ox}}; F(L,\Lambda) = \frac{\dfrac{11}{\Lambda} + 7.3}{100 - 40L}$$

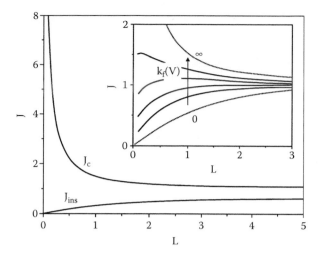

FIGURE 5.8 Approach curves as a function of the heterogeneous reaction rate constant for electron transfer at the substrate, k_f.

where k_f [cm s^{-1}] is the apparent heterogeneous rate constant and D_{ox} is the diffusion coefficient of the active species, in this case the species ox. Thus, the reaction electron transfer rate k_f can be obtained by adjusting the experimental curve to the theoretical Eq. (5.33). Figure 5.8 shows the approach curves for the two limit cases and the transition curves for increasing values of the heterogeneous transfer rate constant (right side insert).

5.2   CHEMICAL AND ELECTRONIC PROPERTIES OF ELECTROCHEMICAL GENERATED INTERFACES ON SILICON

Surface states arising at the Si-SiO$_x$ interface has been a matter of intensive study by using a battery of spectroscopic techniques [17]. The concentration and energy distribution of energy states depends on the number of terminal $\equiv$Si-H, $\equiv$Si-O and $\equiv$Si$^\bullet$ surface species, and on the degree of distortion of Si-Si bonds. It is common to speak about a surface passivation to refer to the reduction of the density of surface states. Passivation is required to reduce the presence of potential recombination centers, which diminish the flux of photoelectrons consumed by the conversion interfacial reaction.

Because of the similar electronegativity of Si and H, the saturation of surface dangling bonds by reaction with H to form Si-H bonds leads to the lowest level of surface state concentration. The weak stability of this type of surface on exposing it to the moist air requires researching other alternatives, as for example, the anchoring of stable organic molecules or the formation of high quality Si-SiO$_2$ interfaces. The growth of tens of oxide monolayers reduces the concentration of interfacial states in comparison with the high defective surface structure arising after the formation of the first oxide monolayers.

Surface states can be regarded as electronic eigenvalues of the Schrödinger equation, which shows a maximum of the probability function $|\psi_{SS}|^2$ at the surface. The origin of the surface state depends on the potential wall at the crystal surface, which in a vacuum results in two types of solutions of the one-dimensional Schrödinger equation. The first one has a Bloch character and drops exponentially in vacuum. The second solution corresponds to the localized state at the surface whose wave function is localized in the surface region, decaying exponentially both into the bulk and the vacuum. When the eigenvalue of the second solution appears within the energy gap of a semiconductor or in localized gaps of the projected band structure of a metal, the solution of the Schrödinger equation is possible by assuming a complex wave vector: $\vec{k} = k_r + ik_{im}$. For a semi-infinite crystal, a possible solution is given by

$$\psi = A \exp(-k_{im}z)\cos\left(\frac{\pi}{2}z+\delta\right) \tag{5.35}$$

inside the crystal and

$$\psi = B \exp\left[-(k_{im} + k_r \tan\delta)z\right] \tag{5.36}$$

outside the crystal. A and B are constants which can be calculated by matching ψ and $d\psi/dz$ at the surface. It should be noted that both solutions (5.35) and (5.36) present decaying eigenfunctions outside (into the vacuum) as well as inside the crystal and hence the state is confined to the surface (Figure 5.9).

This picture is valid by regarding only a wave vector perpendicular to the surface. The parallel component of the wave vector of surface states may cross the projection of the bulk band and overlaps energetically with bulk states. These states are known as surface resonance (see Figure 5.9) and they appear also in the case of

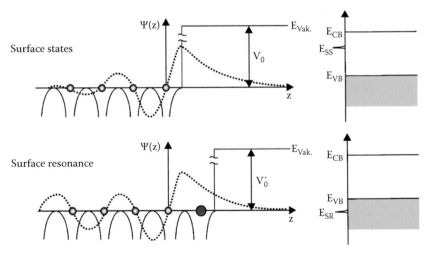

FIGURE 5.9 One-dimensional wave functions for surface states and surface resonance arising from an adsorbed foreign atom; (right) energy band diagrams showing the energy positions of each type of localized states.

adsorption atoms or molecules whose energy levels are within the valence band of the semiconductor.

It is customary to distinguish between intrinsic and extrinsic surface states. The former arise as a result of a break of the crystal lattice symmetry, which alters the boundary condition for the Schrödinger equation. Depending on the binding character of the semiconductor, they can be classified as Tamm states for ionic crystals and as Shockley states for covalent compounds. The former arise from applying the tight-binding approach based in the construction of wave functions from atomic-like orbitals, a method for describing fairly localized electrons. The formation of Tamm states implies a large perturbation of the potential at the surface, a fact that is found in ionic compounds as a result of an asymmetrical Madelung potential. Chemically, it means a heteronuclear splitting of bonding interactions, so that these states adopt energy positions near the band edges. States near the valence band acquire anionic character and are completely filled, while those near the conduction band are empty. Thus, semiconductors with Tamm states, such as ZnO, TiO_2, CdS, and ZnS, generally do not show Fermi level pinning as, for instance, does Si. The approach is also suitable for describing d-electrons in transition metals.

For covalent compounds, it is suitable to describe surface states in the framework of the nearly free electron model, neglecting electron–electron interactions. The resulting solution of the Schrödinger equation receives the name of Shockley states. Shockley states appear due to termination of the crystal symmetry only. Chemically, they correspond to the homonuclear splitting of covalent bonds, which give rise to unsaturated radical states (dangling bonds) at the surface [18]. They are situated close to the center of the band gap and are very reactive. Shockley surface states appear in Si and Ge. Surface states in III-V semiconductors such as GaAs and InP are at the threshold of the classification between Tamm and Shockley states.

Researchers usually refer to extrinsic surface states in semiconductors to indicate the nature of the source of the new electronic states. Extrinsic states arise as a result of strong interactions with adsorbates, metals, and electrolytes in contact with the semiconductor. It must be emphasized that intrinsic and extrinsic surface states exclude each other, because intrinsic states, especially Shockley states, are usually related to surface sites of enhanced reactivity. Adsorbates react preferentially with these states and, hence, new states are formed.

The contact of a silicon surface with aqueous electrolytes, for instance, leads to the formation of an oxide monolayer due to the large reactivity of dangling bonds with water. This presupposes the breakdown of the surface passivation by, for example, hydrogen and the formation of intermediate unsaturated radicals as analyzed below.

The formation of a $Si-SiO_x$ interface brings about a large density of undesired surface states ascribed to the formation of $= Si(-O)_2$, $\equiv Si-O$ and $\equiv Si^{\cdot}$ bounds, on the order of 10^{13}–10^{14} $cm^{-2}eV^{-1}$ [19–21]. Figure 5.10 shows a typical energy distribution reported by Flietner [19] characterized by a U-like form and the appearance of a peak close to the intrinsic Fermi level. The chemical and electronic aspects of states at $Si-SiO_x$ were investigated by using electron paramagnetic resonance (EPR) [22–24]. It was found that the hydrogenation of oxidized silicon leads to the passivation of surface states by formation of H-terminated surfaces. The $Si–SiO_x$ interface is not an abrupt junction but is formed by a transition region of an altered structure

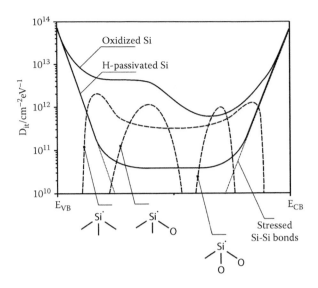

FIGURE 5.10 Solid line: energy distribution of density of interface states at oxidized and passivated silicon surfaces. Dashed line: energy distribution of density of states corresponding to different groups of defects at the Si-SiO$_2$ interface (after Flietner [18,19]).

and stoichiometry. Transmission microscopy and photoemission investigations have shown that this region can vary from 0.3 nm to 3 nm.

The Si 2p$_{3/2}$ core level signal of photoelectron spectroscopy spectrum obtained from silicon covered with a thin oxide layer shows typical peaks, which are shifted by +1, +1.8, +2.5 and +3.6 eV from the line corresponding to bulk silicon [25] (fully coordinated) (see Figure 5.11). These peaks can be ascribed to the presence of

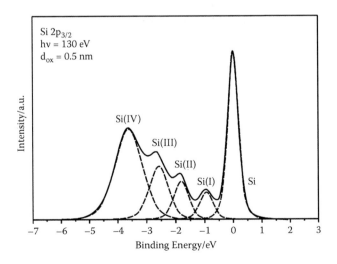

FIGURE 5.11 Si 2p$_{3/2}$ core level signal of Si(100) covered with an ultra-thin oxide layer formed under 2×10^{-5} Torr O$_2$ for 20 s at 750° C (after Himpsel et al. [25]).

intermediate oxidation states of silicon, provided the term oxidation state it is rather understood as the loss of back coordination bonds. The transition region is characterized by a large number of defects consisting of unsaturated dangling bonds such as $\cdot Si \equiv Si_3$ and $\cdot Si \equiv Si_2O$ and distorted bonds. Stressed Si-Si bonds lead to energy distributions that decay exponentially from the band edges toward the middle of the band gap. This type of surface states predominates in well H-passivated surfaces, as can be achieved by chemical or electrochemical methods [26–29]. Best passivated surface can be achieved by electrochemical etching with a minimum density of surface states of 10^{10} cm^{-2} eV^{-1}. H-terminated surface are, however, instable in contact with water and oxidize within a period of time not longer than one hour [30], leading to an increase of the surface states up to 10^{13} cm^{-3}eV^{-1} [31]. A convenient issue is the thermal growth of an oxide layer. This improves the quality of the Si–SiO$_2$ interface by relaxation an elimination of the large amount of defects formed during the first stages of oxide growth. Hattori and Nohira have found that the density of Si$^+$ and Si^{+3} increases during oxide growth until to reach a thickness of 0.5 nm. Upon further growth, they starts to oscillate; a consequence of the inward growth to the oxide. The density of Si^{+2}, considered as the cause of oxide micro-roughness start to decrease at thickness larger than 0.5 nm (see Figure 5.12). The thermal oxide growth is a customary method in the fabrication of conventional crystalline Si solar cells [33]. Thus, the recombination rate is slowed down to values smaller than 10 cm s^{-1}. In an attempt to introduce passivation low temperature methods, plasma-enhanced vapor deposition of SiN$_x$ [34], the anodic oxidation [35], and the adsorption of organic molecules [36,37], among others are being presently investigated.

Studies of electron traps at the Si-SiO$_x$ interface by means of capacity-voltage, electron paramagnetic resonance, and surface photovoltage experiments indicate the presence of rechargeable defects usually referred to as P$_b$-centers [22,24,38]. These traps are associated to determine types of interfacial structures (see Figure 5.13). One can distinguish between P$_{b0}$- and P$_{b1}$-centers ascribed to $\cdot Si \equiv Si_3$ and $\cdot Si \equiv Si_2O$, respectively. The traps are constituted by an unpaired Si sp^3 hybrid orbital along $\langle 111 \rangle$ ($\cdot Si \equiv Si_3$) and $\langle 211 \rangle$ directions ($\cdot Si \equiv Si_2O$). Molecular-orbital analysis also indicates that the unpaired electron in Pb$_1$ resides for 58% in a single unpaired Si hybrid orbital [39]. Electrical measurements demonstrate that P$_b$ centers are the dominant charge trap at the Si-SiO$_2$ interface. Thus, their concentration and energy distribution determine recombination rates and shifts of the semiconductor band edges.

Rechargeable interfacial traps can be detected by means of impedance experiments due to the time-limited charging and discharging processes. A customary representation of the impedance behavior of surface states is given by a series RC-circuit. Figure 5.14 shows commonly used equivalent circuits for the EMOS and MOS junctions. At the frequency region dominated by the capacitive behavior, the total capacitance of the junction is given by

$$\frac{1}{C_T} = \frac{1}{C_{ox}} + \frac{1}{C_{it} + C_{sc}} \tag{5.37}$$

where C_{it} is the capacitance of the interface states. The capacitance of the semiconductor is given by

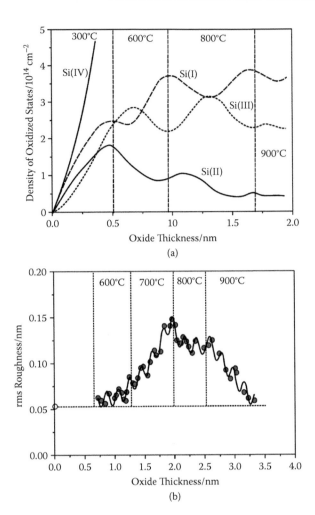

FIGURE 5.12 (a) Dependence of the surface density of intermediate oxidation states on the oxide thickness; (b) variation of the rms surface roughness with progress of oxidation. The solid curve was calculated by assuming a constant amplitude of oscillation of a period of 0.19 nm [after Hattori and Nohira [32]].

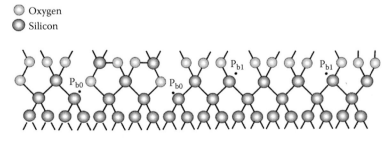

FIGURE 5.13 **(See color insert.)** Schematic depicting the Si-SiO$_2$ atomic structure and the associated electronic defects P$_{b0}$ and P$_{b1}$.

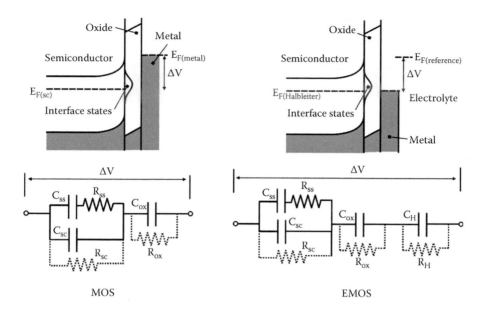

FIGURE 5.14 Energy band diagrams and related customary equivalent circuits for the representation of MOS and EMOS junctions.

$$C_{sc} = \frac{\varepsilon_0 \varepsilon_{sc}}{\left(\dfrac{\varepsilon_0 \varepsilon_{sc} kT}{2Ne^2}\right)^{1/2}} \frac{1 - \exp\left(\pm e\varphi_s / kT\right)}{\sqrt{\left[1 \pm e\varphi_s / kT - \exp\left(\pm e\varphi_s / kT\right)\right]}} \qquad (5.38)$$

where "+" is for p-type and "−" for n-type semiconductors. φ_s represents the semiconductor band bending: $\varphi_s > 0$ for n-type and $\varphi_s < 0$ for p-type semiconductors in depletion. At sufficiently high frequencies, the charging–discharging process at the interface states cannot follow the potential changes and hence, $C_{it} = 0$ in Eq. (5.37).

Figure 5.15 depicts a typical capacitance-voltage (bias) curve of a MOS junction. It should be noted that at high frequencies, the curve presents an S-like course: in accumulation, $C_{sc} \gg C_{ox}$ and $C_T \approx C_{ox}$; in depletion, the semiconductor capacity decreases exponentially up to a capacitance minimum. As the Fermi level of the n-type (p-type) semiconductor reaches the valence band (conduction band) edge, a strong inversion begins. At high frequencies, the generation–recombination processes of minority charge carriers cannot follow the potential changes and the capacitance of the semiconductor is given by the maximal depletion layer. At low frequencies, the accumulation of minority carriers leads to a capacitance increases according to:

$$C_{sc} = \frac{\varepsilon_0 \varepsilon_{sc}}{2\left(\dfrac{\varepsilon_0 \varepsilon_{sc} kT}{2n_{min} e^2}\right)^{1/2}} \exp\left(|e\varphi_s| / 2kT\right) \qquad (5.39)$$

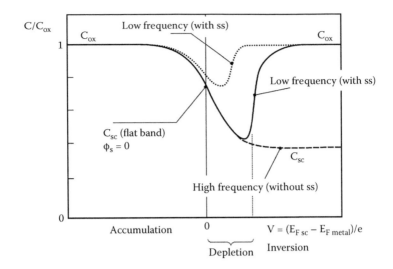

FIGURE 5.15 Typical capacitance-voltage curves for an ideal MOS-junction and for the same junction with interface states.

Thus, the total capacitance reaches again the oxide capacitance. The presence of interface states leads to a stretch-out of the capacitance-voltage curve at low frequencies (dotted line in Figure 5.15). The density of interface states can be calculated by

$$D_{it}(V) = \frac{1}{e^2}\left[\left(\frac{1}{C_{T(f<100Hz)}} - \frac{1}{C_{ox}}\right)^{-1} - C_{sc}(V)\right] \tag{5.40}$$

The relation between the total voltage, V, and the position of the Fermi level in the semiconductor can be found by

$$\varphi_s(V) = \varphi_{s0} + \int_{U_0}^{U}\left[1 - \frac{C_{T(f<100Hz)}(V)}{C_{ox}}\right]dV \tag{5.41}$$

This method was applied by Poindexter et al. [24] for resolving the energy distribution of interface states in thermally produced Si-SiO$_2$ interfaces, which is drawn in Figure 5.16. It can be seen that the distribution of P$_b$ centers presents a double peak structure with maxima at 0.26 eV and 0.84 eV above the valence band. The first peak corresponds to a transition from a positive charged to a neutral defect ($+1 \to 0$), whereas the second peak, near the conduction band, represents the transition of neutral to a negative charged state ($0 \to -1$).

In electrochemical systems, the inversion region is seldom reached, because the accumulation of charge by Fermi level pinning at the band edge is completely absorbed by the double layer capacity, typically of 10–100 μF cm^{-2}. Furthermore, capacitance-voltage curves at frequencies lower than 100 Hz, as required in solid-state

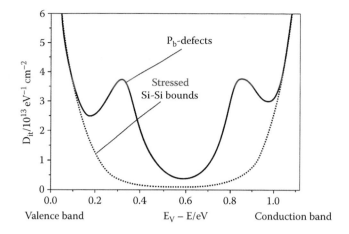

FIGURE 5.16 Energy distribution of the density of interface states at the Si-SiO$_2$ junction formed thermally on p-type silicon (oxide thickness 190 nm) [after Poindexter et al. [24]].

experiments (as used for instance in the experiment in Figure 5.16) are not practicable in EMOS and EOS junctions due to the predominant response of the double layer.

Most studies of the dynamics of the charging–discharging of surface states at E(M)OS junctions were performed in the presence of a thin oxide layer. Hence, $C_{ox} > C_{it}$ and the capacity of the surface states appears as a peak. Let us take a look at the capacitance-voltage curves measured at different frequencies on n-type Si(111) in KCl (Figure 5.17). They show a capacitance peak between −0.6 V and −0.2 V superposed to the exponential decaying capacity of the semiconductor. The peak of −0.35 V at 100 Hz shrinks at higher frequencies. This fact is related with the time constant of the charging processes. The origin of these surface states is related

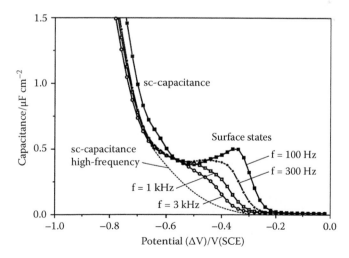

FIGURE 5.17 Capacitance-voltage curves of a freshly etched n-type Si(111) surface recorded at different frequencies in 0.5 M KCl of pH 2.3. $\Delta V = 10$ mV.

to the presence of some oxidized atoms at the otherwise passivated surface. The H-termination was achieved by a treatment in a 40% NH_4F solution. This method allows obtaining a flat atomic surface with a terrace structure. The passivation is, however, not perfect as explained later.

The quantitative connection between experimental impedance measurements and the kinetics of charge trapping can be obtained by a proper formulation of the transfer mechanism, in a similar manner as was proposed in Refs. [40–42].

According to the transfer mechanism schematized in Figure 5.18, the charging rate of surface states can be represented by

$$\frac{dn}{dt} = n_s k_1 [N_{ss} - n] - k_2 n \tag{5.42}$$

where n is the concentration of occupied states, n_s is the electron concentration at the conduction band edge, given by $n_s = n_s^0 \exp[(e/kT)|(V - V_{fb})|$, and N_{ss} is the total number of surface states. The model assumes that the rate constants k_1 and k_2 do not depend on potential. This assumption is justified if one takes into account that transfer processes are controlled by photon–electron interactions, which are determined by the relative energies of bulk and surface states [43]. Thus, the charging current is given by

$$j = e\frac{dn}{dt} \tag{5.43}$$

The harmonic response to the input of an alternating potential of small amplitude to the DC-polarization potential can be described by a linear differential equation, which after Laplace transformation yields

$$\bar{j} = \left(\frac{\partial j}{\partial V}\right)_n \bar{V} + \left(\frac{\partial j}{\partial n}\right)_V \bar{n} \tag{5.44}$$

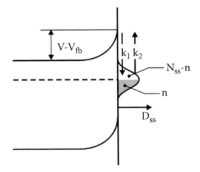

FIGURE 5.18   Energy band diagram of an n-type semiconductor depicting the charging of surface states.

After substituting Laplace transformation of Eq. (5.43) into Eq. (5.44), we have

$$e s \bar{n} = \left(\frac{\partial j}{\partial V} \right)_n \bar{V} + \left(\frac{\partial j}{\partial n} \right)_V \bar{n} \qquad (5.45)$$

where the partial derivatives are referred to as stationary conditions:

$$\left(\frac{\partial j}{\partial V} \right)_n = n_s \frac{e^2}{kT} k_1 (N_{ss} - n); \qquad \left(\frac{\partial j}{\partial n} \right)_V = -e(n_s \quad k_1 + k_2). \qquad (5.46)$$

Solving for $\bar{n}$ in Eq. (5.45) and introducing Eq. (5.46) yields, after a bit of mathematics,

$$\frac{\bar{V}}{j} = Z(i\omega) = \frac{kT}{n_s e^2 k_1 (N_{ss} - n)} - i \frac{kT(n_s k_1 + k_2)}{n_s e^2 k_1 (N_{ss} - n)} \frac{1}{\omega} \qquad (5.47)$$

This impedance expression can be adapted to the customary representation of the surface state by a series RC-circuit. Thus, we have

$$C_{ss}(V) = \frac{n_s(V) e^2 k_1 [N_{ss} - n(V)]}{kT [n_s(V) k_1 + k_2]} \qquad (5.48)$$

$$R_{ss}(V) = \frac{kT}{e^2} \frac{1}{n_s k_1 (N_{ss} - n)} \qquad (5.49)$$

The potential dependence of the capacitance shows a maximum at a potential, for which the following relation applies:

$$\left(\frac{\partial C_{ss}}{\partial V} \right) = \left(\frac{\partial C_{ss}}{\partial n_s} \right) \left(\frac{\partial n_s}{\partial V} \right) + \left(\frac{\partial C_{ss}}{\partial n} \right) \left(\frac{\partial n}{\partial V} \right) = 0 \qquad (5.50)$$

The differentiation Eq. (5.48) with respect to n_s and n, yields

$$\left(\frac{\partial C_{ss}}{\partial n_s} \right) = \frac{e^2}{kT} \frac{k_1 (N_{ss} - n)}{(n_s k_1 + k_2)} - n_s \frac{e^2}{kT} \frac{k_1^2 (N_{ss} - n)}{(n_s k_1 + k_2)^2} \qquad (5.51)$$

$$\left(\frac{\partial C_{ss}}{\partial n} \right) = -\frac{e^2}{kT} \frac{k_1 n_s}{(n_s k_1 + k_2)} \qquad (5.52)$$

Also,

$$\left(\frac{\partial n_s}{\partial V} \right) = n_s \frac{e}{kT} \qquad (5.53)$$

After substituting Eq. (5.51–53) into (5.50) and rearranging, we arrive to

$$(N_{ss} - n)_m = \frac{kT}{e}\left(\frac{\partial n}{\partial V}\right)\left[\frac{n_s k_1 + k_2}{k_2}\right] \tag{5.54}$$

$$C_{ss,m} = (N_{ss} - n)_m \frac{e^2}{kT} - e\left(\frac{\partial n}{\partial V}\right) \tag{5.55}$$

Characteristic parameters $(N_{ss} - n)$, k_1, and k_2 can be calculated from Eq. (5.54) and Eq. (5.55), provided a given distribution of surface states, $(\partial n/\partial V)$, is assumed. A good approach is to adopt a Gaussian distribution function as follows:

$$D_{ss}(E) = \left(\frac{\partial n}{dE}\right) = \frac{N_{SS}}{\sigma\sqrt{2\pi}}\exp\left[-\frac{1}{2}\left(\frac{E - E_0}{\sigma}\right)^2\right] \tag{5.56}$$

where E_0 represents the energy of the distribution center and σ is the distribution width. Accordingly, the potential dependence of surface states is obtained by integration of Eq. (5.56):

$$n(V) = \int_{-\infty}^{E} D_{SS}(E)\,dE = \frac{N_{SS}}{2}\left\{erf\left[\sqrt{1/2}\left(e\frac{V - V_0}{\sigma}\right)\right]+1\right\} \tag{5.57}$$

Figure 5.19 shows the theoretical capacitance-voltage curves as a function of potential calculated by applying Eqs. (5.48), (5.38), and (5.57). It must be noted that the curves

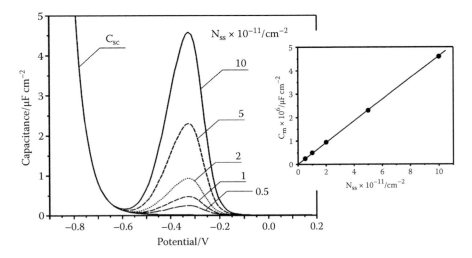

FIGURE 5.19 Calculated capacitance-voltage curves of a semiconductor–electrolyte junction. $n_s = 5 \times 10^{14}$ cm^{-3}, $\sigma = 0.08$ eV, $V_{SS} = -0.40$ V, $V_{fb} = -0.55$ V.

reflect the main characteristics of the experimental measured curves. The integration of the surface states capacitance yields the concentration of surface states as

$$N_{ss} = \frac{1}{e} \int_{-\infty}^{+\infty} C_{ss} \, dV \tag{5.58}$$

Note that the capacitance maximum is directly proportional to the concentration of surface states, as shown in the insert of Figure 5.19.

The etching of Si surfaces in concentrated 40% NH_4F is a customary method to get H-terminated surfaces on Si(111). This etchant leaves a flat morphology characterized by the formation of atomic terraces. The number and morphology of step edges depends on the direction and angle of miscut. This method is preferred instead of immersion for few seconds in a HF solution, because this latter leaves a rather rough surface, in most cases not convenient for microscopy studies. SRPES analysis of surfaces prepared by NH_4F etching indicates, however, that this method does not produce a perfect H-termination, as indicated by capacitance-voltage experiments (see Figure 5.17). The Si 2p core level signal shown in Figure 5.20 indicates the presence of some oxidized surface atoms, which are passivated after short dipping in HF solution [44]. Practically no Si(IV) is detected, while low amounts of Si(I), Si(II), and Si(III) are found, indicating the presence of Si_xO_y in varying stoichiometric

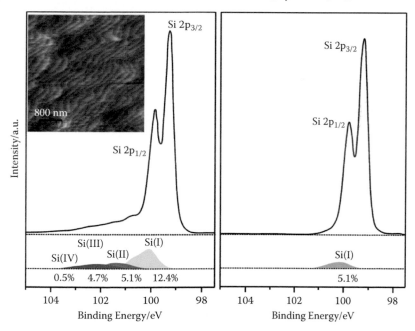

FIGURE 5.20 SRPES Si 2p core level signal of n-type Si(111): (left) etched in NH_4F 40% for 100 s and 10 min with water rinse in between; (right) after additional dipping for 10 s in HF50%. Indicated percentages refer to the integral total signal. Insert: contact-mode AFM picture of the surface after etching in NH_4F and HF.

compositions. Lublow et al. [44] found a low F 1s signal 750 eV excitation energy, suggesting that also such compounds as = Si–H–F and –Si–H–OH have to be considered with two and one Si back bonds, respectively. These contributions result in chemical shifts of about 1 and 1.5 eV, according to earlier density functional theory calculations [45]. A coverage degree of oxidized surface atoms up to 15% can be estimated. These sites are of fundamental importance for the oxidation and deposition mechanisms during the (photo)electrochemical deposition of noble metals on silicon as explained below.

The formation of the first oxide layers on H-terminated silicon is indicated by the appearance of two consecutive anodic peaks in the potentiodynamic current-voltage curves performed in a phthalate buffer solution (see Figure 5.21). This first, narrower peak can be ascribed to the transformation of H- into an OH-termination [46]. The second, wider peak is ascribed to the delayed onset of a high-field oxide growth. It should be noted that passed charge increases linearly with scanning potential, as expected according to the high-field relation:

$$d_{ox} = \beta(V - V_{ox}^f) = \left(\frac{dq}{dV}\right)\frac{\overline{M}_{ox}}{\rho_{ox}zF}(V - V_{ox}^f) \tag{5.59}$$

where β is the growth factor, ρ_{ox} is the oxide density, and $\overline{M}_{ox}$ is the molar mass of SiO_2. V_{ox}^f is the so-called oxide formation potential. Its value is determined to be

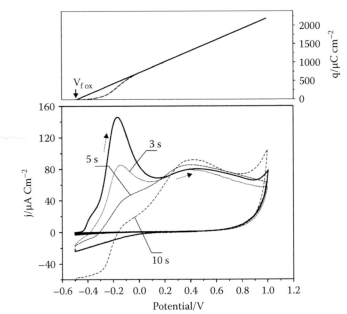

FIGURE 5.21 Current-voltage curves of anodization of n-type Si(111):H in phthalate buffer solution (0.07 M NaHPh + 0.03 M NaOH pH 4.8) at 50 mV s⁻¹ (——). Upper detail: anodic passed charge. Dotted and dashed curves corresponds to those recorded after increasing time lengths of Pt deposition at –0.35 V from 1 mM H₂PtCl₆ + 0.1 M K₂SO₄.

−0.495 V by extrapolation of q vs. V to $q = 0$. Taking $\rho_{ox} \sim 2.1$ g cm^{-3}, typical for an anodic oxide, Eq. (5.59) yields $\beta = 1.09$ nm V^{-1}.

The structure of an oxide film of about 1 nm thickness formed at 0.5 V is revealed by SRPES spectra. Figure 5.22 shows the Si 2p core level signals recorded at different excitation energies. Each signal was deconvoluted by seven components corresponding

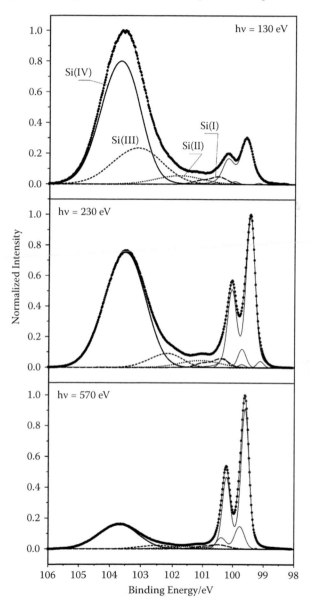

FIGURE 5.22 SRPES Si 2p core level signal of n-type Si(111) covered with a thin anodic oxide layer. The anodization was performed potentiodynamically from −0.5 V to +0.5 V (SCE) in a phathalate buffer of pH 4.8.

to the states Si(I), Si(II), Si(III), Si(IV), and Si(0), and the lines α and β. The α- and β-lines, appearing 0.29 eV above and 0.30 eV below the Si bulk signal, were ascribed by Yazyev and Pasquarello to stretched and compressed Si-Si bonds [47].

Increasing excitation energies provide information of deeper atomic structures, because the attenuation length of photoelectrons emitted from tetra-coordinated silicon increases with the excitation energy. The attenuation length can be obtained experimentally by the relation between the experimentally measured intensity of emitted electrons from the oxide and from the substrate. The flux of photoelectrons arising at the oxide film is given by

$$I_{SiO_2} \sim n_{SiO_2}\sigma_{SiO_2}\int_0^{d_{ox}}\exp\left(-\frac{z}{\lambda_{SiO_2}}\right)dz \tag{5.60}$$

where n_{SiO_2} is the density of Si(IV), and σ_{SiO_2} is the corresponding capture cross section for Si in the oxide. The emission from Si atoms in the substrate is described by the relation

$$I_{Si} \sim n_{Si}\sigma_{Si}\exp\left(-\frac{d_{ox}}{\lambda_{SiO_2}}\right)\int_0^\infty\exp\left(-\frac{z}{\lambda_{Si}}\right)dz \tag{5.61}$$

The combination of Eq. (5.60) and (5.61) yields

$$\lambda_{SiO_2} = \frac{d_{ox}}{\ln\left(\dfrac{I_{SiO_2}}{I_{Si}}\dfrac{I_{Si}^\infty}{I_{SiO_2}^\infty}+1\right)} \tag{5.62}$$

Himpsel et al. [25] measured the oxide thickness by means of TEM and calculated the attenuation length by applying Eq. (5.62). The results are reproduced in Figure 5.23. It can be seen that the dependence of the mean free path with excitation energy has a minimum at 160 eV and increases with a functionality of $\lambda_{ox} \sim (h\nu)^\alpha$. In this way, the relative proportions of Si atoms with different oxidation states of Si obtained from spectra recorded at increasing excitation energies provide information about the structure of the Si-SiO₂ interface. The relative atom density relations are calculated by applying the formula

$$\frac{n_{Si(i)}}{\sum\limits_{i:1..4} n_{Si(i)}} = \frac{I_{Si(i)}/\sigma_{Si(i)}}{\sum\limits_{i:1..4} I_{Si(i)}/\sigma_{Si(i)}} \tag{5.63}$$

where $i:1..4$ refers to the oxidation state and $\sigma_{Si(i)}$ is the corresponding capture cross sections. According to Himpsel et al. [25], these are $\sigma_{Si(I)} = 1.0$, $\sigma_{Si(II)} = 1.1$, $\sigma_{Si(III)} = 1.7$, and $\sigma_{Si(IV)} = 2.2$. Table 5.1 resumes the calculated values. The proportion of Si(I) is practically the same for excitations of 170 and 230 eV. It duplicates its value for an excitation of 570 eV with a corresponding decrease of Si(IV). The proportions of Si(II) and

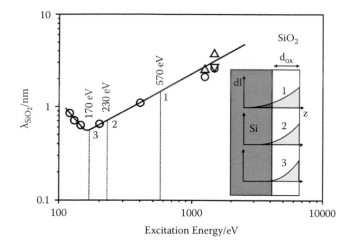

FIGURE 5.23 Attenuation length of photoelectrons in SiO_2 as a function of the excitation energy (from Himpsel et al. [25]).

TABLE 5.1

Relative Abundance of Si Coordination States at the Si/SiOₓ Interface after Anodizing n-Type Si(111) to +0.5 V in a Phthalate Buffer Solution (0.07 M NaHPh + 0.03 M NaOH pH 4.8) at 50 mV s⁻¹ under Illumination

Excitation Energy	Si(I)	Si(II)	Si(III)	Si(IV)
570	0.16	0.11	0.09	0.64
230	0.07	0.09	0.10	0.74
170	0.06	0.14	0.08	0.72

Si(III) remain practically constant for the three excitation energy values. Considering that the anodic oxide has a thickness of 1 nm, one can infer that excitation energies of 170 eV and 230 eV collect information predominantly from the oxide film; the corresponding attenuation lengths are smaller than 1 nm. At 570 eV, information is also collected from the interface. These results indicate the presence of an oxide with a 27% of subcoordinated atoms in the oxide; a characteristic of thin anodic oxides, which contain a large amount of water. The subcoordinated Si atoms are also chain links of a high-field aided hopping transport of O⁼. The thickness of the anodic oxide can be calculated by applying Eq. (5.62), where the oxide intensity of the high defective oxide is calculated by

$$I_{ox} = I_{Si(IV)} + \tfrac{3}{4} I_{Si(III)} + \tfrac{1}{2} I_{Si(II)} + \tfrac{1}{4} I_{Si(I)} \tag{5.64}$$

Thus, an oxide thickness of $d_{ox} \sim 1$ nm can be calculated, in good agreement with the value obtained by the integrated anodic charge; see Eq. (5.59).

The O 1s core level signal can be deconvoluted into two components at 531.8 eV and 532.7 eV (see Figure 5.24). The former can be ascribed to the surface oxide [48].

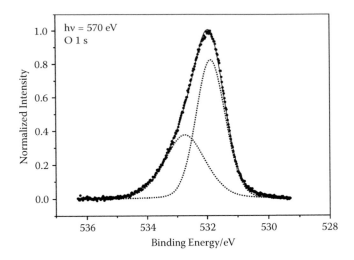

FIGURE 5.24   SRPES O 1s core level signal of n-type Si(111) covered with a thin anodic oxide layer. The anodization was performed potentiodynamically from −0.5 V to +0.5 V (SCE) in a phathalate buffer of pH 4.8.

The latter is to be assigned to adsorbed water. The binding energy of the O 1 s for water-free SiO_2 was determined at 532.3 eV [49]. The shift of 0.5 eV arises from the considerable amount of Si-OH bounds in the anodic oxide.

The formation of nanoislands of catalytic metals by electroreduction of their chloride salts such as H_2PtCl_6, $RhCl_3$, and $IrCl_3$ onto H-terminated silicon brings about side effects: the formation of a thin oxide film and an amorphous Si film due to the inward diffusion of H. The chemical and electronic characteristics of the formed interface delimits the electron transfer process of photoelectrodes. The surface chemical transformations can be followed by SR-PES measurements. The electrodeposition onto Si is selected as a model system to be extrapolated to the study of other systems. The existence of an oxide layer beneath the deposited particles was already discussed with respect to its morphological aspects in Chapter 3. Figure 5.25 shows the Si 2p core level signal of n-type Si(111) recorded with a surface-sensitive excitation energy of 170 eV (low kinetic energy of emitted photoelectrons near the minimum of the attenuation length) before and after the electrochemical deposition of Pt and Ir was carried out *in-system*. Small signals detected at larger binding energies at the foot of the bulk signal at 99.4 eV indicate the presence of some oxidized atoms. The spectral signals obtained after the deposition of a minute amount of metal indicate the formation of an oxide film. The signal was deconvoluted into seven components as in the case of the analysis of anodic oxides. Apart from the tetra-coordinated Si at 99.4 eV and the α- and β-components arising by stretching and compression of Si-Si bonds, lines at 100.4 eV, 101.4 eV, 102.4 eV and 103.4 eV appeared. These were assigned to Si(I), Si(II), Si(III), and Si(IV), respectively.

According to Eq. (5.63), the relative abundance of the different oxidation states can be calculated. The results are shown in Table 5.2. It is worth noting the large abundance of Si(I) and Si(III), which are characteristic of an interface with a large

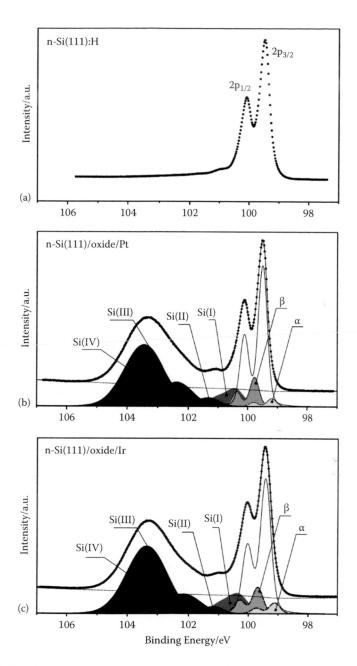

FIGURE 5.25 SRPES Si 2p core level signal of n-type Si(111):H obtained with an excitation energy of hv = 170 eV before (a) and after the deposition of Pt (b) and Ir (c). The metal deposition was carried out potentiodynamically from 0 V up to the first voltammetric peak (−0.35 V for Pt and −0.9 V for Ir) at 10 mV s⁻¹ from 1 mM H_2PtCl_6 + 0.1 M K_2SO_4 and 2 mM $IrCl_3$ + 0.5 M KCl, respectively.

TABLE 5.2

Relative Abundance of Si Coordination States at the Si/SiO$_x$ Interface for an Excitation Energy of 170 eV Obtained after Potentiodynamic Electrodeposition of Pt and Ir from 0 V to −0.35 V SCE (Pt) and −0.9 V SCE (Ir) onto n-Type Si(111)

Coordination State	n-Type Si/Oxide/Pt	n-Type Si/Oxide/Ir
Si(I)	0.206	0.235
Si(II)	0.096	0.071
Si(III)	0.216	0.071
Si(IV)	0.482	0.522

amount of defects arising from dissolution processes taking place during the deposition process.

The oxide thickness calculated by using Eq. (5.62) and taking $\lambda_f \sim 0.65$ nm for $h\nu = 170$ eV yields $d_{ox} = 0.35$ nm after Pt deposition, if one considers $I_{ox} = I_{Si(IV)}$. HRTEM of cross-sectional cuts performed after depositing Pt and Ir (see Figure 3.18) shows, on the other hand, the formation of an interfacial layer of ~ 3 nm. Indeed the amorphous interfacial layer arises mainly as a consequence of inward diffusion of H during the electrochemical deposition.

The thin oxide layer formed during deposition of noble metals is a natural consequence of the position of the redox level of the reduction reactions of their chloride complex below the valence band edge of the semiconductor, as explained later in detail. The large concentration of Si(I) and Si(III) in the very thin oxide film leads inexorably to the formation of a large concentration of interfacial states. Capacitance-voltage curves recorded after the deposition of Pt show a significant increase of the capacitance peak appearing in semiconductor depletion (Figure 5.26). By using Eq. (5.58), one can estimate an increase of the density of surface states by 5.3×10^{12} cm^{-2} after 5 s Pt deposition. The initial density of states is ~ 3.6×10^{11} cm^{-2}. The deposition of Pt also brings about a change in the semiconducting behavior of the substrate, as can be inferred from the steeper increase of the capacitance in accumulation, in comparison with the S-like form of the C-V curve of the n-type Si(111):H surface. The flat band potential was determined by the Mott–Schottky method at high frequencies ($f = 1$–10 kHz) at −0.57 V vs. SCE in sulfate solution. The second capacity peak at −0.85 V probably arises from the energy distribution of surface states. This interpretation implies the assumption of a strong Fermi level pinning and the stretching of the capacitance curve along the potential axis. The potential drop in the oxide is proportional to the amount of charged states. Therefore, the calculation of the density of surface states by integration of the capacitance curve requires, in principle, an estimation of the fraction of potential drop in the oxide. Dotted lines in Figure 5.26 represent the hypothetical case of the response of surface states without Fermi level pinning.

No morphological differences can be noted at first glance in the HRTEM pictures shown for different metals in Figure 3.18. The shape of the capacitance-voltage curves indicates, however, very different behaviors for Rh and Ir (see Figures 5.27

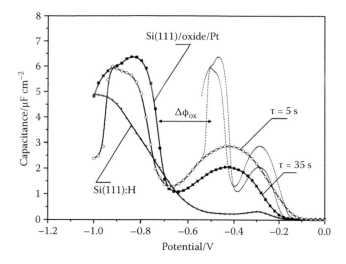

FIGURE 5.26 Capacitance-voltage curves measured at 100 Hz in 0.1 M K_2SO_4 pH 2.3 of n-type Si(111):H before and after electrochemical deposition of Pt. Deposition was performed at -0.35 V from 1 mM $H_2PtCl_6 + 0.1$ M K_2SO_4 for two different pulse lengths.

and 5.28). In both cases the curves do not show a capacitance peak, but rather a sigmoidal shape (experimental curves were corrected by subtracting the electrolyte resistance from the total impedance at high frequencies). The S-like curve of the Si/oxide/Rh interface shifts toward the depletion potential region on decreasing the analysis frequency. In accumulation, the capacitance attains a constant value of about $10\ \mu F\ cm^{-2}$. This value agrees with that expected for an oxide thickness of 0.5 nm, assuming $\varepsilon_{ox} = 6$ (Eq. 2.68). The Si/oxide/Ir interface reaches a constant capacity of $2-3\ \mu F\ cm^{-2}$ in accumulation. An increase of the semiconductor capacitance by accumulation is first observed at -1.0 V, giving strong evidence for Fermi level pinning upon charging of surface states.

The capacitance-voltage curves for the Si/oxide/Rh junctions resemble the typical course of MOS junctions [50]. This behavior cannot be explained by regarding solely the frequency dependence of the charging of interface states and deserves a special analysis. The capacitance-voltage curves at different frequencies are practically identical to that of the ideal MOS. They shift toward more positive potentials upon decreasing the frequency analysis. It should be noted that a shift of about 0.2 V is still observed at 10 kHz! The inverse effect is observed at p-type Si. In contrast to n-type Si, the frequency dispersion is much less and the potential shift amounts to 1.05 V. The potential shift arises in this case as a consequence of the formation of fixed charges at the Si-SiO_2 interface. The concentration of fixed charges can be calculated by the dielectric potential drop in the oxide. Assuming an oxide thickness of 1 nm and a dielectric constant of 6, a concentration of fixed charge $n_{it} = (q_{it}/e) = 0.9 \times 10^{13}\ cm^{-2}$ and $3.46 \times 10^{13}\ cm^{-2}$ for the n-Si/oxide/Rh and p-Si/oxide/Rh junctions, respectively, can be calculated. According to the shift direction observed in the capacitance-voltage curves, the formation of interfacial fixed charges must be positive for n-type and negative for p-type silicon. A possible explanation can

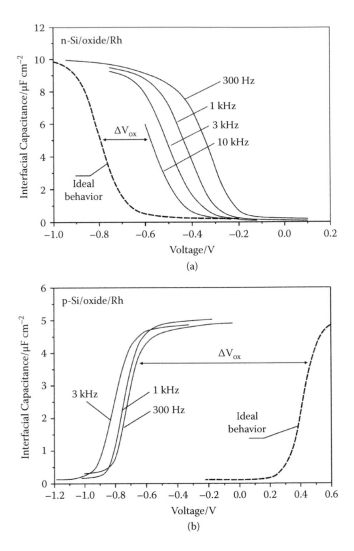

FIGURE 5.27 Capacitance-voltage curves of (a) Rh/oxide/n-Si junction and (b) Rh/oxide/p-Si recorded at different frequencies in 0.1 M K_2SO_4 pH 2.3. Dashed lines represent the ideal response of the Si/oxide junction for a donor concentration of 10^{15} cm^{-3} and flat band potentials of −0.55 V for n-type Si and 0.15 V for p-type Si.

be formulated in terms of the reaction of P_b centers with hydrogen diffusing into the silicon. This reaction was extensively investigated, motivated in part by the so-called negative bias temperature instability arising in field effect transistors (MOSFET) [51]. The formation of fixed interfacial charge traps by hydrogen was observed after annealing of Si-SiO$_2$ interfaces in a hydrogen atmosphere, a standard technique used to reduce the concentration of dangling bonds. Experiments performed by Afanas'ev and Stesmans [52] on annealing Si-SiO$_2$ interfaces in a hydrogen atmosphere at 450–800°C indicate the formation of a large density of positively charged centers (about

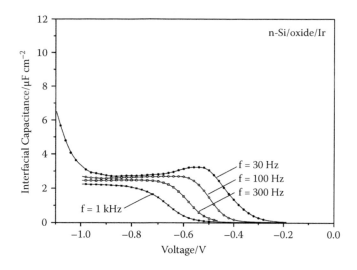

FIGURE 5.28 Capacitance-voltage curves of Ir/oxide/n-Si junction in 0.5 M KCl pH 2.3 recorded at different frequencies. Ir particles were deposited electrochemically from an $IrCl_3$ solution at −0.9 V.

10^{13} cm^{-2}) not correlated with the density of dangling bonds. Various models have been proposed to describe the formation of charge traps [53–57]. One of them [57,58] considers the conversion of E′ centers, a positively charged oxygen vacancy with a Si dangling bond, into a H-related interface trap. Although the mechanism of formation of fixed charges at the Si-SiO_2 interface is not fully understood, it is reasonable to assume that there is a charge exchange between interface states and the nearest oxide defects. In the case of n-type Si/oxide junctions, a proton bridge is formed by insertion of atomic hydrogen. As a result, the dangling bond is passivated and a positive charge is created by formation of an over-coordinated O atom. On the other hand, the density of states of P_b centers is reduced, and thus also the band bending generated by charge exchange between interfaces states and the conduction band. For p-type Si/oxide interfaces, the valence band character of interface states neutralizes the nearest E′ centers by insertion of atomic hydrogen. As the available interface states for charge exchange with the valence band are exhausted by reaction with H and E′ centers, the band bending retrogrades. The accumulated positive charge at the Si-interface leads to dielectric potential drop at the oxide. These processes are represented schematically in Figure 5.29. The reason why this effect is especially observed for Si/oxide/Rh junction remains unclear. A possible explanation may be the particularly strong atomic hydrogen adsorption onto Rh surfaces and its subsequent inward diffusion to the oxide covered film.

The especially high concentration of interface states formed by surface oxidation during the first instants of electrochemical deposition leads to a strong Fermi level pinning reflected in the SRPES valence band spectra recorded at a surface-sensitive excitation energy of 150 eV (minimal inelastic mean free path) (Figure 5.30). The linear extrapolation of the rising part of the signal to zero intensity for a surface sensitive excitation energy indicates the position of the Fermi level at the investigated surface.

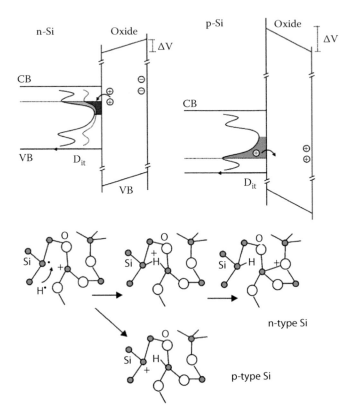

FIGURE 5.29 Energy band diagrams and corresponding atomic structures representing the charging processes at n-type/oxide and p-type/oxide junctions.

It can be observed that the Fermi level of the n-type Si(111) surface, fresh etched in 40% NH_4F, is placed 0.195 eV below the surface conduction band edge. According to the resistance of the used wafer of 5–7 Ωcm, a donor concentration of 6–8 × 10^{14} cm^{-3} is calculated. This corresponds to a difference $E_{CB} - E_F = 0.26$ eV, and means that the n-type Si(111):H is in accumulation. This effect is stronger on so-called *step-bunch* surfaces, which present a stair-type structure arising from a stacking of dissolving terraces by a layer-by-layer dissolution type mechanism [59,60]. The charge accumulation in the case of atomic terraces is explained in terms of spatial redistribution related with different numbers of terminal H-bonds on the terraces and at their edges.

After the electrodeposition of Pt from a H_2PtCl_6 solution within the depletion potential region, the Fermi level shifts to a new position 0.45 eV below the conduction band; the semiconductor is now in depletion with a band bending of 0.2 V. The semiconductor is, however, in its flat band after the anodic oxidation at 0.5 V in a phthalate solution. This fact seems to indicate an improvement of the quality of the Si-SiO$_2$ interface by a reduction of the density of traps upon relaxation of the junction by formation of a thicker oxide. However, additional capacitance-voltage studies do not show precisely such an effect. The capacitance-voltage curves carried out after anodization at increasing potentials show that the capacitance peak does not increase

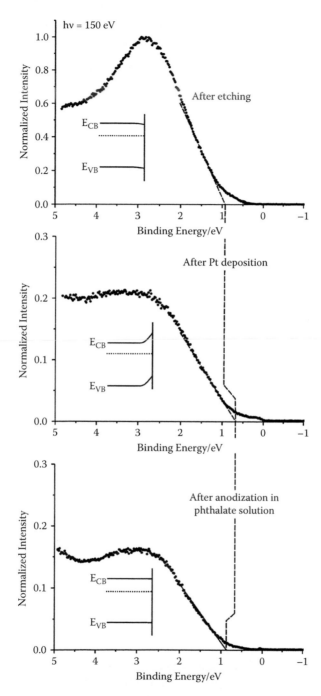

FIGURE 5.30 SRPES valence band spectra taken with an excitation energy of 150 eV on n-type Si(111) (a) after etching in 40% NH_4F, (b) after (a) and deposition of Pt, and (c) after (a) and potentiodynamic anodization up to 0.5 V in phthalate solution. Pt deposition was performed potentiodynamically at 10 mV s^{-1} from 0 V to -0.35 V from 1 mM + 0.1 M K_2SO_4.

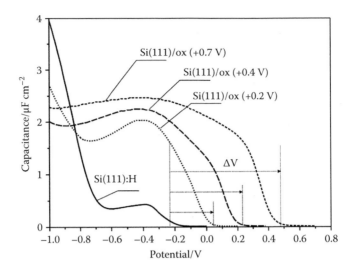

FIGURE 5.31 Capacitance-voltage curves of n-type Si(111):H before and after anodization at increasingly anodic limits in 0.1 M sodium phthalate solution under illumination.
Frequency: 100 Hz. Illumination: W-I lamp with an intensity of 100 mW cm⁻².

significantly between 0.2 V and 0.7 V, that is, during the formation of the first oxide
monolayers. It should be noted, however, that the capacitance peak widens in a linear
relationship with the anodic potential.

This effect can be interpreted as a direct consequence of the dielectric potential
drop in the formed oxide; it also implies a charge exchange between interface states
and electron traps in the oxide. Let us assume a constant density of interface states
D_{it}. By a charge exchange at the Si-SiO₂ interface, a potential drop is established in
the oxide, given by

$$\Delta V = \frac{\int_{E_{ss}}^{E_F} D_{it}(E)\,dE}{C_{ox}} = \frac{q_{it}}{C_{ox}} = \frac{d_{ox}q_{it}}{\varepsilon_{ox}\varepsilon_0} = q_{it}\frac{\beta(V_a - V_{ox}^f)}{\varepsilon_{ox}\varepsilon_0} \qquad (5.65)$$

where β is the growth factor, V_a is the anodic voltage, and V_{ox}^f is the oxide formation potential.

The etching of Si(111) surfaces in 40% NH₄F leaves a surface structure characterized by atomic terraces with zigzag edges and a height of 0.314 nm, as can be
seen in the tapping-mode AFM images depicted in Figure 5.32. This type of surface
structure exposes H-terminated surface atoms with different number of Si-Si back
bonds. The atoms on the triangular terraces show single dangling bonds. Atoms at
the tip of terraces and edges parallel to the $\langle \bar{1}12 \rangle$ direction are coordinated with two
H-atoms. These atoms are especially susceptible to a nucleophilic attack of water
molecule as the first step in the formation of Si-OH bonds with participation of two
holes. The breakdown of the H-termination by the formation of –OH terminal bonds

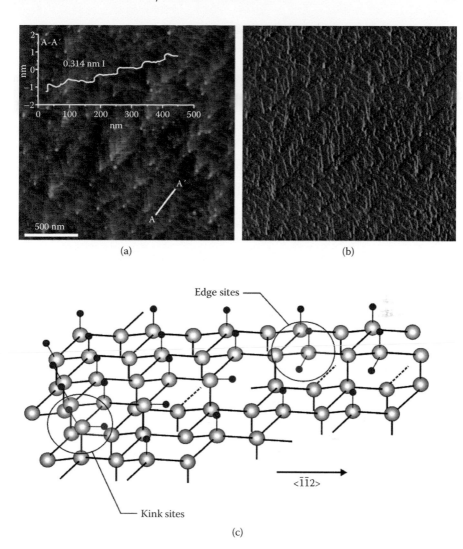

FIGURE 5.32 (a) and (b) Tapping-mode AFM pictures of an n-type Si(111) surface fresh etched in 40% NH$_4$F. (a) height-mode; (b) modulation mode; (c) atomic structure of silicon (111) surface. Double coordinated atoms are pointed out.

at the tip of the terraces constitutes the trigger for the formation of the first oxide layers in silicon. The mechanism of oxide formation is depicted in Scheme 5.1. The formation of polarized Si-OH bonds weakens the Si-Si back bonds, making them susceptible to further attack of water. This latter reaction leaves as products a surface atom with two back bonds and an atom three-fold coordinated with –OH groups and one Si-Si back bond. In a subsequent step two adjacent Si-OH bonds condense to form a Si-O-Si bridge. The process repeats until the oxide film covers the whole surface, extending in a direction perpendicular to the $\langle \bar{1}\bar{1}2 \rangle$ direction. Further growth after the formation of the first oxide monolayer proceeds as already discussed in

Chapter 2 by high-field aided transport of O^{2-} from the electrolyte to the surface and counter transfer of holes.

The electrodeposition of noble metals such as Pt, Ir, Rh, and Os is carried out from solutions of their chloride complexes. The thermodynamic standard reduction potentials for the different chloride complexes are much more positive than the flat band potential of n-type silicon (Eqs. 5.66–69) [61,62]. In the vacuum scale, the corresponding redox levels are situated below the valence band:

$$PtCl_6^{2-} + 2\ e^- \rightarrow PtCl_4^{2-} + 2\ Cl^- \qquad V^0 = 0.708\ V\ (SCE) \qquad (5.66)$$

$$PtCl_4^{2-} + 2\ e^- \rightarrow Pt + 2\ Cl^- \qquad V^0 = 0.535\ V\ (SCE) \qquad (5.67)$$

$$IrCl_6^{3-} + 3\ e^- \rightarrow Ir + 6\ Cl^- \qquad V^0 = 0.61\ V\ (SCE) \qquad (5.68)$$

$$RhCl_6^{3-} + 3\ e^- \rightarrow Rh + 6\ Cl^- \qquad V^0 = 0.26\ V\ (SCE) \qquad (5.69)$$

The current-voltage curves depicted in Figure 5.33 indicate that the deposition of Pt and Rh requires a large overpotential, over 0.5 V. Current flow is observed as the semiconductor is in depletion. Because of the high energy barrier imposed by the semiconductor band bending, charge transfer is permitted by injection of holes from the empty states of the redox level to the valence band or to interface states [63]. The electron transfer is represented by energy band diagrams at three different potential values in Figure 5.33. The energy of the empty electronic states in the solution is represented by the Gaussian distribution function according to the Marcus–Gerischer Theory (see Chapter 4):

$$D_{ox}(E) = \frac{1}{\sqrt{4\pi\lambda_{ox}kT}} \exp\left[-\frac{(E - E_{redox} + \lambda_{ox})^2}{4\lambda_{ox}kT} \right] \qquad (5.70)$$

The fact that deposition is observed just as the Fermi level of the semiconductor is driven above the neutrality level of surface states provides experimental evidence supporting the supposition of hole trapping by surface states. The electroreduction of a metal cation requires the trapping of four holes in the case of Pt and three holes for Rh and Ir. Trapping of holes in interface states has several implications. We have already mentioned, that interface states arise in surface oxidized atoms. Due to the increased susceptibility of kink and edge atoms to be oxidized, it is expected that the low number of oxidized atoms in the uncompleted H-termination achieved by chemical etching in 40% NH_4F are preferentially located at these sites. Initial oxidized atoms trigger further oxidation by hole supply. Oxidized surface atoms at kink and edge sites play a decisive role in the nucleation process during electrodeposition, because they are the sites of enhanced reactivity: chemical and electronic. This reactivity is due in part to the enhanced radial electric field around these sites given by $E(r) = \Delta\varphi/r$, where $\Delta\varphi$ is the potential drop at the Helmholtz layer and r is the radius of the site. Accordingly, the capacity at the spiky sites increases inversely proportional to the radius as

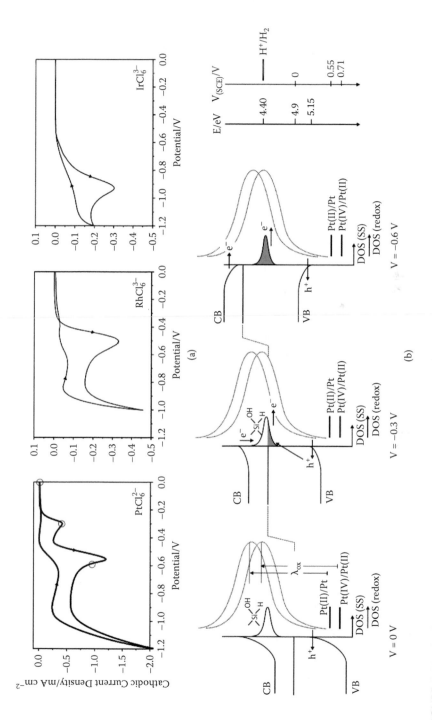

FIGURE 5.33 (a) Current-voltage curves of the electroreduction of chloride complexes of noble metals onto n-type Si(111):H surfaces. Scan rate: 10 mVs^{-1}. Solutions: 1 mM H$_2$PtCl$_6$ + 0.1 M K$_2$SO$_4$, 1 mM RhCl$_3$ + 0.5 KCl and 2 mM IrCl$_3$ + 0.5 M KCl; (b) energy band diagram representing different stages of the electroreduction of PtCl$_6^{2-}$. A work function of 4.9 eV was assumed for the SCE electrode.

$$C_{H(r)} = \frac{\varepsilon_H \varepsilon_0}{r}\left(1 + \frac{r}{d_H}\right) \qquad (5.71)$$

where d_H is the width of the planar Helmholtz layer. The localized increase of the interfacial electric field and the enhanced electron density around oxidized atoms related to surface states brings about a spatial selectivity of deposition reflected in a preferential formation of crystallites at the tip of zigzag-like terraces. Figure 5.34 shows the surface morphology before and after a few seconds of Pt electrodeposition from $H_2PtCl_6^{2-}$ onto n-type Si(111):H at -0.35 V (i.e., in depletion). It can be noted that in addition to the formation of some metal particles, a step-bunch type of

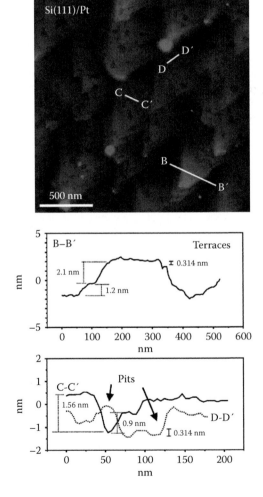

FIGURE 5.34 Tapping-mode AFM pictures of an n-type Si(111) after 3 s Pt deposition at -0.35 V from 1 mM H_2PtCl_6 + 0.1 M K_2SO_4. Some representative cross sections are also shown.

dissolution has led to the formation of large triangular terraces with a height of some monolayers, as clearly shown by the cross sections performed at some selected sites of the AFM picture. The appearance of some pits of 1–2 nm depth allows us to infer the formation of an anodic oxide film with enhanced dissolution at some points.

The enhanced reactivity of oxidized surface atoms in electrocrystallization of noble metals is reflected by comparing current-voltage curves performed on n-type Si(111), the surface of which was chemically etched (i) with 40% NH_4F, with (ii) the additional a second etching step with 50% HF (Figure 5.35). It is assumed that the additional etching step hydrogenates the remaining oxidized atoms from the first step. It is clearly seen that the elimination of these preferential nucleation sites leads to an increase of the nucleation work pointed out by a larger overpotential.

The electrocrystallization selectivity of oxidized atoms is also related to a surface complex chemical process involving the formation of transitory chemical bonds.

An SRPES Pt 4f core level signal obtained at the very first instant of electrodeposition onto n-type Si(111):H shows a line at 71.85 eV, shifted 0.95 eV from the $4f^{7/2}$ line corresponding to metal Pt (see Figure 5.36). This line is also observed after deconvolution of the Pt 4f signal obtained after some metal deposition. The appearance of a line shifted by about 1 eV suggests the formation of a Pt-O bond, if one takes into account chemical shifts of 0.3 eV for PtO and 4.1 eV for PtO_2 reported in the literature. The formation of silicides can be excluded at low temperatures, especially on oxidized surfaces. A value of 72.3 eV for the Pt $4f^{7/2}$ line on PtO has been reported [64–66].

Similarly, the SRPES Rh 3d core level signal indicates the presence of two oxidation states after electrodeposition of Rh onto n-type Si(111):H (see Figure 5.37). The predominant signal is observed at a binding energy of 307.2 eV, which can be ascribed to metallic Rh. The spectrum shows a shoulder indicative of a second line shifted +0.85 eV. In principle, this line can be ascribed to the formation of Rh_2O_3 according to data reported in the literature [65]. The formation of a bulk oxide can,

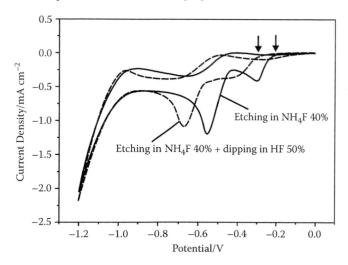

FIGURE 5.35 Current-voltage curve of n-type Si(111) in 1 mM H_2PtCl_6 + 0.1 M K_2SO_4 after etching in 40% NH_4F and 40% NH_4F + dipping in 50% HF. Scan rate: 10 mV s^{-1}.

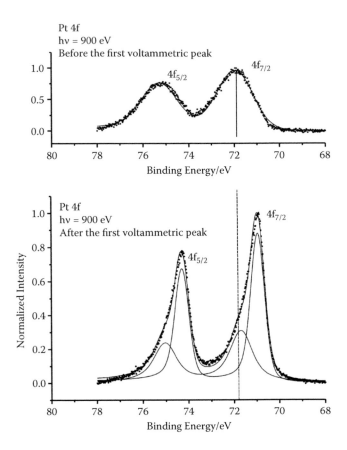

FIGURE 5.36 SRPES Pt 4f core level signal of an n-type Si(111) surface after potentio-dynamic deposition of Pt: (a) sample taken out of the solution before the first current-peak at −0.35 V; (b) sample taken out after the current peak ($V < -0.35$): refer to j-V curve shown in Figure 5.33a. A Shirley-type background and asymmetry factor of $\alpha = 0.33$ for the line at 70.9 eV were used.

however, be disregarded considering the standard oxidation potential of Rh: +0.7 V (NHE). The formation of an Rh-O bond linking the metal particles to the oxidized surface is here also assumed. Interestingly, the formation of a Rh(III) oxide mono-layer with a corundum structure was found to show a chemical shift of +0.85 eV [67].

In light of these experimental findings, it is acceptable to postulate an interaction between the sp^3 hybrid orbital of a surface Si atom and the d^2sp^3 orbital of the halogenide metal complex resulting in an O-bridge, which facilitates the electron transfer of the reduction process. Figure 5.38 shows a reaction scheme representing the reduction process.

The ligand orbitals of the metal complex are partially occupied by water molecules [68,69]. The Pt(IV) complex presents six coordination sites, forming an octahedron. First, the complex is attracted to the hydroxylated site by hydrogen bridge bonds. A proton transfer occurs by a Grottus-like mechanism. This weakens the Si-O bond

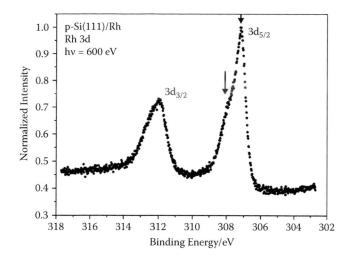

FIGURE 5.37   SRPES Rh 3d core level signal of a p-Si(111) surface after 10 s of Rh deposition at −0.4 V from 2 mM RhCl$_3$ + 0.5 M NaCl.

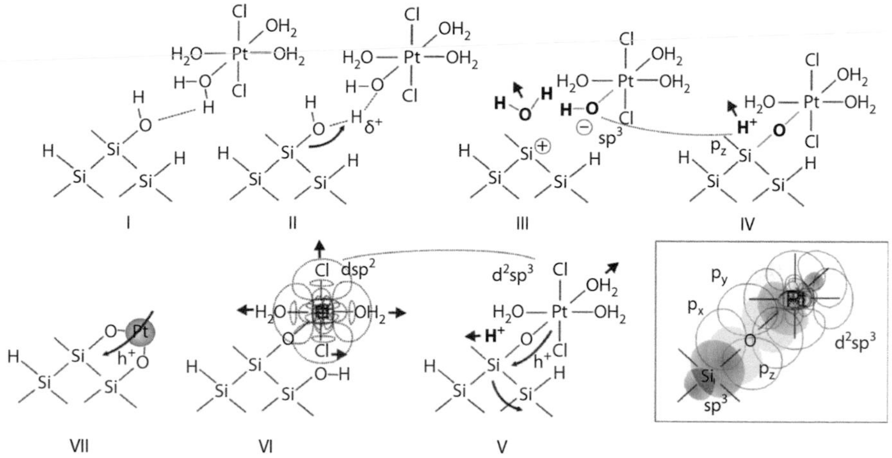

FIGURE 5.38   Reaction mechanism for the nucleation of crystallites of noble metals onto silicon.

and leads to the expulsion of a water molecule. The electrostatic attraction of the positively charged surface atom and the hydroxyl group results in the formation of a Pt-O-Si bridge with the release of a proton by breakdown of the sp^3 hybridization and the formation of more stable bond by coupling of a p-orbital with one of the sp^3 orbitals of the silicon atom. This bond is thought to accelerate the transfer of holes from the metal cation to the semiconductor. The metal complex reduces at first to Pt(II) characterized by a planar complex structure by formation of a dsp^2 hybrid orbital. Holes injected into the semiconductor through the O-bridge trigger the oxidation of Si by the mechanism described in Scheme 5.1. The reduction of the adsorbed Pt(II)

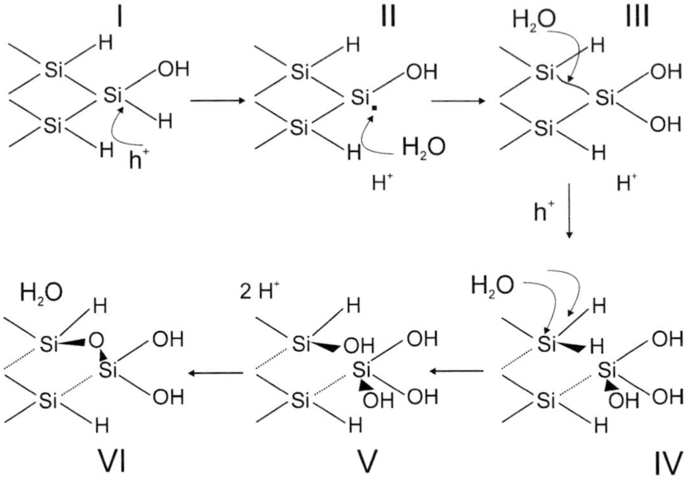

SCHEME 5.1   Reaction mechanism for the formation of the first oxide layers.

complex may lead to the formation of a second Pt-O bond by a substitution reaction of the coordination water with the –OH group of the adjacent Si surface atom. As a result, the first Pt ad-atom forms onto the underlying oxide monolayer.

At 0 V, the n-type Si is in depletion and the Fermi level is below the neutrality level of surface states. Under these conditions, holes from empty redox levels can only be injected into the valence band. Accumulation of holes triggers the oxidation of the surface in the same way as it occurs by photogeneration of holes under illumination. The presence of a thin oxide layer can be detected by SRPES, but not the presence of Pt. Here, the oxidation process is coupled to the first reduction step of Pt: Pt(IV) → Pt(II) and it is probably nonspecific. The nucleation of Pt requires filled surface states.

Because holes are the majority charge carrier, the electrocrystallization of noble metals onto p-type Si surfaces is inherently associated with the presence of an oxide film, independently of the reduction process. The thickness of the oxide film depends on the initial potential applied in the deposition process. Reduction theoretically requires electrons, which should be provided by illumination of the p-type semiconductor. The electroreduction of $RhCl_6^{3-}$ complex also occurs in the dark, as indicated by the voltammetry shown in Figure 5.39. The current-voltage curve shows an initial anodic current due to the surface oxidation. Electroreduction takes place at potentials V < –0.2 V, that is, as the Fermi level is positioned above the neutrality level of surface states and the localized states becomes hot spots with an excess of electron density. Illumination brings about solely an increase of about 20% of the current peak. This experimental evidence supports the reduction mechanism by injection of holes into interface states. A direct injection of holes into the valence band is prohibitive due to the large energy barrier imposed by the band bending. Illumination brings about an increase of the density of occupied interface states. The presence of two correlated cathodic peaks and two anodic peaks is related to the hydrogen adsorption on deposited Rh particles as discussed in Chapter 4.

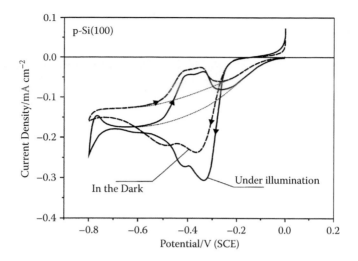

FIGURE 5.39 Current-voltage curves for the electrodeposition of Rh onto H-terminated p-type Si(100) from 1 mM $RhCl_3$ + 0.5 KCl in the dark and under illumination. Scan rate: 10 mV s^{-1}. Illumination: W-I lamp with an incident power of 100 mW cm^{-2}.

TM-AFM pictures shown in Figure 5.40 resume the collateral transformations of the n-type Si(111) substrate occurring during electroreduction of noble metal complexes. The evolution of the morphology depends on the chemistry of the solution. One common aspect, however, is the preferential deposition of particles at edge and kink sites, pointed out by AFM pictures taken at the first instant of deposition.

The supply of holes by reducing complexes maintains the oxidation of the substrate. The resulting morphology depends on the solution chemistry, which more or less influences the dissolution processes. A step-bunch dissolution mechanism is observed for the deposition of Pt from sulfate solutions. Kink and edge sites play an important role in the morphological evolution of the silicon substrate due to their special reactivity. Thus, different morphologies appear according to the extent and direction of the miscut. It is interesting to note, for instance, the effect of the addition of i-propanol on the final substrate morphology developed during the deposition of Ir from its chloride complex [70].

Several theories have been formulated to explain the course of the topographical changes in fluoride solutions [71–73]. For instance, the dissolution of n-type Si(111) in diluted NH_4F solutions starts with the removal of a kink site with two dangling bonds and continues perpendicular to the $\langle \overline{11}2 \rangle$ direction. As a consequence, an equalization of the zigzag terraces occurs.

The dissolution process accompanying the formation of metallic nanojunctions is arrested by the formation of an oxide film. The initial dissolution of Si in fluoride free solutions is aided in the initial stages of oxide formation by chloride ions. It was observed that the low frequency interfacial capacitance of n-type Si(111) increases linearly with time when the surface is polarized at −0.30 V (SCE) in a 0.5 M KCl solution of pH 2.3. On the contrary, the capacitance remains constant in sulfate solutions. This experiment indicates that chloride ions trigger the oxidation of the surface

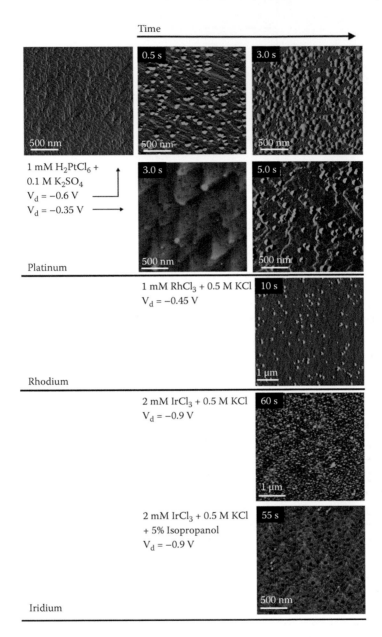

FIGURE 5.40 Tapping-mode AFM pictures of n-type Si(111) after potentiostatic electrode-position of different noble metals from their chloride complexes.

and hence the related increase of surface state concentration. The effect of chloride is reflected in another experiment consisting of applying a triangular potential program as shown in Figure 5.41. The current, initially cathodic, adopts anodic values from -0.1 V on. The anodic process persists at the reverse scan until complete inhibition at -0.5 V: a customary induction-like process observed in pitting corrosion processes

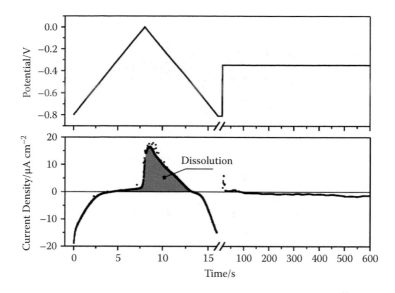

FIGURE 5.41 Potentiodynamic investigation of chloride-assisted dissolution of n-type Si(111) in 0.5 M KCl, pH = 2.3.

initiated by halogenides. A subsequent return to −0.35 V shows practically no anodic current, indicating the formation of a dissolution barrier: the oxide film. The passed anodic charge amounts to 46.8 μC cm⁻², which would correspond to the electrosorption of ¼ of a monolayer. The fact that this process is not observed in sulfate solutions lets us infer that the electrosorption of Cl⁻ with the supply of holes on reactive surface sites accelerates the oxidation process by strong polarization of Si-Si back bonds and subsequent solvolytic attack of water to form new Si-OH bonds [74]. The electrosorption of halogenides was the argument used by other researchers to explain the experimentally observed shift of the flat band potential of silicon in concentrated HX solutions (X: I, Br, Cl) [75]. Direct evidence for electrosorption could only be found for I⁻ by means of XPS experiments. The absence of a Cl or Br spectral signals was explained in terms of an exchange of adsorbed X⁻ with OH⁻ during rinsing of the sample [76].

The surface dissolution of n-type Si(111) silicon in depletion assisted by electrosorption of Cl⁻ leads to the formation of large terraces with zigzag edges of a height of about 1 nm as indicated by the TM-AFM picture shown in Figure 5.42. In light of the experimental evidence, the mechanism depicted in Scheme 5.2 can be formulated. Cl⁻ adsorbs onto an already oxidized kink site by consuming a hole (the formulation of this step was experimentally found by the observation of an anodic current). The trapping of a hole leads to the breakdown of the Si-Si back bond and the formation of a Si⁻ and a Si-Cl bond. Si⁻ reacts with water to form a Si-OH with release of a proton and the injection of one electron into the conduction band. This latter is consumed by the proton reduction reaction to form H₂. The formation of hydrogen bubbles is experimentally observed after prolonged polarization time. Further solvolytic water attack results in the formation of soluble chlorsilicates. It should be noted, that steps I and III represent

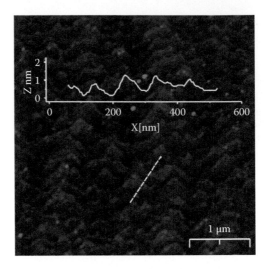

FIGURE 5.42 Tapping-mode AFM picture of an n-type Si(111) surface after holding it for one hour at −0.35 V in 0.5 M KCl, pH 2.3.

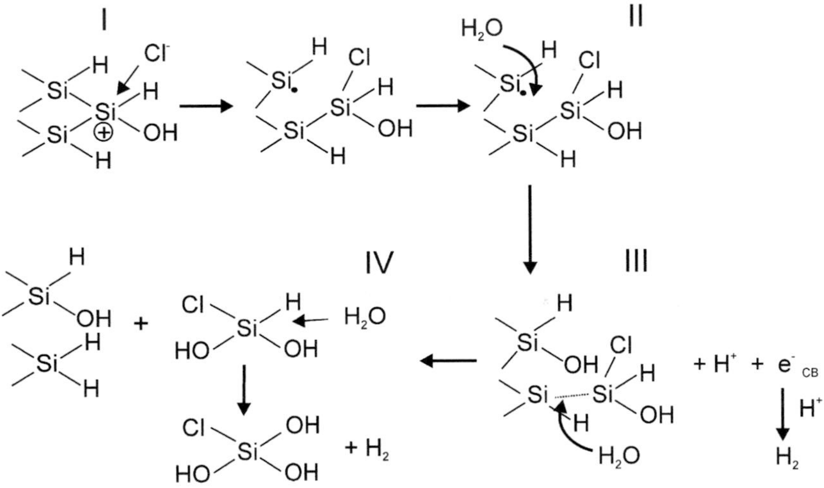

SCHEME 5.2 Reaction mechanism of chloride-assisted Si dissolution.

anodic and cathodic counterparts of the corrosion processes. The substrate dissolution aided by electrosorption of Cl^- competes with the formation of an oxide film. This gives rise to the formation of pits during the first instants of deposition.

5.3 MODULATION OF THE INTERFACIAL POTENTIAL IN NANODIMENSIONED EMOS CONTACTS

There is a lot of experimental evidence for a local modification of the band bending of the semiconductor upon depositing nanodimensioned metal islands. This results in an efficient collection of photogenerated charge carriers when modified semiconductors are used in photoelectrochemical cells in the photovoltaic mode [77–80]. Nakato et al. [80,81] explained the efficient collection in terms of the change of the semiconductor band bending under the metal islands as a consequence of formation of mirror charges at the semiconductor–metal contact. This immature model takes into account neither the chemical transformation of the surface during deposition nor the presence of surface states.

From the general expression of the semiconductor capacitance, the following relation can be derived for the semiconductor in depletion:

$$\frac{1}{C_{sc}^2} = \frac{2}{(e\varepsilon_{HL}\varepsilon_0 N)}\left(\Delta\varphi_{sc} - \frac{kT}{e}\right) \tag{5.72}$$

This expression is known as the Mott–Schottky equation. The representation of the experimentally measured capacitance as $1/C^2$ vs. V shows a linear relationship, the slope of which is inversely proportional to the density of impurity concentration. The extrapolation of the linear part of the curve toward the zero ordinate yields the flat band potential. It is customary to measure Mott–Schottky plots at frequencies $f > 3$–10 kHz, where the response of surface states is suppressed. Experimentally, it is observed that the electrodeposition of noble metals brings about a drop of the slope of the Mott–Schottky plot of n-type and p-type Si. Figure 5.43 shows this for Ir and Pt deposits onto n-type Si(111). It should be noted that in the case of Ir, the slope drops by a factor of 0.64 and the flat band potential shifts by about 0.3 V toward more negative potentials. A similar effect, but to a lesser extent, is observed after Pt deposition. Several causes can be addressed here: (i) a modification of the doping concentration of the semiconductor near the interface due to inward diffusion of hydrogen during deposition [82–84], (ii) a reduction of the thickness of the space charge layer under the nanodimensioned MOS junctions, and (iii) a potential drop in the oxide film due to charging of interface states. The effects of the inward diffusion of H were discussed in Chapter 3. In n-type Si, interstitial H brings about, in principle, a compensation of the doping concentration and the formation of an np-type junction. The drop of Mott–Schottky slope is also observed to decrease by further deposition. Hence, the effects of nanodimensioned MOS junctions seem to play a minor role here.

The charging of surface states is compensated by a dielectric response of the thin oxide film. In an ideal MOS junction, the total capacitance is given by $C_T^{-1} = C_{sc}^{-1} + C_{ox}^{-1} + C_H^{-1}$. Now, let us assume an anodic silicon oxide of 1 nm thickness and a dielectric constant of 6 [85]. Thus, we have $C_{ox} = \varepsilon_{ox}\varepsilon_0 / d_{ox} = 5.31\ \mu\text{F cm}^{-2}$. This value is much larger than the semiconductor capacitance in depletion: $C_{sc} < 0.1\ \mu\text{F cm}^{-2}$. This means that in such a case, the capacitance is still dominated by the semiconductor. In the presence of interface states, $C_{sc} + C_{it} > C_{ox}$, and hence the oxide influences the capacitance behavior. The total potential drop splits into two parts: one at the semiconductor and one at the oxide, the relation of which is given by $\Delta V_{ox}/\Delta V_{sc} = [C_{sc} + C_{it}]/C_{ox}$.

Electrochemically prepared planar MOS junctions allow analysis of the electronic structure of the Si-SiO$_2$ interface without the influence of the electrolyte and a better

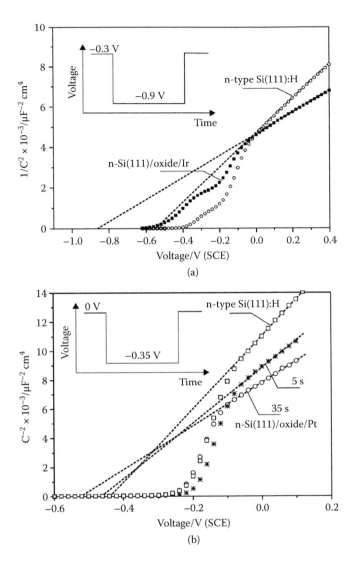

FIGURE 5.43 Mott–Schottky plots of n-type Si(111) recorded at 3 kHz before (hydrogen-terminated surface) and after the electrodeposition of Ir (a) and Pt (b) in 0.5 M KCl and 0.1 M K_2SO_4, both at pH 2.3, respectively. Deposition times: 55 s for Ir and 5 and 35 s for Pt. Inserts indicate the voltage-time program used for deposition.

definition of the potential drop at the oxide–semiconductor interface. The current-voltage curves performed with Pt–oxide–Si(111) junctions prepared electrochemically by longtime deposition (30 min) in depletion and in accumulation show a diode-type behavior. The *J-V* characteristics can be described by the Schottky equation (Figure 5.44):

$$j = A^*T^2 \exp\left(-\frac{e}{kT}\Phi_B\right)\exp\left(-\frac{e}{kTn}V\right)\left[\exp\left(\frac{e}{kT}V\right)-1\right] \qquad (5.73)$$

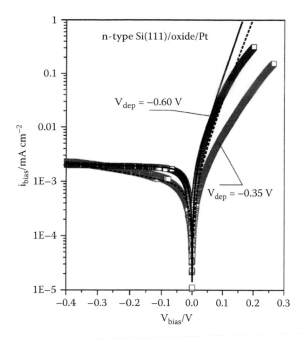

FIGURE 5.44 Current-voltage characteristics measured on solid-state MOS junctions made electrochemically by longtime deposition of Pt.

where Φ_B is the energy barrier of the junction and n is the ideal factor. The experimental curves can be approximated to Eq. (5.73), assuming ideal factors of 1.01 and 1.07 for contacts obtained by deposition in accumulation, $V = -0.6$ V (SCE), and in depletion, $V = -0.35$ V (SCE), respectively. In reverse bias, interface states are completely discharged and the MOS junction behaves as ideal. The larger deviation from ideality observed in forward bias arises as a consequence of the charging of surface states. The reverse current densities are adjusted by assuming junction barriers of 0.755 eV and 0.775 eV for MOS obtained at -0.6 V and -0.35 V, respectively. The Mott–Schottky plots obtained for both MOS junctions shows a linear behavior in reverse bias (Figure 5.45). From the extrapolation to the zero ordinate, one can calculate a flat band potential $V_{fb} = 0.475$ V. Hence, the junction barrier is $\Phi_B = 0.725$ eV. This value differs from those calculated from the J-V characteristic, indicating that the transfer mechanism is not purely controlled by thermionic emission. A donor concentration of 1.90×10^{14} cm^{-3} can be calculated from the slope for the device prepared in accumulation. This value contrasts with 1.35×10^{15} cm^{-3} obtained on n-type Si(111):H in 0.5 KCl (pH 2.3), this being a consequence of a potential drop distribution between semiconductors and Helmholtz layer. The collection efficiency of a semiconductor decorated with metal islands implies the formation of an enhanced electric field underneath the small MOS junctions. In principle, the lateral distribution of the solid–electrolyte capacitance can be disregarded as the origin of such electrical surface modulation. The capacitance of the metal–electrolyte interface in noble metals typically shows values of 20–40 μF cm^{-2} [86,87]. The SiO$_x$–electrolyte interface, on the other hand, shows values of 100 to

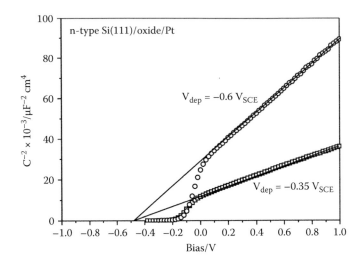

FIGURE 5.45 Mott–Schottky plots of the MOS junctions in (a) obtained at 10 kHz.

200 µF cm^{-2} in concentrated electrolytes (>0.1 M) [88]. Therefore, the potential drop for the whole conditioned photoelectrode is governed by the Si-SiO$_x$ interface. Because the oxide film covers the surface homogeneously, no inherent modulation of the electric field is expected. A lateral modulation indeed appears under illumination. The following explanation is based on an n-type Si surface (based on energy band diagrams shown in Figure 5.46). The explanation can be extended for the case of p-type Si. In the first picture, the photoelectrode semiconductor is in depletion. Let us assume, as is usually the case, that the density of interface states is large enough so that Fermi level pinning arises. Hence, each change of the applied potential in the dark results in a modification of the electric field in the oxide, given by $\Delta\varphi_{ox}/d_{ox}$, due to its dielectric nature. Now we turn the light on, holding the applied potential. The band bending under the areas free of metal draws back due to the charge accumulation until the generation rate equals the recombination in the bulk (k_V) and at interface states (k_{rec}). At constant potential, the position of the Fermi level remains unchanged and the reduction of the band bending can only occur by an increase of the potential drop in the oxide. This occurs by occupation of interface states. Under the metal, photogenerated charges are further transported to the electrolyte and are consumed by the electrochemical reaction. Thus, an accumulation of charge at interface states is limited, and hence the retraction of the band banding. The modulation of the thickness of the space charge layer is governed by the rate of charge consumption at the catalytic centers. The space charge layer protrusions generated by catalytic centers extend the catching area of photogenerated charges which reach the lateral modulated electric field by diffusion. In the case of Si, the passivating oxide constitutes a barrier for electron transfer to the electrochemical reaction centers. But a thicker oxide provides better passivation. The so-called nanoemitter cell concept circumvents this drawback by separating the passivated from the electroactive zones.

The scanning electrochemical microscopy (SECM) technique provides a convenient method to study the electron transfer via the surface states under semiconductor depletion in the dark. This information is extracted from the current-distance

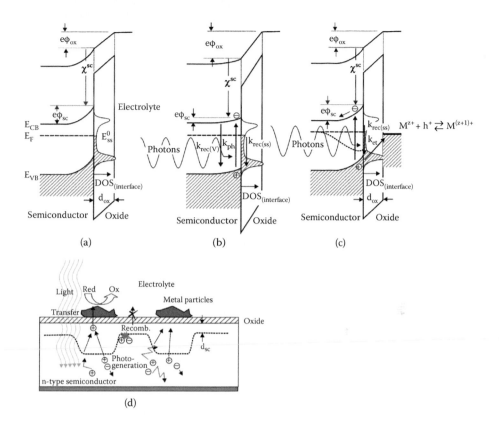

FIGURE 5.46 Energy band diagrams representing the EMOS junction of reduced dimensions formed by electrodeposition in the dark (a) and under illumination (b) and (c). (b) and (c) represent the EOS and EMOS junctions, respectively. The schematic in (d) represents the lateral modulation of the space charge layer arising from different charge transfer rates at the metal-covered and metal-free areas of the photoelectrode.

curves recorded during the approach of the Pt microelectrode to the semiconductor surface in a $Fe(CN)_6^{-3/-4}$ solution. In this electrode arrangement, the oxidation of Fe(II) species at the Pt-probe is set under diffusion control, whereas the resulting Fe(III) species are reduced back to Fe(II) at the substrate. For an irreversible heterogeneous reaction occurring at the substrate, the transfer rate constant can be calculated by fitting the experimental current distance curves to Eqs. (5.31–35). The potential dependence of the rate constant measured on MOS junctions formed after deposition of Pt and Rh shows that the electron transfer is up to ten times faster in the latter than in the former case (see Figure 5.47). In principle, the low transfer rate constant observed in the n-type Si(111)/Pt interface in comparison with that in n-type Si(111)/Rh can be ascribed to a thicker oxide film, which limits the tunnel process. This is not unexpected in view of the larger hole injection driven by the more positive potential of the couple Pt(IV)/Pt(II) than Rh(III)/Rh(0).

The conductive properties of the anodic oxide formed on InP lead to a very different picture of the MOS junctions than that we have seen for the Si-SiO$_2$–metal

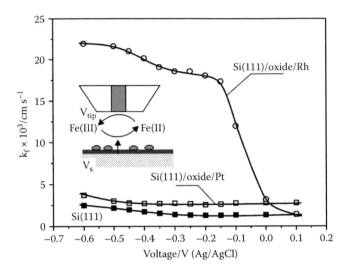

FIGURE 5.47 Dark voltage dependence of the effective heterogeneous rate constant for the electron transfer at n-type Si(111) before and after the coverage with Pt and Rh particles. $V_{tip} = 0.60$ V. Solution: 10 mM Fe(CN)$_6^{-4}$ + 0.1 M K$_2$SO$_4$. Diameter of Pt tip: 25 μm.

system. The transfer of electrons ejected from the conduction band through the oxide takes place by resonance tunneling by conduction band states introduced by local In$_2$O$_3$-like chemical environments. Tubbsing et al. [89] arrived at such an interpretation based on photoelectrochemical characterization of p-InP in different redox-couples. They also observed a positive shift of the flat band potential under illumination of 0.3 V for p-type InP and 0.18 V for n-type InP in 1 M HCl solution. This experimental finding was interpreted in terms of a trapping of light-excited electrons in the surface states, which leads to a shift of the semiconductor band edges toward higher energies (more negative values in the electrochemical scale). The addition of a redox couple with a standard potential more negative than the standard potential for hydrogen evolution reaction was observed to shift the onset of photocurrent on p-type InP toward more positive potentials. The faster transfer of light-excited electrons to the redox system depresses the charging of surface states and the consequent band shift.

The formation of a thin oxide layer on InP was observed to reduce the leakage current and to increase the Schottky barrier by about 0.3 eV in MOS Au/oxide/n-type InP junctions [90,91]. Wada et al. have found that the growth of an oxide film is connected with the formation of interface states and negative charges in the oxide near the oxide–semiconductor interface. Assuming a flat distribution of surface states, the barrier height is expected to change with potential according to

$$\Phi_B = c_3(\Phi_m - \chi) + (1 - c_3)(E_g - \varphi_0) + c_3 \frac{d_{ox}eN_{ox}}{\varepsilon_0\varepsilon_{ox}} + \frac{c_0}{2} \tag{5.74}$$

$$- \left[c_0 \left(\Phi_B^0 + \frac{c_0}{4} - V_{bias} - E_F - \frac{kT}{e} + c_3 \frac{d_{ox}eN_{ox}}{\varepsilon_0\varepsilon_{ox}} \right) \right]^{1/2}$$

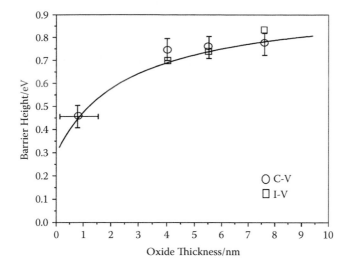

FIGURE 5.48   Barrier height of a solid-state Au/oxide/n-type InP junction as a function of the oxide thickness. The solid line was calculated by Eq. (5.74) by taking $D_{it} = 6.0 \times 10^{12}$ cm^{-2}eV^{-1} and $N_{ox} = 2.8 \times 10^{12}$ cm^{-2}. The other parameters are as indicated in the text. Experimental data taken from Wada et al. [92].

where

$$c_0 = \frac{2c_2^2 e N_D}{\varepsilon_0 \varepsilon_{sc}}; c_2 = d_{0x} \frac{\varepsilon_0 \varepsilon_{sc}}{\varepsilon_0 \varepsilon_{ox} + e^2 d_{ox} D_{is}}; c_3 = \frac{\varepsilon_0 \varepsilon_{ox}}{\varepsilon_0 \varepsilon_{ox} + e^2 d_{ox} D_{is}}$$

N_{ox} and d_{ox} are the surface density of oxide charges and the oxide thickness, respectively. φ_0 is the surface Fermi level of the semiconductor before contact measured above the valence band edge. This was assumed to be 0.82 eV. E_F is the Fermi level measured from the conduction band. Taking the parameters $\chi = 4.40$ eV, $\Phi_m = 4.70$ eV, $E_g = 1.34$ eV, $\varepsilon_{sc} = 12.35$, $\Phi_B^0 = 0.82$, $N_D = 2 \times 10^{15}$ cm^{-3}, $E_F = 0.12$ eV, and $\varepsilon_{ox} = 3$, values of $D_{is} = 6 \times 10^{12}$ eV^{-1}cm^{-2} and $N_{ox} = 2.8 \times 10^{12}$ cm^{-2} were found to reproduce the experimental values as shown in Figure 5.48 [92].

The chemical and structural origin of interface states and oxide negative charges is not totally clear. They are ascribed to P vacancies at the InP–oxide interfaces. The charging of surface states by trapping of photoinduced electrons at p-type InP/oxide interface decreases the maximal theoretical achievable photopotential in EMOS junctions for light-induced hydrogen evolution produced electrochemically by cycling in HCl and subsequent photoelectrodeposition of Rh.

REFERENCES

1. Hüfner, S. 2003. *Photoelectron Spectroscopy: Principles and Applications*, 3rd ed. Berlin: Springer-Verlag.
2. Koopmans, T. 1933. Über die Zuordnung von Wellenfunktionen und Eigenwerten zu den einzelnen Elektronen eines Atoms. *Physica* 1:103.

3. The symbol $|\psi\rangle$ is known as *ket* and stems from the so-called *braket* notation introduced by P. Dirac. It refers to the state described by the wave function ψ without introducing any specific mathematical representation. The projection of the wave function upon a determined basis, as for instance the position vector $|\vec{r}\rangle$, is represented in quantum mechanics as $\langle\vec{r}|\psi\rangle = \psi(\vec{r})$.

4. Hüfner, S. and G.K. Wertheim. 1975. Core-line asymmetries in the x-ray photoemission spectra of metals. *Phys. Rev. B* 11:678.

5. Hüfner, S., Wertheim, G.K., Buchanan, D.N.E. and K.W. West. 1974. Core line asymmetries in XPS-spectra of metals. *Phys. Lett.* 46A:420.

6. Doniach, S. and M. Sunjic. 1979. Many-electron singularity in X-ray photoemission and X-ray line spectra from metals. *J. Phys. C* 3:285.

7. Seah, M.P. 1986. Data compilations: Their use to improve measurement certainty in surface analysis by aes and xps. *Surf. Interf. Anal.* 9:85.

8. Tamura, S., Powell, C.J. and D.R. Penn. 1991. Calculations of electron inelastic mean free paths II. Data for 27 elements over the 50–2000 eV range. *Surf. Interface Anal.* 17:911.

9. Tamura, S., Powell, C.J. and D.R. Penn. 1991. Calculations of electron inelastic mean free paths III. Data for 15 anorganic compounds over the 50–2000 eV range. *Surf. Interface Anal.* 17:927.

10. Copper, J.W. 1962. Photoionization from outer atomic subshells. A model study. *Phys. Rev.* 128:681.

11. Yeh, J.J. and I. Lindau. 1985. Atomic subshell photoionization cross sections and asymmetry parameters: 1<Z<103. *Atomic Data and Nuclear Data Tables* 32:1.

12. Lewerenz, H.J. 1997. Surface scientific aspects in semiconductor electrochemistry. *Chem. Soc. Rev.* 26:239.

13. Bard, A.J. and L.R. Faulkner. 1980. *Electrochemical Methods: Fundamentals and Applications*. New York: John Wiley & Sons.

14. Scholz, F. (Ed.). 2002. *Electroanalytical Methods: Guide to Experiments and Applications*. Heidelberg: Springer.

15. Mirkin, M. V. and B. R. Horrocks. 2000. Electroanalytical measurements using the scanning electrochemical microscope. *Anal. Chim. Acta* 406:119.

16. Horrocks, B.R., Mirkin, M.V. and A.J. Bard. 1994. Scanning electrochemical microscopy. Application to investigation of the kinetics of heterogeneous electron transfer at semiconductor (WSe_2 and Si) electrodes. *J. Phys. Chem.* 98:9106.

17. Chabal, J.Y. 2001. *Fundamental Aspects of Silicon Oxidation*. Berlin: Springer.

18. Flietner, H. 1988. Spectrum and nature of defects at interfaces of semiconductors with predeominant homopolar bonding. *Surf. Sci.* 200:463.

19. Flietner, H. 1974. Spectrum and nature of surface states. *Surf. Sci.* 46:251.

20. Angermann, H., Kliefoth, K., Füssel, W. and H. Flietner. 1995. Defect generation at silicon surfaces during etching and initial stage of oxidation. *Microelect. Eng.* 28:51.

21. Lam, Y.W. 1971. Surface-state density and surface potential in MIS capacitors by surface photovoltage measurements. I. *J. Phys. D: Appl. Phys.* 4:1370.

22. Stesmans, A., Braet, J. and J. Witters. 1984. X and K band ESR study of the Pb interface centres in thermally oxidized p-type (001)Si wafers at low temperatures and influence of medium-dose As^+ ion implantation. *Surf. Sci.* 141:255.

23. Poindexter, E.H., Caplan, P.J., Deal, B.E. and R.R. Razouk. 1981. Interface states and electron spin resonance centers in thermally oxidized (111) and (100) silicon wafers. *J. Appl. Phys.* 52:879.

24. Poindexter, E.H., Gerardi, G.J., Rueckel, M.-E., Caplan, P.J., Johnson, N.M. and D.K. Biegelsen. 1984. Electronic traps and Pb centers at the Si/SiO2 interface: Band-gap energy distribution. *J. Appl. Phys.* 56:2844.

25. Himpsel, F.J., McFeely, F.R., Taleb-Ibrahimi, A. and J.A. Yarmoff. 1988. Microscopic structure of the SiO2/Si interface. *Phys. Rev. B* 38:6084.
26. Kern, W. 1990. The evolution of silicon wafer cleaning technology. *J. Electrochem. Soc.* 137:1887.
27. Higashi, G.S., Chabal, Y.J., Trucks, G.W. and K. Raghavachari. 1990. Ideal hydrogen termination of the Si(111) surface. *Appl. Phys. Lett.* 56:656.
28. Bitzer, T., Gruyters, M., Lewerenz, H.J. and K. Jacobi. 1993. Electrochemically prepared Si(111) 1×1-H surface. *Appl. Phys. Lett.* 63:397.
29. Rauscher, S., Dittrich, Th., Aggour, M., Rappich, J., Flietner, H. and H.J. Lewerenz. 1995. Reduced interface state density after photocurrent oscillations and electrochemical hydrogenation of n-Si(111): a surface photovoltage investigation. *Appl. Phys. Lett.* 66:3018.
30. Boonekamp, E.P., Kelly, J.J., van de Ven, J. and A.H.M. Sondag. 1994. The chemical oxidation of hydrogen-terminated silicon (111) surfaces in water studied in situ with Fourier transform infrared spectroscopy. *J. Appl. Phys.* 75:8121.
31. Angermann, H. 2002. Characterization of wet-chemically treated silicon interfaces by surface photovoltage measurements. *Anal. Bioanal. Chem.* 374:676.
32. Hattori, T and H. Nohira. 2001. Oxidation of H-terminated silicon in Chabal, J.Y. (Ed.), *Fundamental Aspects of Silicon Oxidation.* Berlin: Springer.
33. Aberle, A.G. 2000. Surface passivation of crystalline silicon solar cells: a review. *Prog. Photovolt: Res. Appl.* 8:473.
34. Hezel, R. and K. Jäger. 1989. Low-temperature surface passivation of silicon for solar cells. *J. Electrochem. Soc.* 136:518.
35. Aggour, M., Skorupska, K., Stempel Pereira, T., Jungblut, H., Grzanna, J. and H.J. Lewerenz. 2007. Photoactive silicon-based nanostructure by self-organized electrochemical processing. *J. Electrochem. Soc.* 154:H794.
36. Rappich, J., Zhang, X., Chapel, S., Sun, G. and K. Hinrichs. 2009. Passivation of Si surfaces by hydrogen and organic molecules investigated by in-situ photoluminescence techniques. *Solid State Phenomena* 156–158:363
37. Royea, W.J., Juang, A. and N.S. Lewis. 2000. Preparation of air-stable, low recombination velocity Si(111) surfaces through alkyl termination. *Appl. Phys. Lett.* 77:1988.
38. Poindexter, E.H., Caplan, P.J., Deal, B.E. and R.R. Razouk. 1981. Interface states and electron spin resonance centers in thermally oxidized (111) and (100) silicon wafers. *J. Appl. Phys.* 52:879.
39. Stesmans, A., Nouwen, B. and V.V. Afanas'ev. 1998. Pb_1 interface defect in thermal (100)Si/SiO2:^{29}Si hyperfine interaction. *Phys. Rev. B* 58:15801.
40. Hoffmann, P.M., Oskam, G. and P.C. Searson. 1998. Analysis of the impedance response due to surface states at the semiconductor/solution interface. *J. Appl. Phys.* 83:4309.
41. Oskam, G., Hoffmann, P.M. and P.C. Searson. 1996. In situ measurements of interface states at silicon surfaces in fluoride solutions. *Phys. Rev. Lett.* 76:1521.
42. Oskam, G., Hoffmann, P.M., Schmidt, J.C. and P.C. Searson. 1996. Energetics and kinetics of surface states at n-type silicon surfaces in aqueous fluoride solutions. *J. Phys. Chem.* 100:1801.
43. Zeiser, A., Bücking, N., Götte, J., Förstner, J., Hahn, P., Schmidt, W.G. and A. Knorr. 2004. Dynamics of the phonon-induced electron transfer between semiconductor bulk and surface state. *Phys. Stat. Sol. (b)* 241:R60.
44. Lublow, M., Stempel, T., Skorupska, K., Muñoz, A.G., Kanis, M. and H.J. Lewerenz Lublow. 2008. Morphological and chemical optimization of ex situ NH4F 40% conditioned Si(111) 1×1:H. *Appl. Phys. Lett.* 93:062112.

45. Lewerenz, H. J., Aggour, M., Murrell, C., Kanis, M., Jungblut, H., Jakubowicz, J., Cox, P.A., Campbell, S.A., Hoffmann, P. and D. Schmeisser. 2003. Initial stages of structure formation on silicon electrodes investigated by photoelectron spectroscopy using synchrotron radiation and in situ atomic force microscopy. *J. Electrochem. Soc.* 150:E185.
46. Muñoz, A.G., Moehring, A. and M.M. Lohrengel. 2002. Anodic oxidation of chemically hydrogenated Si(100). *Electrochim. Acta* 47:2751.
47. Yazyev, O.V. and A. Pasquarello. 2006. Origin of fine structure in Si photoelectron spectra at silicon surfaces and interfaces. *Phys. Rev. Lett.* 96:157601.
48. Wagner, C.D., Naumkin, A.V., Kraut-Vass, A., Allison, J.W., Powell, C.J. and J.R. Rumble Jr. In *NIST X-Ray Photoelectron Spectroscopy Database.*
49. Verdaguer, A., Weis, C., Oncins, G., Ketteler, G., Bluhm, H. and M. Salmeron. 2008. Growth and structure of water on SiO2 films on Si investigated by Kelvin probe microscopy and in situ X-ray spectroscopies. Lawrence Berkeley National Laboratory, 2008.
50. Sze, S.M. 1981. *Physics of Semiconductors Devices*, 2nd ed. New York: John Wiley & Sons.
51. Schroder, D.K. 2007. Negative bias temperature instability: What do we understand? *Microelectron. Reliab.* 47:841.
52. Afanas'ev V.V. and A. Stesman. 1998. *Appl. Phys. Lett.* 72:79; *Phys. Rev. Lett.* 23:5176.
53. Ushio, J. and T. Maruizumi. 2007. Transfer of positive fixed charge between Si and SiO2 in Si/SiO2 interface with hydrogen migration. *ECS Trans.* 6:27.
54. Rashkeev, S.N., Fleetwood, D.M., Schrimpf, R.D. and S.T. Pantelides. 2001. Defect generation by hydrogen at the Si-SiO2 interface. *Phys. Rev. Lett.* 87:165506.
55. de Nijs, J.M.M., Druijf, K.G., Afanas'ev, V.V., van der Drift, E. and P. Balk. 1994. Hydrogen induced donor-type Si/SiO2 interface states. *Appl. Phys. Lett.* 65:2428.
56. Vanheusden, K. and A. Stesmans. 1994. Positive charging of buried SiO_2 by hydrogenation. *Appl. Phys. Lett.* 64:2575.
57. Warren, W.L., Vanheusden, K.,. Schwank, J.R, Fleetwood, D.M., Winokur, P.S. and R.A.B. Devine. 1996. Mechanism for anneal-induced interfacial charging in SiO_2 thin films on Si. *Appl. Phys. Lett.* 68:2993.
58. Godet, J., Giustino, F. and A. Pasquarello. 2007. Proton-induced fixed positive charge at the $Si(100)$-SiO_2 interface. *Phys. Rev. Lett.* 99:126102.
59. Skorupska, K., Lublow, M., Kanis, M., Jungblut, H. and H.J. Lewerenz. 2005. Electrochemical preparation of a stable accumulation layer on Si: A synchrotron radiation photoelectron spectroscopy study. *Appl. Phys. Lett.* 87:262101.
60. Garcia, S.P., Bao, H. and M.A. Hines. 2004. Etchant anisotropy controls the step bunching instability in KOH etching of silicon. *Phys. Rev. Lett.* 93:166102.
61. Llopis, J.F. and F. Colom. 1976. Platinum, in *Encyclopedia of Electrochemistry of the Elements*, Vol. 6, Ch. VI-4, A.J. Bard, (Ed.). New York: Marcel Dekker.
62. Rao, C.R.K. and D.C. Trivedi. 2005. Chemical and electrochemical depositions of platinum group metals and their applications. *Coord. Chem. Rev.* 249:613.
63. Muñoz A.G. and H.J. Lewerenz. 2008. Platinum deposition onto Si single crystal surfaces I. Chemical, electronic and morphological surface processes under depletion conditions. *J. Electrochem. Soc.* 155:D527; 2009. Platinum deposition onto silicon single crystal surfaces II: Mechanistic aspects of the growth of Pt islands. *J. Electrochem. Soc.* 156:D242.
64. Hecq, M., Hecq, A., Delrue, J.P. and T. Robert. 1979. Sputtering deposition, XPS and X-ray diffraction characterization of oxygen-platinum compounds. *J. Less-Common Met.* 64:P25.
65. NIST, XPS Database. 2004. Web. Edn., National Institute of Standards and Technology, Gaitherburg, MD.

66. Blackstock, J.J., Stewart, D.R. and Z. Li. 2005. Plasma-produced ultra-thin platinum-oxide films for nanoelectronics: physical characterization. *Appl. Phys. A* 80:1343.
67. Dri, C., Africh, C., Esch, F., Comelli, G., Dubay, O., Köhler, L., Mittendorfer, F., G. Kresse, Dudin, P. and M. Kiskinova. 2006. Initial oxidation of the Rh(110) surface: Ordered adsorption and surface oxide structures. *J. Chem. Phys.* 125:094701.
68. Benguerel, E., Demopoulos, G.P. and G.B. Harris. 1996. Speciation and separation of rhodium(III) from chloride solutions: A critical review. *Hydrometallurgy* 40:135.
69. Spieker, W.A., Liua, J., Miller, J.T., Kropf, A.J. and J.R. Regalbuto. 2002. An EXAFS study of the co-ordination chemistry of hydrogen hexachloroplatinate(IV) 1. Speciation in aqueous solution. *Appl. Cat. A: General* 232:219.
70. Muñoz, A.G. and H.J. Lewerenz. 2009. Electroplating of iridium onto single crystal silicon: Chemical and electronic properties of n-Si(111)/Ir nano-junctions. *J. Electrochem. Soc.* 156:D184.
71. Lewerenz, H.J., Jakubowicz, J. and H. Jungblut. 2004. Metastable stage in porous silicon formation: the role of H-terminated low index faces. *Electrochem. Commun.* 6:1243.
72. Hines, M.A. 2001. *Fundamental Aspects of Silicon Oxidation*, Y.J. Chabal (Ed.), p. 12. Berlin: Springer Verlag.
73. García, S., Bao, H. and M.A. Hines. 2004. Effects of diffusional processes on crystal etching: Kinematic theory extended to two dimensions. *J. Phys. Chem. B* 108:6062.
74. Stempel, T., Muñoz, A.G., Skorupska, K., Lublow, M., Kanis, M. and H.J. Lewerenz. 2009. Surface chemistry and nanotopography of step-bunched silicon surfaces: In-system SRPES and SPM investigations. *ECS Trans.* 19:403.
75. Fujitani, M., Hinogami, R., Jia, J.G., Ishida, M., Morisawa, K., Yae, S. and Y. Nakato. 1997. Modulation of Flachbandpotential and increase in photovoltage for n-Si electrodes by formation of halogen atom terminated surface bonds. *Chem. Lett.* 26:1041.
76. Zhou, X., Ishida, M., Imanishi, A. and Y. Nakato. 2000. Reactions of Si-H to Si-X (X = Halogen) bonds at H-terminated Si(111) surfaces in hydrogen halide solutions in the presence of oxidants. *Electrochim. Acta* 45:4655.
77. Nakato, Y. and H. Tsubomura. 1992. Silicon photoelectrodes modified with ultrafine metal islands. *Electrochim. Acta* 37:897.
78. Nakato, Y., Ueda, K., Yano, H. and H. Tsubomura. 1988. Effect of microscopic discontinuity of metal overlayers on the photovoltages in metal-coated semiconductor-liquid junction photoelectrochemical cells for efficient solar energy conversion. *J. Phys. Chem.* 92:2316.
79. Jia, J., Fujitani, M., Yae, S. and Y. Nakato. 1997. Hole diffusion length and temperature dependence of photovoltages for n-Si electrodes modified with LB layers of ultrafine platinum particles. *Electrochim. Acta* 42:431.
80. Yae, S., Kitagaki, M. and T. Hagihara, Miyoshi, Y., Matsuda, H., Parkinson, B.A. and Y. Nakato. 2001. Electrochemical deposition of fine Pt particles on n-Si electrodes for efficient photoelectrochemical solar cells. *Electrochim. Acta* 47:345.
81. Allongue, P. 1992. Physics and Applications of Semiconductor Electrodes Covered with Metal Clusters. *Modern Aspects of Electrochemistry*, Vol. 23, B.E. Conway, J.O'M. Bockris, R.E. White (Eds.). New York: Plenum Press.
82. Rizk, R., de Mierry, P., Ballutaud, D. and M. Aucouturier. 1991. Hydrogen diffusion and passivation processes in p- and n-type crystalline silicon. *Phys. Rev. B.* 44:6141.
83. Allongue, P., Henry de Villeneuve, C., Pinsard, L. and M.C. Bernard. 1995. Evidence for hydrogen incorporation during porous silicon formation. *Appl. Phys. Lett.* 67:941.
84. Pearton, S.J., Corbett, J.W. and T.S. Shi. 1987. Hydrogen in crystalline semiconductors. *Appl. Phys. A* 43:153.
85. Muñoz, A.G. and M.M. Lohrengel. 2002. Kinetics of oxide growth and oxygen evolution on p-Si in neutral aqueous electrolytes. *J. Solid State Electrochem.* 6:513.

86. Pajkossy, T. and D.M. Kolb. 2001. Double layer capacitance of Pt(111) single crystal electrodes. *Electrochim. Acta* 46:3063.
87. Pajkossy, T. and D.M. Kolb. 2009. The interfacial capacitance of Rh(111) in HCl solutions. *Electrochim. Acta* 54:3594.
88. Bouse, L. and P. Bergveld. 1983. On the impedance of the silicon dioxide/electrolyte interface. *J. Electroanal. Chem.* 152:25.
89. Tubbesing, K., Meissner, D., Memming, R. and B. Kastening. 1986. On the kinetics of electron transfer reactions at illuminated InP electrodes. *J. Electroanal. Chem.* 214:685.
90. Cardwell, M.J. and R.F. Peart. 1973. Measurement of carrier-concentration profiles in expitaxial indium phosphide. *Electron. Lett.* 9:88.
91. Wada, O. and A. Majerfeld. 1978. Low leakage nearly ideal Schottky barriers to n-InP. *Electron. Lett.* 14:125.
92. Wada, O., Majerfeld, A. and P.N. Robson. 1982. InP Schottky contacts with increased barrier height. *Solid-State Electronics* 25:381.

6 Optical Effects: Current Approaches and Future Directions

The atomic structure and the related electronic properties of the MOS junction made by electrochemical deposition define the adsorption spectrum of the photoelectrode and, hence, the practical efficiency of sunlight-induced conversion processes. As a model system, let us analyze the photoelectrochemical behavior of a p-type Si(111) surface onto which Rh particles were photoelectrochemically deposited. As already discussed in the foregoing chapters, p-type Si/oxide/Rh junctions of reduced dimensions are obtained. Figure 6.1a shows the quantum yield recorded at constant incident frequencies in 1 M HClO$_4$ solution. For the investigated frequencies the onset of photocurrent can be observed at potentials more negative than −0.15 V (SCE). This clearly indicates a loss of photopotential of 0.21 eV from a maximum theoretically achievable photopotential of about 0.3 eV. This effect is expected by virtue of a shift of the band edge toward more negative potentials, as observed by capacitance measurements performed on this system (see Figure 5.27). Quantum yield values over 40% are attained, however, with a maximum of 2.3 eV. The dependency of quantum efficiency with the energy of incident photons is shown for different potentials in Figure 6.1b. The quantum yield is calculated as

$$\eta = \frac{j_{ph} \times h\nu}{W(h\nu) \times e} \tag{6.1}$$

The onset of light absorption is observed at photons of energy larger than the band gap. The efficiency increase at larger energies reflects the reduction of the adsorption length, and hence the length of the diffusion path of light-induced charge carriers in the field-free region toward the space charge layer. Figure 6.2 shows the dependency of the absorption coefficient for Si with the incident wavelength. The adsorption distance is reduced from some tens of millimeters at 1.1 eV up to some hundreds of nanometers at 400 nm. It must be noted that below 1.2 eV the adsorption distance becomes larger than the wafer thickness, usually 300–400 µm, and the charge collection reduces practically to zero.

Photoelectrons generated very near to the Si-SiO$_x$ interface, on the other hand, have a low mobility and hence a large probability to recombine. As a consequence, the efficiency decreases again to zero. The normalized quantum efficiency curves taken at −0.26 V and −1.0 V do not show significant differences (Figure 6.1c). This indicates that there the semiconductor presents a Fermi level pinning at the potential region

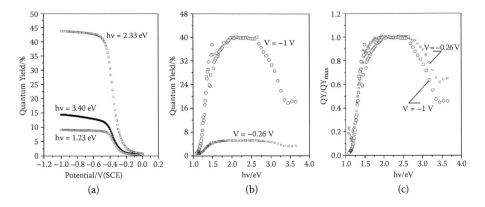

FIGURE 6.1 Quantum yield of light-induced hydrogen evolution on electrochemically prepared p-type Si/oxide/Rh photoelectrodes in 1 M HClO$_4$: (a) as a function of potential for different photon energies; (b) as a function of incident photon energy at two different applied potentials; (c) normalized quantum yield as a function of applied potential.

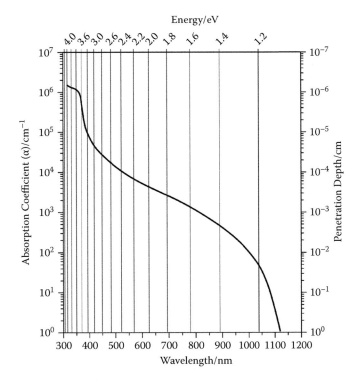

FIGURE 6.2 Absorption coefficient and penetration depth of light into Si as a function of the wavelength.

for photoinduced hydrogen evolution; that is, the relative relation quantum efficiency-to-absorption length does not change. The maximal achieved efficiency, on the other hand, is controlled by the rate of proton reduction at the metal–electrolyte interface.

As mentioned at the end of Chapter 5, the separation of active and passive zones at the photoelectrodes by the construction of metal fingers imbedded in a passivated matrix leads to a modulation of the space charge layer under illumination. The deposition of noble metals such as Rh or Pt in a structured surface leads to a photopotential increase. Figure 6.3 shows the respective current-voltage curves and compares them with the corresponding current-voltage curve for Rh in the same acid solution and that for an ideal EMOS contact. In an idealized performance, the current increases rapidly to its saturation value near the open circuit potential. The real efficiency, however, presently represents only a fifteenth of the theoretical maximal output (compare the shadowed rectangle for both cases), pointing to losses by interfacial recombination and ohmic losses. The incorporation of an interfacial dipole layer by insertion of a dielectric film, for instance by atomic layer deposition or by surface chemical modification, could help to increase the expected efficiency. In this case, the current–voltage characteristic should shift to the right by a voltage difference close to the affinity change $\Delta\chi/e$.

The conversion efficiency is also affected by the size and distribution of pores. The maximal distance of separation among the metal charge collectors, d, is defined by the relation

$$d \leq \sqrt{2} \times L_{diff,h,e} + W + r_{pore} \qquad (6.2)$$

where $L_{diff,h,e}$ is the diffusion length of holes or electrons, W is the thickness of the space charge layer, and r_{pore} is the pore radius. In the particular case of crystalline

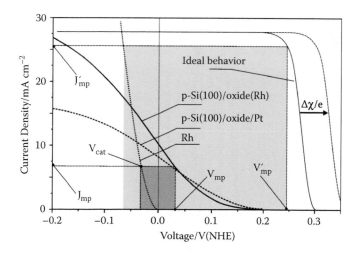

FIGURE 6.3 Photocurrent-voltage characteristics of p-type Si/porous silicon-oxide/metal photoelectrodes in 1M HClO$_4$. Light intensity: 100 mWcm^{-2}. The theoretical maximal achievable photocurrent-voltage characteristic and the hypothetical dipole shift of the semiconductor bands are also included.

Si, with a diffusion length of several hundred micrometers, the metal coverage of a photoelectrode structure fulfilling condition (6.2) may be insufficient to maintain the large photoinduced flux of the minority charge carrier; thus, a kinetic limitation arises.

In another approach, an efficient harvesting of photoinduced charge carriers can be achieved by a rodlike structure, as shown schematically in Figure 6.4. One of the main features of this type of structure is the orthogonalization of light absorption and carrier separation, which allow an increase of the red sensitivity and an increase of the metal-covered surface for the catalytic reaction. Rodlike structures can be prepared, for instance, by lithographic techniques combined with reactive ion etching [1], or by using a copper-catalyzed, vapor-liquid-solid-growth process, with $SiCl_4$ and BCl_3 as precursors [2,3]. In principle, the potential for high efficiency stems from the short radial distances, which photogenerated charge carriers have to traverse before they reach the catalytic centers. Based on this concept, some energy photovoltaic and photoelectrochemical conversion devices including highly ordered, vertically oriented Si wire arrays, have been developed [4–6]. Figure 6.5 shows SEM pictures of rodlike structures fabricated using ion milling and vapor-liquid-solid-growth techniques. The improved light absorption in this type of array, however, is overshadowed by the large concentration of surface states due to the reaction of the naked silicon surface with the electrolyte and the large damages

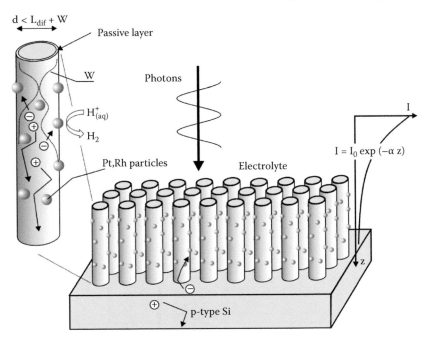

FIGURE 6.4 (See color insert.) Schematic of a rodlike photoelectrocatalytic structure where photon absorption and excess carrier separation occur in an orthogonalized configuration; the catalytically active nanoparticles are supposed to form rectifying junctions with the p-type semiconductor upon contact with the respective redox electrolyte with a semispherical space charge region modulated by the curvature of the rod circumference; $I(z)$: absorption profile, α: absorption coefficient, d: diameter of the rods, W: length of space charge layer.

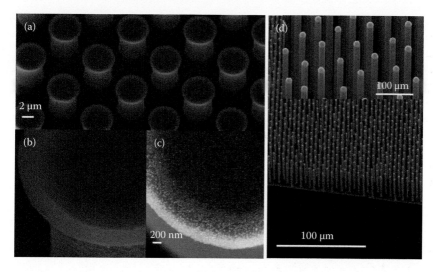

FIGURE 6.5   SEM pictures of rodlike structures: (a,b,c) obtained by ion milling technique; (a and b) before Rh deposition; (c) after Rh deposition; (d) Si-rods made by vapor-liquid-solid-growth techniques (after Kayes et al. [4], with permission of AIP).

introduced particularly in the ion-milling method. This leads to Fermi level pinning which limits the achievable photopotential as indicated, for instance, in the photocurrent-voltage obtained after Rh deposition on a rodlike structured surface (see Figure 6.6). Hence, the search of an adequate passivation method constitutes a new challenge for future research in this area.

A third approach to improve the conversion efficiency of devices based on metal–semiconductor junctions exploits the enhanced light absorption generated by high-ordered arrays of metal particles of reduced dimensions. The enhanced light absorption arises as a result of two related phenomena: light scattering and generation of surface plasmons. The latter consists of coherent oscillations of conducting electrons in metals induced by the electric field of the incident light [7]. This optical property forms the basis of some spectroscopy techniques such as surface enhanced Raman spectroscopy (SERS) and second harmonic generation (SHG).

If particles are arranged with a particular regularity, coupled plasmon–polariton modes may be set up by near-field dipole interactions that lead to coherent propagation of energy along the array. The field enhancement at the particle surface is confined to the near-field region of the particle over distances much smaller than the wavelength of light. If a second metal nanoparticle is placed in this near-field region, the interaction strength between the particles is greatly enhanced, as shown by the simulation reported by Maier et al. (Figure 6.7) [8].

The interaction of light with nanodimensioned particles can be quantitatively described by the solution of Maxwell's equations in spherical coordinates. This task was undertaken by Mie in 1908 [9]. Let I_0 be the intensity of the incident light onto a small particle of radius a. The intensity of the scattered light I at a distance r from the particle is given by

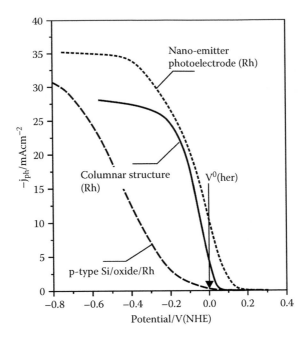

FIGURE 6.6 Photocurrent-voltage curve of rodlike p-type Si after deposition of 9.26 mC cm^{-2} of Rh in 1 M HClO$_4$. Comparison is made with a corresponding nanoemitter-type and plane p-type Si photoelectrode decorated with Rh. I-W lamp with an intensity of 1200 mW cm^{-2}.

$$I = I_0 \frac{f(\theta,\phi)}{k^2 r^2} \tag{6.3}$$

where k is the wave number defined as $k = 2\pi/\lambda$, and f is a dimensionless function of the direction. According to Eq. (6.3), the energy of the dispersed light in all directions can be equal to the energy of the incident light falling on the area σ_{sca}, given by (see scheme in Figure 6.8)

$$\sigma_{sca} = \frac{1}{k^2} \int f(\theta,\phi)\sin\theta\, d\theta\, d\phi \tag{6.4}$$

Part of the incident light is absorbed by the particle, and hence the corresponding energy can be represented by the energy incident on the area σ_{abs}. Thus, according to the law of conservation of energy, we have

$$\sigma_{ext} = \sigma_{sca} + \sigma_{abs} \tag{6.5}$$

The quantities $\sigma_{ext,sca,abs}$ are called the cross sections of the particle for extinction, scattering, and absorption, respectively. The light dispersion is customarily expressed in terms of efficiency factors defined as $Q_{ext,sca,abs} = \sigma_{ext,sca,abs}/\pi a^2$. For particles smaller

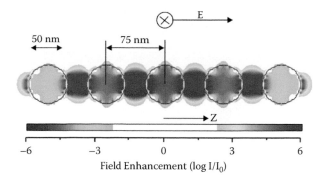

Field Enhancement (log I/I_0)

FIGURE 6.7 **(See color insert.)** Distribution of the component of the electric field parallel to the z-direction for an array of 5 Au nanoparticles in air excited at the collective resonance frequency by a propagating plane wave with the electric field polarized in the z-direction (adapted from Maier et al. [8]).

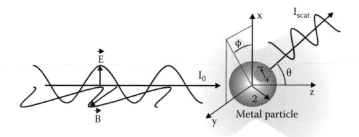

FIGURE 6.8 Schematic depicting the spherical light scattering by small particles.

than the wavelength of incident light embedded in a nonabsorbing dielectric medium, the Mie theory predicts

$$Q_{scat} = \frac{2}{3\pi}\lambda^2\alpha^6 \left| \frac{m^2 - \varepsilon_m}{m^2 + 2\varepsilon_m} \right|^2 \qquad (6.6)$$

$$Q_{abs} = -\frac{\lambda^2}{\pi}\alpha^3 \, \text{Im}\left(\frac{m^2 - \varepsilon_m}{m^2 + 2\varepsilon_m} \right) \qquad (6.7)$$

where m is the complex diffraction index of the particle material, $m = n - ik$, and ε_m is the dielectric permittivity of the surrounding medium. α is a dimensionless parameter given by $\alpha = 2\pi a/\lambda$. For nonmagnetic materials, $m = n = \varepsilon^{1/2}$ and $Q_{abs} = 0$. Hence, an extinction peak appears at $\varepsilon(\omega) \approx -2\,\varepsilon_m$ [10]. The frequency at which this occurs is called the Fröhlich frequency. The spectral characteristics of the extinction cross section, such as position, damping, and strength of the dipole, as well as of the higher-order plasmon resonances of single metal nanoparticles depend on the particle, the geometry, and the dielectric function of the surrounding host [11,12]. For

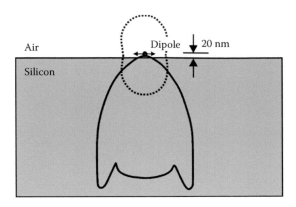

FIGURE 6.9 Radiation pattern for a point dipole oriented parallel to the surface at 20 nm from the air-silicon interface. The radiation pattern for a dipole radiating in the free space is also shown as a reference.

the sake of analysis, the large variety of synthesized shapes of metal particles is often approximated to spheres or spheroids.

The Fröhlich frequency of some metals, such as Ag and Au, appears within the visible region. Therefore, they can be used for light harvesting in the design of photovoltaic and photoelectrochemical devices. There are some ways to do this. First, metal nanoparticles can trap propagating plane waves from sunlight and couple them with the absorbing semiconductor film by folding the light into it. The scattering behavior of a radiating Hertzian dipole with a dipole moment placed at 20 nm above a Si surface is depicted in Figure 6.9. For the sake of comparison, the radiation pattern for a dipole in free space is also shown. It should be noted that the majority of the radiation is entering the Si substrate and only a tiny fraction of radiation is back scattered into air [13].

The deposition of Au particles onto semiconductor substrates is considered a reliable alternative to increase the light adsorption in photovoltaic and photoelectrochemical systems. Figure 6.10 shows the absorbance of colloidal Au nanoparticles in solution for different sizes. It should be noted that they typically adsorb around 530 nm (2.3 eV) and are also known to be influenced by the shape of the particles. The HRSEM image of a p-type Si sample with deposited Au nanoparticles shown as an inset of Figure 6.10 indicates spherical shapes with a diameter of ~40 nm and a population of 3.14×10^{10} cm^{-2}.

Colloidal solution of Au can be obtained by reduction of refluxing $HAuCl_4$ with sodium citrate or by sol-gel methods, as reported in Ref. [14], where Au nanoparticles are stabilized by a surrounding layer of adsorbed phosphodecanomolybdenic acid. The subsequent adsorption on the Si surface is performed by potential scanning between −0.1 V and −0.25 V in the dark, after adding an aliquot of the colloidal solution to the water-containing cell. A constant current of 45 μAcm^{-2} was observed during the potential scan, attributed to the reduction of Mo(VI) to Mo(V) in the structure of phosphomolybdic acid adsorbed on the surface of nanoparticles [15]. At the deposition potential, the p-type silicon surface is covered by an ultra-thin oxide layer. Therefore, there is a large probability that the attachment of Au nanoparticles occurs by hydrogen bonding of the adsorbed reduced phosphomolybdic acid with

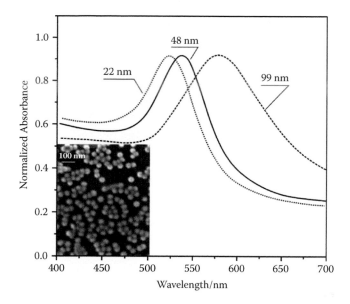

FIGURE 6.10   Typical optical absorption spectra for spherical Au nanoparticles. Inset: SEM picture of Au nanoparticles deposited onto p-type Si.

the oxidized silicon surface. Also, some agglomeration of these particles into larger units can be discerned.

The deposition of Au nanoparticles without a stabilizing molybdenate shell onto p-type Si electrodes shows a photocurrent in acidic solutions comparable to that observed for the nanoemitter photoelectrode type decorated with Pt or Rh. Presently, it is not possible however to attribute the improvements unambiguously to the influence of localized surface plasmons. It is known that Au nanoparticles are catalytically considerably more active than layer-type films or single crystals [16–18]. A direct influence of surface plasmon excitation on catalytic reactions has not yet been confirmed. For light-induced hydrogen evolution, a direct interaction of the surface-plasmon polariton with the vibrational excitations of metal–hydrogen bonds could result in a direct influence. The typical localized surface plasmon resonances of noble metal nanoparticles are energetically located in the high frequency visible part of the spectrum, having frequencies in the 10^{15} Hz range (500 nm ~6×10^{14} Hz). The Si–H bending mode around 2100 cm^{-1} corresponds to a wavelength of 4.7 μm and a frequency of 6.3×10^{13} Hz. Thus, the plasmon resonance is about one order of magnitude faster than the vibrational excitations. One might envisage a possible direct interaction when comparing the frequencies and time scales in Heitler–London van der Waals interactions that result in substantial influence on interatomic bonding [19]. For an atom, the orbiting frequency on the Bohr radius is 3×10^{15} Hz, which defines the frequency range of an atomic stroboscopic dipole that is the basis of high frequency dipole-induced dipole interactions. If it becomes possible to tune surface plasmon polariton resonances to vibronic excitations, a direct influence on a photocatalytic reaction by increased desorption, for instance, might also become possible.

In another approach, plasmonic particles can be imbedded several tens of nanometers in the semiconductor to act as sub-wavelength antennas by coupling of the near field to the semiconductor. A third approach consists of the insertion of a corrugated metallic film on the back surface of a thin semiconducting absorber that couples sunlight by generation of surface plasmon–polariton modes with travel laterally at the metal-semiconductor interface [20]. Surface plasmon–polaritons are electromagnetic waves travelling along the contact of an absorber layer and a metal back contact. Near the plasmon resonance frequency, an evanescent electromagnetic surface plasmon–polariton field is confined at the interface by dimensions much shorter than the wavelength [21,22]. Incident light can thus be trapped and guided laterally on the photoelectrode. Figure 6.11a depicts the dispersion of surface plasmon–polaritons generated at Ag/Si interfaces.

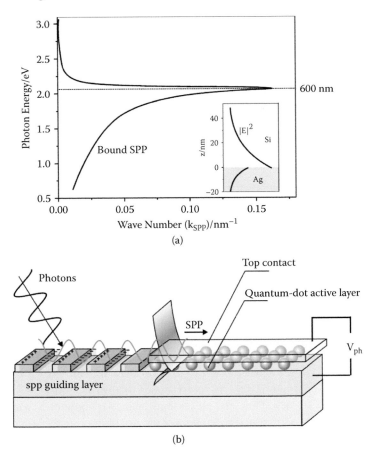

(a)

(b)

FIGURE 6.11   **(See color insert.)** (a) Dispersion diagram showing the relationship between frequency of incident light and wave vector ($2\pi/\lambda$) of surface plasmon–polaritons (SPP) on a Ag/Si interface. The bound SPP mode appears at energies below the surface plasmon resonance energy of 2.07 V for Ag. The inset schematizes the profile of the SPP mode along the Si/Ag interface; (b) plasmonic quantum-dot solar cells designed for enhanced light absorption in ultra-thin quantum-dot layers by coupling to SPP modes propagating in the plane of the interface Ag/Si.

In an advanced design, trapping of light by generation of surface-plasmon polaritons is thought to solve absorption problems in quantum-dot solar cells (see Figure 6.11b). This type of cell requires thick quantum-dot layers to yield an effective light absorption. This, in turn, limits the carrier transport. It was demonstrated that a 20 nm thick layer of CdSe quantum dots deposited on Ag film can absorb light by surface-plasmon polaritons within a decay length of 1.2 μm at a photon-energy above 2.3 eV [23].

The enhanced absorption of light by using ordered arrangements of nanosized metal particles opens new horizons in photoelectrocatalysis. In principle, metal thin films or adequate arrangement of nanoparticles can be applied to increase the optical path length in thin active photovoltaic layers, thus enhancing the overall light absorption [24,25]. Recently, evidence has been presented that confirms that electrons can be exchanged from an excited surface of metal particles to chemisorbed molecules, resulting in photochemistry processes. For instance, colloidal Ag nanocrystals stabilized by sodium citrate build up a photovoltage under visible light excitation that is caused by irreversible "hot holes" photo-oxidation of adsorbed citrate anion. This creates a driving force for photochemical transformation of round shaped Ag seeds of 8 nm into 70 nm single crystal disk prisms under room light in a Ag^+ solution in a novel type of light-driven Ostwald ripening. In another experimental approach, it was found that the acceleration of electrochemical reactions at the surface of illuminated plasmonic particles arises due to conversion of plasmonic energy into heat that raises the medium temperature [27–29]. A local temperature increase of up to 30 K was detected for Au suspensions with a concentration of 2 mM (Au) on illumination with a 514 nm laser with an intensity of 3 Wcm^{-2} [29]. The general expression for the absorption cross section of a nanoparticle illuminated by a plane wave is

$$\sigma_{abs} = \frac{2\pi n}{\lambda_0} \frac{1}{\varepsilon_0 \left|\vec{E}_0\right|} \int_{np} Im[\varepsilon(\omega)] \left|\vec{E}(\vec{r})\right|^2 d\vec{r} \qquad (6.8)$$

where n is the optical index of the surrounding medium, $\varepsilon(\omega)$ is the permittivity of the nanoparticle material, and $\vec{E}(\vec{r})$ is the total electric field amplitude of incident light. The integral is calculated over the volume of the particle. Thus, the power of heat generation is given by

$$P = \sigma_{abs} I_0 = \sigma_{abs} \frac{n \times c \times \varepsilon_0}{2} \left|\vec{E}_0\right|^2 \qquad (6.9)$$

where I_0 is the incident irradiance. Figure 6.12 reproduces the calculation results of the heat dissipation by Au nanoparticles at the water/polymer interface [30,31] for an incident light power of 10^4 Wcm^{-2}. It should be noted that the heat generation increases with the number of absorbing nanoparticles. Hence, it is expected that similar changes of temperature are reached for a dense arrangement of Au nanoparticles by illumination with light of terrestrial sunlight intensities.

The construction of high ordered arrangements with plasmonic properties is limited to metals such as Ag and Au. The former presents a high chemical reactivity against

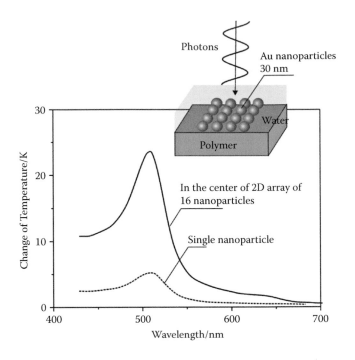

FIGURE 6.12 Temperature increase upon illumination of a single Au nanoparticle and an array of 16 nanoparticles at a polymer/water interface as a function of the wavelength of the incident light for an intensity of 10^4 Wcm^{-2}. $\varepsilon_{water} = 1.8$ (after Govorov and Richardson [30]).

some solution compositions; the latter is, in comparison with transition metals, a bad catalyst for hydrogen evolution. An elegant solution would be the deposition of a monolayer of a catalytic active metal, such as Pt, Rh, Ir, or Pd onto Au nanoparticles.

Within the inorganic concepts of photoelectrodes based on EMOS junctions, the fabrication of thin film photoelectrodes consisting of a stacking of a thin p-InP homoepitaxial layer deposited by MOVPE, an ultra-thin interfacial oxide, and a network of interlinked Rh nanoparticles constitutes a promising approach. Technical applications become possible by using liftoff processes [32] for thin films on substrate wafers. Details about the fabrication of this type of layer were offered in Chapter 3. Figure 6.13a depicts a scheme of the photoelectrode structure and its working mode in a three-electrode arrangement. The HR-TEM picture in Figure 6.13b shows how this structure looks. Although the currently achieved performance of this type of junction can still be improved (see the current-voltage characteristic in Figure 6.13c) by, for instance, modifying the oxide formation parameters, this concept offers a reliable way to be followed. Beyond the performance improvements which can be achieved by enlarging the dipolar shift of semiconductor band edges caused by the adsorption of halogenides, it is also important to point out the role played by the metal film. First, a closed particulate film avoids the contact of the aggressive electrolyte with the semiconductor, thus introducing operational stability. Additionally, a particulate metal film may bring beneficial effects with respect to the light absorption because of the light scattering. This is exemplified by calculations performed

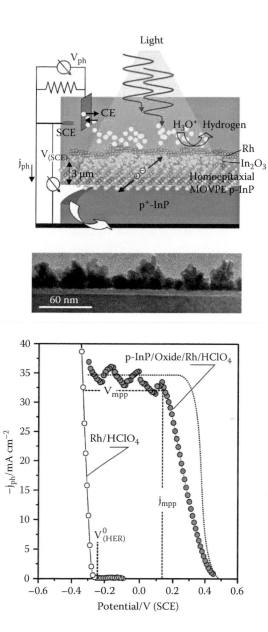

FIGURE 6.13 **(See color insert.)** Schematic of a light conversion EMOS junction in 1 M HClO4 and the output power characteristic formed electrochemically on thin homoepitaxial films of p-InP. Light intensity: 100 mWcm^{-2} (W-I lamp); open circles: Rh I-V curve; V$_{mpp}$, j$_{mpp}$ denote voltage and current at the maximum power point. (.....) expected I-V characteristic for improved charge transfer.

using the Mie theory. For the calculations, a Gauss distribution of deposited particles was assumed with a variance of 20% and mean size of 60 nm (Figure 6.14a). This was selected in agreement with the experimental observations of electrodeposited particles (see Figure 6.13b). Particle-particle interactions were ignored as well as the influence of the layered substrate comprising the InP bulk material and the top-surface oxide layer. To determine the scattering and absorption efficiencies, Q_{sca} and Q_{abs}, the Mie coefficients were calculated using spherical Bessel functions. An implementation of the equations for a computer-based evaluation can be found, for example, in ref. [33]. Q_{sca} results from the integration of the scattered radiation power over all directions and is related to the scattering cross section, σ_{sca}. While Q_{sca} characterizes the scattered far-field, Q_{abs} depends on the internal field within the particle. Energy conservation requires that any loss of radiation is related either to scattering or absorption: $Q_{ext} = Q_{sca} + Q_{abs}$ where Q_{ext} denotes the so-called extinction efficiency. Of particular interest for a photoactive device is the distinction between forward and backward scattering. Forward scattering can increase the amount of light-induced excess minority carriers and therefore the efficiency of the device. Therefore, the respective contributions, Q_{fwd} and Q_{back}, to the total scattering $Q_{sca} = Q_{fwd} + Q_{back}$ were calculated as well. In Figure 6.14b, the photon energy dependent forward scattering behavior is shown for the nanoparticles with the assumed size distribution. In Figure 6.14c, the different scattering behavior for the particles at photon energies of 2 and 4 eV is shown in detail by the intensity distribution versus scattering angle.

Figure 6.14d compares the reflectance data of InP bulk and those of the Rh/oxide/InP junction. A considerably lower reflectance of the device by about 30% indicates effective forward scattering into the absorber. Figure 6.14e depicts the calculated scattering efficiency $Q_{sca} = Q_{fwd} + Q_{back}$ as the sum of electromagnetic radiation scattered in the forward (fwd) and, respectively, backward (back) direction. The absorption efficiency is denoted by Q_{abs}. It follows that scattering of light is most efficient in the forward direction and that the overall scattering efficiency Q_{sca} decreases for lower photon energies. Near ~3.1 eV, the Rh/oxide junction starts to behave like a planar multilayer structure. Consequently, multiple reflections at the respective Rh/oxide/InP interfaces lead to interference patterns and an increasing reflectance signal (Figure 14da, dotted curve and magnification therein). The optical properties of the system are therefore sensitively dependent on the ratio of wavelength and particle cross section.

Combined, often named hybrid-systems joints the proven stability and light absorption of some inorganic semiconductors with the catalytic properties of some natural enzymes. In particular, the attachment of the photosystem I as the catalytic element of photoelectrodes for light-induced hydrogen evolution is being pursued [34,35]. The design of hybrid photoelectrodes requires overcoming one of its major challenges: the stability of complex proteic systems and the cross linking of organic groups with the semiconductor surface. For the latter issue, some experience is extracted from research about immobilization of medically relevant biomolecules with a structure similar to that of photosynthesis [36,37]. Molecular complexes such as the reverse transcriptase of the retrovirus HIV 1 (human immunodeficiency virus) and the AMV (avian myoblastosis virus) were able to be attached to Si surfaces with relative success [38].

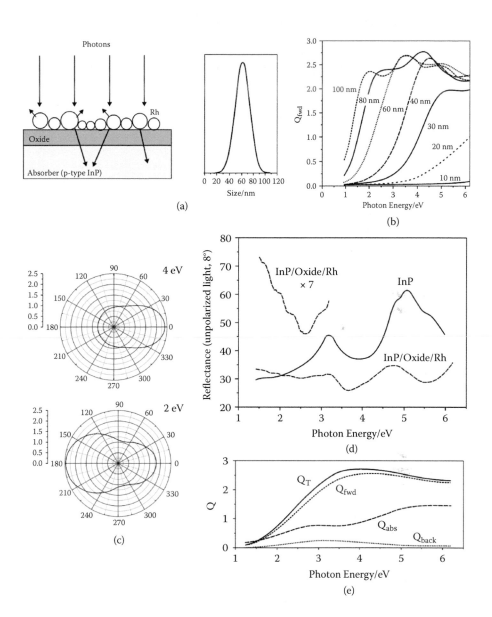

FIGURE 6.14 Light scattering in film particulate films of Rh on oxidized p-type InP thin films. (a) Particle size distribution; (b) forward scattering efficiency factors as a function of the incident photon energy for particles of various sizes; (c) intensity of the scattered light versus scattering angle of Rh particles with the size distribution shown in (a). The scattering behavior is exemplified for two different photon energies, 2 eV and 4 eV, respectively; (d) reflectance of InP (solid line) and the Rh/oxide/InP junction (dashed line). The inset shows a magnification of interference patterns observed for the reflectance of the device; (e) calculated efficiency factors as a function of incident photon energy.

It was found that DLVO (Derjaguin–Landau–Verwey–Overbeek) forces [39] and non-DLVO forces [40,41] play an essential role in the attachment process. Attachment by means of DLVO forces results from a balance of attracting van der Waals forces and electrostatic forces, usually repulsive, arising from the overlap of the charge double layers formed at the solid–electrolyte interface. The application of the DLVO-theory for a general case of two particles of different radii and surface potentials yields the following expression for the total interaction energy [42]:

$$U(d) = -\frac{A}{6}\frac{r_1 r_2}{r_1 + r_r}\frac{1}{d} + 64\pi\varepsilon_r\varepsilon_0\gamma_1\gamma_2\frac{r_1 r_2}{r_1 + r_r}\left(\frac{kT}{ve}\right)^2 \exp(-\kappa d) + U_B$$

where A is the Hamaker constant, which depends on the dielectric properties of medium, substrate, and attaching material. d is the separation between both spheres or radius $r_{1,2}$. The factors $\gamma_{1,2}$ are given by $\gamma_i = \tanh(ve\varphi_i / 4kT)$ where φ_i is the surface potential and v is the charge number. κ is the Debye length, the inverse of which gives the length of the double layer:

$$\kappa = \sqrt{F^2 \sum z_i^2 c_{i,0} / RT\varepsilon_0\varepsilon_r}$$

U_B represents the Born repulsion forces. The curve for stable colloids presents a secondary minimum that represents the distance between colloidal particles in a stable dispersion. Depending on the charge of particle and surface (in the limit $r_2 \to \infty$) and the ionic strength of the medium, the colloidal biological molecule can attach to the surface if the electrostatic term becomes smaller than the van der Waals term.

In bio-inspired photoelectrochemical systems, the isoelectric point (IP) of the protein-based enzymes plays a decisive role in the immobilization process onto the semiconductor surface. For instance, for the reverse transcriptase (RT) of HIV 1, the IP is between 8.5 and 9. Hence, for physiological solutions with a customary pH of about 7, the RT carries an overall positive charge. It would thus be attracted to negatively charged surface sites, a fact which has indeed been observed on defects on MoTe$_2$ surfaces using scanning tunneling microscopy imaging [43–45], or on step-bunched type Si surfaces, where a negative charge along the step edges [45] results in electrostatic attraction. Figure 6.15a shows a tapping-mode AFM picture showing the adhesion of bio-molecules (RT of the AMV) onto a step-bunch structured silicon surface. Here, one can see that the large coiled chains are mainly localized at the step edges at a height of about 20 nm (see cross section in Figure 6.15b). The potential profile for this cross section obtained by Kelvin probe microscopy indicates that a negative charge is accumulated at these sites (see Figure 6.15c). The origin of this charge accumulation is still not clear. It is probably related to a predominance of double hydrogen-coordinated Si atoms at the wall of steps.

The immobilization of protein-based macromolecules on VI layered transition metal dichalcogenides is also of particular interest due to the exposure of metal d-states with a large density of states at atomic terraces when layers perpendicular to the c-axis are exposed to the electrolyte. These states are especially reactive and may

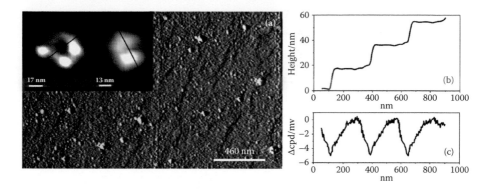

FIGURE 6.15   Immobilization of proteins onto semiconductor surfaces. (a) Tapping-mode AFM image of reverse transcriptase of AMV deposited onto a step-bunch structure n-type Si(111) surface. Inset: enlarged images of the adsorbed protein at two common positions; (b) and (c) cross section showing the stepped structure of silicon substrate and the related variation of the contact potential difference measured with a Kelvin probe microscope.

lead to the formation of surface dipoles by forming bonds with species in solution. For instance, it was observed that the flat band potential of a steeped surface WSe_2 varies with a linear relationship with the log $[I^-]$, a fact that was also observed on Si surfaces in HI solutions [46].

Non-DLVO hydration forces generated by the formation of a bound hydration layer around the protein units of enzymes yields a repulsion component. It is thought that at a distance much shorter than the Debye length, positively charged ammonium units of amino acids (lysine, arginine, histidine) $R-NH_3^+$ form part of the double layer at the host semiconductor, replacing solvated cations at the Helmholtz layer.

RT molecules immobilized onto structured surfaces appear to be in an upright, kidney-shaped position, as suggested by tapping-mode AFM pictures (see Figure 6.15a). For the charge carrier movement from the semiconductor to the tip in STM imaging experiments, a solvation-induced detrapping of electrons on the outer polypeptides, where so-called bio-water exists, has been invoked [47,48]. This mechanism also explains the findings of Guckenberger and coworkers on moisture-induced tunneling from insulating substrates through DNA to the STM tip [49]. For charge transfer from metals or semiconductors to enzymes, this finding seems to indicate that injected electrons are not likely to easily reach the reactive sites, which are typically inside the molecules, via tunneling and hopping. Therefore, it would be worthwhile to attempt EET using a long-range resonance process, as provided by the Förster mechanism. The direct use of enzymes, such as water oxidase and hydrogenase, even if they are genetically modified in terrestrial solar fuel generating systems, however, is unlikely to be successful due to their low stability. The synthesis of more stable macromolecular units appears to be a more promising route, particularly in conjunction with conducting electrodes [50].

The use of enzymes as electrocatalysts in reliable photoelectrodes requires mimicking the active part by eliminating the inherently unstable protein part of natural photosynthetic components. This in turn would contribute to facilitation of the electron

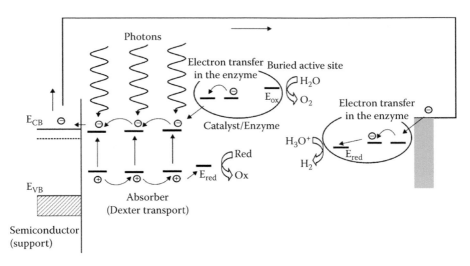

FIGURE 6.16 Envisaged water splitting system with nanoparticle absorbers on a semiconductor, Dexter exciton transport, excitation energy transfer to the reactive site of an enzyme or macromolecule, and successive water oxidation at the anode; hydrogen evolution at the cathode occurs as a dark reaction in this scheme.

transfer process. Architectures should consist of mono- and hetero-metallic cores where the metal atoms are chelated by ligands or organic molecules as a source of free electrons. Artificial photosynthetic systems are presently designed on the basis of Co- and Ni-diimine-dioxime and Ni-Ru complexes [51], Ni-Fe di-thiolatehydrides [52], as well as Ru complexes [53], all being classified as excellent catalysts in the dark.

Figure 6.16 shows a schematic of an envisaged bio-inspired catalytic system. Here, nanoparticle absorbers are assumed that have been deposited on a semiconducting support that would allow rectification, thus using the system in an analogous manner to a photodiode-type solar cell. The transport along the absorber particles is taken as exciton hopping as in the Dexter mechanism. The next step is excitation energy transfer from the surface nanoparticles to the macromolecular catalyst center where the reaction occurs at the reaction center. This could occur by classical Marcus-type charge transfer but here, the catalytic reaction center has been assumed to be too far away for efficient tunneling and the excitation energy transfer process could take place via Förster transfer, for example. The photoprocess has been assumed to occur at the anode side where water is oxidized. The reduction to hydrogen takes place at a conducting (metallic) counter electrode that has been modified with an appropriate catalyst, too. Although the process works in solution, it has not yet been possible to prepare electrodes where current flows to a photoelectrochemical solar cell arrangement. Molecules with catalytic centers have to be immobilized on semiconductor surfaces in a way so that the energetic position of their HOMO and LUMO levels match with the conduction or valence band in the case of photoreduction or photo-oxidation, respectively. As an absorber, the use of AlGaInN is envisaged due to the possibility of tuning the band gap between that of InN (0.7 eV) and that of AlN (6 eV). This allows the change of the energetic position of the conduction band by a 60% of the band gap change [54].

Another approach is the construction of photocathodes by adsorption of hydrogenase onto nanoparticles of TiO_2 [55,56]. Other, more complex approaches attempt to mimic the natural green algae by deposition of chloroplasts onto semiconductors and their coupling to hydrogenase by means of a redox couple [57,58]. Such a system involves the electron transfer from the semiconductor to the active center of the protein, consisting of a NiFe or FeFe complex. The transfer takes place over FeS clusters, which are separated by a regular distance (see Figure 6.17). It is not totally clear what mechanism governs the electron transfer processes in enzymatic systems. The theory of electron transfer in catalytic electrochemistry of enzymes was analyzed in recent years [59–61]. In general, the catalytic cycle of NiFe hydrogenase can be broken down into two parts: (i) interfacial electron transfer steps, involving the oxidation or reduction of the surface exposed [4Fe4S] cluster, and (ii) intramolecular

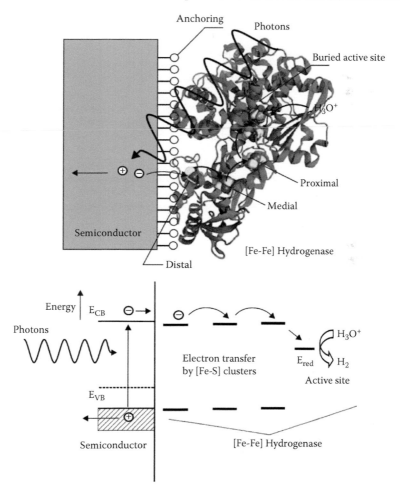

FIGURE 6.17　(**See color insert.**) Electron transfer at a photocathode constituted by a semiconducting absorber covered by electrocatalytic hydrogenase attached to the surface by an organic anchor.

electron transfer along the chain of FeS clusters together with H_2 diffusion and binding, active site chemistry, and proton transfers. The rate of interfacial electron transfer can be described by a Marcus-type mechanism. Hence, its rate depends exponentially on the electrode potential. Recently, Madden et al. [62] reported electrochemical experiments performed onto [Fe-Fe]-hydrogenase immobilized onto Au electrons by anchoring it by carboxylate terminated SAM layers. Their results show that these enzymes present a catalytic activity comparable to that shown by platinum for the reduction of water in neutral media.

From the numerous photoelectrode concepts based on inorganic compounds, hybrid systems, and those mimicking the photosynthesis processes, it is presently impossible to define a research mainstream. A lot of designs are being intensively investigated, and surely only some of them will lead to industrial applications.

REFERENCES

1. Lewerenz, H.J., Skorupska, K., Muñoz, A.G., Stempel, T., Nüsse, N., Lublow, M., Vo-Dinh, T. and P. Kulesza. 2011. Micro- and nanotopographies for photoelectrochemical energy conversion. II: Photoelectrocatalysis—Classical and advanced systems. *Electrochimica Acta* 56:10726.
2. Boettcher, S.W, Spurgeon, J.M., Putnam, M.C., Warren, E.L., Turner-Evans, D.B., Kelzenberg, M.D., Maiolo, J.R., Atwater, H.A. and N.S. Lewis. 2010. Energy-conversion properties of vapor-liquid-solid–grown silicon wire-array photocathodes. *Science* 327:185.
3. Plass, K.E., Filler, M.A., Spurgeon, J.M., Kayes, B.M., Maldonado, S., Brunschwig, B.S., Atwater, H.A. and N.S. Lewis. 2009. Flexible polymer-embedded Si wire arrays. *Adv. Mater.* 21:325.
4. Kayes, B.M., Filler, M.A., Putnam, M.C., Kelzenberg, M.D., Lewis, N.S. and H.A. Atwater. 2007. *Appl. Phys. Lett.* 91:103110.
5. Kelzenberg, M.D., Turner-Evans, D.B., Kayes, B.M., Filler, M.A., Putnam, M.C., Lewis, N.S. and H.A. Atwater. 2008. *Nano. Lett.* 8:710.
6. Maiolo, J.R., Kayes, B.M., Filler, M.A., Putnam, M.C., Kelzenberg, M.D., Atwater, H.A. and N.S. Lewis. 2007. *J. Am. Chem. Soc.* 129:12346.
7. Maier, S.A. and H.A. Atwater. 2005. *J. Appl. Phys.* 98:011101.
8. Maier, S.A., Kik, P.G., Brongersma, M.L., Atwater, H.A., Meltzer, S., Requicha, A.A.G. and B.E. Koel. 2002. Observation of coupled plasmon-polariton modes of plasmon waveguides for electromagnetic energy transport below the diffraction limit. *Mater. Res. Soc. Symp. Proc.* 722:L6.2.1.
9. Mie, G. 1908. Beiträge zur Optik trüber Medien, speziell kolloidaler Metalllösungen. *Ann. Phys.* 25:445.
10. Dragoman, D. and M. Dragoman. 2002. *Optical Characterization of Solids.* Berlin: Springer-Verlag.
11. Kreibig, U. and M. Vollmer. 1995. *Optical Properties of Metal Clusters.* Berlin: Springer.
12. Hao, E. and G.C. Schatz. 2004. *J. Chem. Phys.* 120:357.
13. Catchpole, K.R. and A. Polman. 2008. Plasmonic solar cells. *Opt. Express* 16:21793.
14. Ernst, A.Z., Sun, L., Wiaderek, K., Kolary, A., Zoladek, S., Kulesza, P.J. and J.A. Cox. 2007. *Electroanalysis* 19:2103.
15. Whittingham, M.S. 1978. *Prog. Solid State Chem.* 12:41.
16. Perez, J., Gonzalez, E. and H.M. Villullas. 1998. *J. Phys. Chem. B* 102:10931.
17. Merga, G., Saucedo, N., Class, L.C., Puthussery, J. and D. Meisel. 2010. *J. Phys. Chem. C* 114:14811.

18. Mirdamadi-Esfahani, M., Mostavi, M., Kelta, B., Nadjo, L., Kooyman, P. and H. Remlta. 2010. Bimetallic Au-Pt nanoparticles synthesized by radiolysis: Application in electro-catalysis. *Gold Bull.* 43:49.
19. Israelachvili, J.N. 1992. *Intermolecular and Surface Forces*, 2nd ed. London: Academic Press.
20. Atwater, H.A. and A. Polman. 2010. Plasmonics for improved photovoltaic devices. *Nature Materials* 9: 202.
21. Berini, P. 2000. Plasmon–polariton waves guided by thin lossy metal films of finite width: Bound modes of symmetric structures. *Phys. Rev. B* 61:10484.
22. Berini, P. 2001. Plasmon–polariton waves guided by thin lossy metal films of finite width: Bound modes of asymmetric structures. *Phys. Rev. B* 63:125417.
23. Pacifici, D., Lezec, H. and H.A. Atwater. 2007. All-optical modulation by plasmonic excitation of CdSe quantum dots. *Nature Photon* 1:402.
24. Nakayama, K., Tanabe, K. and H.A. Atwater. 2008. Plasmonic nanoparticle enhanced light absorption in GaAs solar cells. *Appl. Phys. Lett.* 93:121904.
25. Chang, T.H., Wu, P.H., Chen, S.H., Chan, C.H., Lee, C.C., Chen, C.C. and Y.K. Su. 2009. Efficiency enhancement in GaAs solar cells using self-assembled microspheres. *Opt. Express* 17:6519.
26. Brus, L. 2008. Noble metal nanocrystals: Plasmon electron transfer photochemistry and single-molecule Raman spectroscopy. *Acc. Chem. Res.* 41:1742.
27. Yen, C.W. and M.A. El-Sayed. 2009. Plasmonic field effect on the hexacyanoferrate (III)-thiosulfate electron transfer catalytic reaction on gold nanoparticles: electromagnetic or thermal?. *J. Phys. Chem. C* 113:19585.
28. Hu, M., Chen, J., Li, Z.-Y., Au, L., Hartland, G.V., Li, X., Marquez, M. and Y. Xia. 2006. Gold nanostructures: engineering their plasmonic properties for biomedical applications. *Chem. Soc. Rev.* 35:1084.
29. Jain, P.K., Huang, X., El-Sayed, I.H. and M.A. El-Sayed. 2008. Noble Metals on the Nanoscale: Optical and Photothermal Properties and Some Applications in Imaging, Sensing, Biology, and Medicine. *Acc. Chem. Res.* 41:1578.
30. Govorov, A.O. and H.H. Richardson. 2007. Generating heat with metal nanoparticles. *Nanotoday* 2(1):30.
31. Govorov, A.O., Zhang, W., Skeini, T., Richardson, H., Lee, J. and N.A. Kotov. 2006. Gold nanoparticle ensembles as heaters and actuators: Melting and collective plasmon resonances. *Nanoscale Res. Lett.* 1:84.
32. Schermer, J.J., Bauhuis, G.J., Mulder, P., Haverkamp, E.J., van Deelen, J., van Niftrik, A.T.J. and P.K. Larsen. 2006. Photon confinement in high-efficiency, thin-film III–V solar cells obtained by epitaxial lift-off. *Thin Solid Films* 511–512:645.
33. Kreibig, U. and M. Vollmer. 1995. *Optical Properties of Metal Clusters*. Berlin: Springer-Verlag.
34. Golbeck, J.H. 2006. *Photosystem I (The Light-Driven Plastocyanin: Ferredoxin Oxidoreductase)*. Dordrecht: Springer.
35. Hambourger, M., Gervaldo, M., Svedruzic, D., King, P.W., Gust, D., Ghirardi, M., Moore, A.L. and T.A. Moore. 2008. [FeFe]-Hydrogenase-catalyzed H_2 production in a photoelectrochemical biofuel cell. *J. Am. Soc.* 130:2015.
36. Cosnier, S. 1999. Biomolecule immobilization on electrode surfaces by entrapment or attachment to electrochemically polymerized films: A review. *Biosens. Bioelectron.* 14:443.
37. Yamamoto, T. and T. Fujii. 2007. Active immobilization of biomolecules on a hybrid three-dimensional nanoelectrode by dielectrophoresis for single-biomolecule study. *Nanotechnology* 18:495503.
38. Skorupska, K., Lewerenz, H.J. and T. Vo-Dinh. 2009. Scanning probe characterization of enzymes deposited onto step-bunched silicon nanostructures. *Phys. Scr.* 79:065801.

39. Derjaguin, B.V. and L.D. Landau. 1941. Theory of the stability of strongly charged lyophobic sols and of the adhesion of strongly charged particles in solution of electrolytes. *Acta Phys. Chim. USSR* 14:633.

40. Grasso, D., Subramanian, K., Butkus, M., Strevett, K. and J. Bergendahl. 2002. A review of non-DLVO interactions in environmental colloidal systems. *Rev. Environ. Sci. Biotechnol.* 1:17.

41. Petsev, D.N. and P.G. Vekilov. 2000. Evidence for non-DLVO hydration interactions in solutions of the protein apoferritin. *Phys. Rev. Lett.* 84:1339.

42. Gregory, J. 1975. Interaction of unequal double layers at constant charge. *J. Colloid Interface Sci.* 51:44.

43. Jungblut, H., Campbell, S.A., Giersig, M., Müller, D.J. and H.J. Lewerenz. 1992. *Faraday Discuss.* 94:183.

44. Lewerenz, H.J. 2008. Enzyme–semiconductor interactions: Routes from fundamental aspects to photoactive devices. *Phys. Status Solidi B* 245:1884.

45. Skorupska, K., Lublow, M., Kanis, M., Jungblut, H. and H.J. Lewerenz. 2005. Electrochemical preparation of a stable accumulation layer on Si: A synchrotron radiation photoelectron spectroscopy study. *Appl. Phys. Lett.* 87:262101.

46. Skorupska, K., Lewerenz, H.J., Smith, J.R., Kulesza, P.J., Mernagh, D. and S.A. Campbell. 2011. Macromolecule–semiconductor interfaces: From enzyme immobilization to photoelectrocatalytical applications. *J. Electroanal. Chem.* 662:169.

47. Campbell, S.A., Smith, J.R., Jungblut, H. and H.J. Lewerenz. 2007. Protein imaging on a semiconducting substrate: a scanning tunneling microscopy investigation. *J. Electroanal. Chem.* 599:313.

48. Lewerenz, H.J., Skorupska, K., Smith, J.R. and S.A. Campbell. 2009. Surface chemistry and electronics of semiconductor–nanosystem junctions II: Enzyme immobilization, charge transport aspects and scanning probe microscopy imaging. *J. Solid State Electrochem.* 13:195.

49. Guckenberger, R., Heim, M., Cevc, G., Knapp, H. F., Wiegrabe, W. and A. Hillebrand. 1994. Scanning tunneling microscopy of insulators and biological specimens based on lateral conductivity of ultrathin water films. *Science* 266:1538.

50. Dempsey, J.L., Esswein, A.J., Manke, D.R., Rosenthal, J., Soper, J.D. and D.G. Nocera. 2005. Molecular chemistry of consequence to renewable energy. *Inorg. Chem.* 44:6879.

51. Oudart, Y., Artero, V., Pécaut, J. and M. Fontecave. 2006. [Ni(xbsms)Ru(CO)$_2$Cl$_2$]: A bioinspired nickel–ruthenium functional model of [NiFe] hydrogenase. *Inorg. Chem.* 45:4334.

52. Barton, B.E., Matthew Whaley, C., Rauchfuss, Th.B. and D.L. Gray. 2009. Nickel–iron dithiolato hydrides relevant to the [NiFe]-hydrogenase active site. *J. Am. Chem. Soc.* 131:6942.

53. Hiraishi, T., Kamachi, T. and I. Okura. 1999. Photoinduced hydrogen evolution with viologen-linked ruthenium(II) complexes and hydrogenase. *J. Mol. Cat. A: Chem.* 138:107.

54. Shah, P., Mitin, V., Grupen, M., Song, G.H. and K. Hess. 1996. Numerical simulation of wide band-gap AlGaN/InGaN light-emitting diodes for output power characteristics and emission spectra. *J. Appl. Phys.* 79:2755.

55. Cuendet, P., Rao, K.K., Grätzel, M. and D.O. Hall. 1986. Light induced H$_2$ evolution in a hydrogenase-TiO$_2$ particle system by direct electron transfer or via rhodium complexes. *Biochimie* 68:217.

56. Renugopalakrishnan, V. et al. 2009. Nanomaterials for energy conversion applications, Nalwa, H.S. (Ed.). *Nanomaterials for Energy Storage Applications.* Valencia, CA: American Scientific Publishers.

57. Benemann, J.R., Berenson, J.A., Kaplan, N.O. and M.D. Kamen. 1973. Hydrogen evolution by a chloroplast-ferredoxin-hydrogenase system. *Proc. Nat. Acad. Sci. USA* 70:2317.

58. Rao, K.K., Gogotov, I.N. and D.O. Hall. 1978. Hydrogen evolution by chloroplast-hydrogenase systems: Improvements and additional observations. *Biochimie* 60:291.

59. Léger, C., Lederer, F., Guigliarelli, B. and P. Bertrand. 2006. Electron flow in multi-center enzymes: Theory, applications, and consequences on the natural design of redox chains. *J. Am. Chem. Soc.* 128:180.

60. Armstrong, F.A. 2005. Recent developments in dynamic electrochemical studies of adsorbed enzymes and their active sites. *Curr. Op. Chem. Biol.* 9:110.

61. Reda, T. and J.J. Hirst. 2006. Interpreting the catalytic voltammetry of an adsorbed enzyme by considering substrate mass transfer, enzyme turnover, and interfacial electron transport. *J. Phys. Chem. B* 110:1394.

62. Madden, C., Vaughn, M.D., Díez-Perez, I., Brown, K.A., King, P.W., Gust, D., Moore, A.L. and T.A. Moore. 2012. Catalytic turnover of [FeFe]-hydrogenase based on single-molecule imaging. *J. Am. Chem. Soc.* 134:1577.

Index